METHODS IN MOLECULAR BIOLOGY

For further volumes:
https://www.springer.com/series/7651

For over 35 years, biological scientists have come to rely on the research protocols and methodologies in the critically acclaimed Methods in Molecular Biology series. The series was the first to introduce the step-by-step protocols approach that has become the standard in all biomedical protocol publishing. Each protocol is provided in readily-reproducible step-by step fashion, opening with an introductory overview, a list of the materials and reagents needed to complete the experiment, and followed by a detailed procedure that is supported with a helpful notes section offering tips and tricks of the trade as well as troubleshooting advice. These hallmark features were introduced by series editor Dr. John Walker and constitute the key ingredient in each and every volume of the Methods in Molecular Biology series. Tested and trusted, comprehensive and reliable, all protocols from the series are indexed in PubMed.

Manuscript Instructions https://drive.google.com/file/d/1K4Az5vpAFVzYadJ3jKx2-QSo7Dqa_SNA/view?usp=sharing

Root Development

Methods and Protocols

Edited by

Steffen Vanneste

Department Plants and Crops - HortiCell, Ghent University, Ghent, Belgium

Inge Verstraeten

Department Plants and Crops - HortiRoot, Ghent University, Ghent, Belgium

Humana Press

Editors
Steffen Vanneste
Department Plants and Crops - HortiCell
Ghent University
Ghent, Belgium

Inge Verstraeten
Department Plants and Crops - HortiRoot
Ghent University
Ghent, Belgium

ISSN 1064-3745 ISSN 1940-6029 (electronic)
Methods in Molecular Biology
ISBN 978-1-0716-5602-0 ISBN 978-1-0716-5603-7 (eBook)
https://doi.org/10.1007/978-1-0716-5603-7

This Humana imprint is published by the registered company Springer Science+Business Media, LLC, part of Springer Nature.
The registered company address is: 1 New York Plaza, New York, NY 10004, U.S.A.

Preface

The plant's root system fulfills many essential functions that support growth and development. Roots are responsible for the uptake of water and nutrients, anchoring the plant to the substrate, and providing a dynamic interface for interactions with the soil microbiome. The root system's relatively simple anatomy, combined with the spatial separation of developmental processes, has made it an attractive model for a wide range of plant scientists. This interest has been accompanied by the continuous development and refinement of tools to investigate diverse aspects of root biology.

In this volume, we present a collection of recently developed protocols and experimental approaches to study and manipulate roots. Each protocol is described in detail by specialists who routinely apply these methods in their own research, ensuring both reliability and practical relevance.

Recognizing that "Seeing is believing," a substantial part of this book focuses on imaging-based approaches, including anatomical analyses, live-cell imaging, cell biological techniques, and the visualization of gene expression patterns. In addition, a range of omics approaches is covered. The volume also provides detailed methodologies to measure and manipulate key root parameters.

The first part of the book is dedicated to the study of roots in non-*Arabidopsis* species. It includes protocols for members of the *Cucurbitaceae* family, soybean, white lupin, the fern *Ceratopteris richardii*, the grass *Brachypodium distachyon*, and maize. These chapters address plant cultivation, functional analyses, nutrient uptake assays, methods to reveal internal anatomy, microfluidic approaches for real-time analysis and manipulation of root growth, as well as automated image analysis of roots and root hairs.

The second part presents diagnostic methodologies and experimental strategies to manipulate root processes such as the localized control of water availability and pH, the analysis of osmotic pressure, auxin transport rates, membrane protein dynamics, the distribution of labile iron species, and cell proliferation within the root meristem.

The final part focuses on workflows and protocols to assess gene expression at the transcript and (phospho)proteome levels. These include highly sensitive *in situ* hybridization techniques, RNA sequencing of sorted cell populations, single-cell RNA sequencing, membrane protein analysis, and high-throughput (phospho)proteomics.

Together, the contributions in this volume aim to advance root research across diverse plant species by providing detailed, step-by-step descriptions of novel methodologies, as well as improved and adapted versions of established protocols. Each chapter includes practical guidance and technical insights to facilitate successful implementation. We anticipate that the availability of these protocols will promote their broad adoption and stimulate further innovation in root biology research.

Ghent, Belgium *Steffen Vanneste*
Ghent, Belgium *Inge Verstraeten*

Contents

Contributors

KELVIN ADEMA • *Department of Plant Sciences, Cluster of Plant Developmental Biology, Laboratory of Cell and Developmental Biology, Wageningen University, Wageningen, The Netherlands*
HAIYUE AI • *Institute of Science and Technology Austria, Klosterneuburg, Austria*
CARINE ALCON • *IPSiM, Univ Montpellier, CNRS, INRAE, Montpellier, France*
ALEJANDRO ARAGÓN-RAYGOZA • *Molecular and Developmental Complexity Laboratory, Advanced Genomics Unit, Cinvestav, México*
JASON BANDA • *Plant and Crop Sciences, School of Biosciences, University of Nottingham, Leicestershire, UK*
JASIM BASHEER • *King Abdullah University of Science and Technology (KAUST), Plant Science Program, Biological and Environmental Science and Engineering Division (BESE), Thuwal, Saudi Arabia*
TOM BEECKMAN • *Department of Plant Biotechnology and Bioinformatics, Ghent University, Ghent, Belgium; Center for Plant Systems Biology, VIB, Ghent, Belgium*
KEVIN BELLANDE • *Laboratory of Molecular and Cellular Biology, Institute of Biology, University of Neuchâtel, Neuchâtel, Switzerland; IPSiM, University of Montpellier, CNRS, INRAE, Institut Agro, Montpellier, France*
MALCOLM J. BENNETT • *Plant and Crop Sciences, School of Biosciences, University of Nottingham, Leicestershire, UK*
IWONA BERNACKA-WOJCIK • *Laboratory of Organic Electronics, Department of Science and Technology, Linköping University, Norrköping, Sweden*
LAURENT BROTTIER • *Institut des Sciences des Plantes de Montpellier (IPSiM), UMR 5004 CNRS/INRAE/Institut Agro, Université de Montpellier, Montpellier, Cedex, France*
VASYL BRYKOV • *Institute of Experimental Botany of the Czech Academy of Sciences, Prague, Czech Republic*
JODI BIANCA CALLWOOD • *Department of Genetics, Development, and Cell Biology, Iowa State University, Ames, IA, USA*
LIUDMILA CHELYSHEVA • *Université Paris-Saclay, INRAE, AgroParisTech, Institute Jean-Pierre Bourgin for Plant Sciences, Versailles, France*
DANIELLE COLYER • *Plant and Crop Sciences, School of Biosciences, University of Nottingham, Leicestershire, UK*
ALFREDO CRUZ-RAMÍREZ • *Molecular and Developmental Complexity Laboratory, Advanced Genomics Unit, Cinvestav, México*
KIRILL N. DEMCHENKO • *Laboratory of Cellular and Molecular Mechanisms of Plant Development, Komarov Botanical Institute, Russian Academy of Sciences, Saint Petersburg, Russia*
RICCARDO DI MAMBRO • *Department of Biology, University of Pisa, Pisa, Italy*
JAMES DONALDSON • *Plant and Crop Sciences, School of Biosciences, University of Nottingham, Leicestershire, UK*
JULIE DUCLA • *Umeå Plant Science Centre (UPSC), Department of Forest Genetics and Plant Physiology, Swedish University of Agricultural Sciences, Umeå, Sweden*

Annie Espinal-Centeno • *Molecular and Developmental Complexity Laboratory, Advanced Genomics Unit, Cinvestav, México*
Matyáš Fendrych • *Institute of Experimental Botany of the Czech Academy of Sciences, Prague, Czech Republic*
Yulin Gan • *MOE Key Laboratory for Cellular Dynamics, Center for Advanced Interdisciplinary Science and Biomedicine of IHM, Hefei National Laboratory for Physical Sciences at the Microscale, Division of Life Sciences and Medicine, University of Science and Technology of China, Hefei, China*
Mélanie Gasser • *Institut des Sciences des Plantes de Montpellier (IPSiM), UMR 5004 CNRS/INRAE/Institut Agro, Université de Montpellier, Montpellier, Cedex, France*
Kris Gevaert • *Department of Biomolecular Medicine, Ghent University, Ghent, Belgium; Center for Medical Biotechnology, VIB, Ghent, Belgium*
Clarice Felix Gonzales • *Department of Genetics, Development, and Cell Biology, Iowa State University, Ames, IA, USA*
Sofie Goormachtig • *Center for Plant Systems Biology, VIB, Ghent, Belgium; Department of Plant Biotechnology and Bioinformatics, Ghent University, Ghent, Belgium*
Carolin Grones • *Department of Plant Sciences, Cluster of Plant Developmental Biology, Laboratory of Cell and Developmental Biology, Wageningen University, Wageningen, The Netherlands*
Peter Grones • *Cluster of Plant Developmental Biology, Cell and Developmental Biology, Wageningen University & Research, Wageningen, The Netherlands*
Jakub Hajný • *Laboratory of Growth Regulators, Czech Academy of Sciences, Institute of Experimental Botany and Palacky University, Olomouc, Czech Republic*
Elena L. Ilina • *Laboratory of Cellular and Molecular Mechanisms of Plant Development, Komarov Botanical Institute, Russian Academy of Sciences, Saint Petersburg, Russia*
Tilman Jacob • *Institute of Agricultural and Nutritional Sciences, Martin-Luther-University Halle-Wittenberg, Halle (Saale), Germany*
Cristovão De Jesus Vieira Teixeira • *Laboratory of Molecular and Cellular Biology, Institute of Biology, University of Neuchâtel, Neuchâtel, Switzerland*
Joris Jourquin • *Department of Plant Biotechnology and Bioinformatics, Ghent University, Ghent, Belgium; Center for Plant Systems Biology, VIB, Ghent, Belgium; UGent TechTransfer Office, Present address: Ghent University, Ghent, Belgium*
Dior Rose Kelley • *Department of Genetics, Development, and Cell Biology, Iowa State University, Ames, IA, USA*
Alexey S. Kiryushkin • *Laboratory of Cellular and Molecular Mechanisms of Plant Development, Komarov Botanical Institute, Russian Academy of Sciences, Saint Petersburg, Russia*
Wouter Kohlen • *Department of Plant Sciences, Cluster of Plant Developmental Biology, Laboratory of Cell and Developmental Biology, Wageningen University, Wageningen, The Netherlands; Department of Plant Sciences, Cluster of Plant Developmental Biology, Laboratory of Molecular Biology, Wageningen University, Wageningen, The Netherlands*
Dominik Kralík • *Institute of Experimental Botany of the Czech Academy of Sciences, Prague, Czech Republic; Department of Chemical Engineering, University of Chemistry and Technology, Prague, Czech Republic*
Vsevolod S. Kukushkin • *Laboratory of Cellular and Molecular Mechanisms of Plant Development, Komarov Botanical Institute, Russian Academy of Sciences, Saint Petersburg, Russia*

LEV LEVENETS • *Institute of Experimental Botany of the Czech Academy of Sciences, Prague, Czech Republic; Department of Experimental Plant Biology, Charles University, Prague, Czech Republic*
PETRA MARHAVA • *King Abdullah University of Science and Technology (KAUST), Plant Science Program, Biological and Environmental Science and Engineering Division (BESE), Thuwal, Saudi Arabia*
PETER MARHAVÝ • *King Abdullah University of Science and Technology (KAUST), Plant Science Program, Biological and Environmental Science and Engineering Division, Thuwal, Kingdom of Saudi Arabia*
LAURENCE MARQUÈS • *Institut des Sciences des Plantes de Montpellier (IPSiM), UMR 5004 CNRS/INRAE/Institut Agro, Université de Montpellier, Montpellier, Cedex, France*
POONAM MEHRA • *Plant and Crop Sciences, School of Biosciences, University of Nottingham, Leicestershire, UK*
DOMINGO MÉNDEZ-ALVAREZ • *Molecular and Developmental Complexity Laboratory, Advanced Genomics Unit, Cinvestav, México*
HANS MOTTE • *Department of Plant Biotechnology and Bioinformatics, Ghent University, Ghent, Belgium; Center for Plant Systems Biology, VIB, Ghent, Belgium*
SOPHIA MÜLLER • *Department of Plant Sciences, Cluster of Plant Developmental Biology, Laboratory of Cell and Developmental Biology, Wageningen University, Wageningen, The Netherlands*
JEAN-CHRISTOPHE PALAUQUI • *Université Paris-Saclay, INRAE, AgroParisTech, Institute Jean-Pierre Bourgin for Plant Sciences, Versailles, France*
KATHARINA PAWLOWSKI • *Department of Ecology, Environment and Plant Sciences, Stockholm University, Stockholm, Sweden*
YAKUN PENG • *MOE Key Laboratory for Cellular Dynamics, Center for Advanced Interdisciplinary Science and Biomedicine of IHM, Hefei National Laboratory for Physical Sciences at the Microscale, Division of Life Sciences and Medicine, University of Science and Technology of China, Hefei, China*
BENJAMIN PÉRET • *Institut des Sciences des Plantes de Montpellier (IPSiM), UMR 5004 CNRS/INRAE/Institut Agro, Université de Montpellier, Montpellier, Cedex, France*
ROMAN PLESKOT • *Institute of Experimental Botany, Czech Academy of Sciences, Prague, Czech Republic*
FENG QU • *Department of Plant Pathology, The Ohio State University, Wooster, OH, USA*
MARCEL QUINT • *Institute of Agricultural and Nutritional Sciences, Martin-Luther-University Halle-Wittenberg, Halle (Saale), Germany*
MARK ROOSJEN • *Laboratory of Biochemistry, Wageningen University, Wageningen, The Netherlands*
ELENA SALVI • *Department of Biology, University of Pisa, Pisa, Italy*
AMBER VAN SETERS • *Department of Plant Sciences, Cluster of Plant Developmental Biology, Laboratory of Cell and Developmental Biology, Wageningen University, Wageningen, The Netherlands*
WOUTER SMET • *Department of Plant Biotechnology and Bioinformatics, Ghent University, Ghent, Belgium; Center for Plant Systems Biology, VIB, Ghent, Belgium; Present address: Biological and Environmental Science and Engineering (BESE) Division, King Abdullah University of Science and Technology (KAUST), Thuwal, Saudi Arabia*
MARTINA ŠPUNDOVÁ • *Department of Biophysics, Faculty of Science, Palacky University, Olomouc, Czech Republic*

ELENI STAVRINIDOU • *Umeå Plant Science Centre (UPSC), Department of Forest Genetics and Plant Physiology, Swedish University of Agricultural Sciences, Umeå, Sweden; Laboratory of Organic Electronics, Department of Science and Technology, Linköping University, Norrköping, Sweden*
SYLWIA STRUK • *Center for Plant Systems Biology, VIB, Ghent, Belgium; Department of Biomolecular Medicine, Ghent University, Ghent, Belgium*
SHUTANG TAN • *MOE Key Laboratory for Cellular Dynamics, Center for Advanced Interdisciplinary Science and Biomedicine of IHM, Hefei National Laboratory for Physical Sciences at the Microscale, Division of Life Sciences and Medicine, University of Science and Technology of China, Hefei, China*
ALESSIO TERENZI • *Department of Biology, University of Pisa, Pisa, Italy*
KENZO VEREECKEN • *Department of Plant Biotechnology and Bioinformatics, Ghent University, Ghent, Belgium; Center for Plant Systems Biology, VIB, Ghent, Belgium; Present address: Eastman, Ghent, Belgium*
JOOP E. M. VERMEER • *Laboratory of Molecular and Cellular Biology, Institute of Biology, University of Neuchâtel, Neuchâtel, Switzerland*
JAN VERWAEREN • *Department of Data Analysis and Mathematical Modelling, Ghent University, Ghent, Belgium*
DOLF WEIJERS • *Laboratory of Biochemistry, Wageningen University, Wageningen, The Netherlands*
GIJS WILBRINK • *Cluster of Plant Developmental Biology, Cell and Developmental Biology, Wageningen University & Research, Wageningen, The Netherlands*
JANA WITMER • *Department of Plant Sciences, Cluster of Plant Developmental Biology, Laboratory of Cell and Developmental Biology, Wageningen University, Wageningen, The Netherlands*
TOU CHEU XIONG • *IPSiM, Univ Montpellier, CNRS, INRAE, Montpellier, France*

Part I

Going Beyond Arabidopsis

Check for updates

Chapter 1

Composite *Cucurbitaceae* Plants as a Tool to Study Physiological and Molecular Mechanisms of Root Development

Elena L. Ilina, Alexey S. Kiryushkin, Vsevolod S. Kukushkin, Katharina Pawlowski and Kirill N. Demchenko

Abstract

Composite plants are a convenient model for obtaining a whole plant with undisturbed regulatory links between the shoot and root system, while the roots carry the properties specified by the researcher. This approach allows expressing transgenes only in roots, which in some cases may be preferable over the expression of transgenes in all organs of a non-chimeric transgenic plant. Composite plants have been widely used to study the nitrogen fixation process, mycorrhizal symbioses, and the interaction of plant roots with various pathogens. The current protocol for *Agrobacterium rhizogenes*-mediated transformation of *Cucurbitaceae* species, cucumber (*Cucumis sativus*) and squash (*Cucurbita pepo*), allows obtaining multiple transgenic roots within 4 weeks. These transgenic roots can contain different modifications such as promoter–reporter fusions, altered gene expression levels, and CRISPR/Cas9-edited genomes. Additionally, different selectable and visually screenable markers of transgenicity, as well as the potential cross-activation between promoters in vectors containing multiple expression cassettes, are described.

Keywords Plant transformation, *Agrobacterium rhizogenes*, Composite plants, Root, Cucumber, *Cucurbitaceae*

1 Introduction

To address a number of fundamental and applied scientific issues related to the root system, such as the molecular and hormonal mechanisms of root development, the interaction between the root system and pathogens or symbionts, plant mineral nutrition, and phytoremediation, it is necessary to have plants of interest with experimentally defined root features. Collections of mutants are accessible for a limited number of species, but reverse genetics offers a powerful approach for manipulations of plant genes, such as overexpression or inhibition of expression by RNA-interference, CRISPR/Cas9-mediated genome editing, and promoter–reporter fusions for

Steffen Vanneste and Inge Verstraeten (eds.), *Root Development*, Methods in Molecular Biology, vol. 3086,
https://doi.org/10.1007/978-1-0716-5603-7_1,

localization of gene expression. The main requirement in this case is an established transformation technique. Producing transgenic plants through *Agrobacterium tumefaciens*-mediated transformation is typically a lengthy, labor-intensive process with low efficiency (*A. tumefaciens* was revised as *Rhizobium radiobacter* but has reverted to its original name [1, 2]). The main problem in many cases is not the virulence of *A. tumefaciens* itself, but the regeneration of plants from calli, which seems to be a bottleneck. Moreover, it was determined that regeneration and transformation efficiency are two different genetically controlled factors [3]. These parameters vary greatly between different genotypes, so researchers are restricted to the use of those with established methods of *A. tumefaciens*-mediated transformation. Good examples are the "Desiree" cultivar of potato [4], although here, meanwhile, the number of transformable cultivars has increased [5], or the "Golden Promise" cultivar of barley [6], although here, anther culture methods promise increased applicability [7]. In spite of the recent advances for the majority of genotypes, *A. tumefaciens*-based transformation methods still have to be adapted.

To date, many species or specific cultivars remain recalcitrant both to Agrobacterium inoculation and to subsequent plant regeneration, despite modern approaches that include the use of morphogenesis regulators to facilitate transformant regeneration [8]. Another issue is the acquisition of new, undesirable traits in regenerated plants due to the phenomenon of somaclonal variation [9].

To overcome these problems, tissue culture-free protocols (*e.g.*, cut–dip–budding (CDB) delivery system; the regenerative activity–dependent in planta injection delivery (RAPID)) have been developed [10,11]; these avoid the callus induction step and are intended for direct transgenic plant regeneration. Currently, the application of such protocols is mainly restricted to species with root-suckering capability [10] or the ability to form tubers with high regenerative capabilities [11].

An alternative method for introducing new properties into the root system is the inoculation of plants with *Agrobacterium rhizogenes* (*Rhizobium rhizogenes*), which induces the development of agravio-tropic "hairy roots" due to the transfer of T-DNA from root-inducing plasmid (pRi) to the nuclear genome of plant cells [12]. To obtain transgenic roots with desired traits, a binary vector, harboring the gene of interest within its T-DNA region, is incorporated into agrobacterial cells. Co-transformed transgenic roots develop from the inoculation site if transfer of T-DNA from the binary vector occurs cooperatively with the transfer of T-DNA from pRi. Often, investigations are carried out on hairy root cultures induced on isolated explants [13]. Hairy root cultures show indefinite growth without added phytohormones and represent good expression systems for the production of secondary metabolites or plant proteins [14].

However, studies on plant development require a model that retains the regulatory connections between the root system and the shoot. Composite plants, which consist of a wild-type shoot and a semi-transgenic root system, meet this condition [15]. Composite plants can be obtained if the *A. rhizogenes* inoculation site, i.e., the wounding site, is located at the base of the seedling hypocotyl. So, co-transformed transgenic roots will develop a wild-type root system instead.

The problem of the hairy root phenotype and altered phytohormone levels [16] can be overcome by using moderate laboratory strains of *A. rhizogenes*. Previously, we demonstrated for squash that co-transformed transgenic roots induced by *A. rhizogenes* R1000 do not differ morphologically or in hormonal status from wild-type roots, as evidenced by their normal tissue organization and the distribution of cellular auxin response maxima [17]. As a rule, *A. rhizogenes* is characterized by a broader host specificity and increased virulence compared to *A. tumefaciens* [18]. Thus, generating composite plants may be accessible even for recalcitrant species. For example, a highly effective protocol exists for obtaining composite plants of pea [19], whereas *A. tumefaciens*-mediated transformation of pea remains genotype-dependent, and efficient methods have only begun to be developed recently [20, 21]. In addition, the time frame for producing composite plants is much shorter than the full cycle required to obtain whole transgenic plants [22].

In this chapter, we present a method developed for *Arabidopsis thaliana* and *Medicago truncatula* [23] and adapted for two species of the *Cucurbitaceae* family—squash (*Cucurbita pepo*) [17] and cucumber (*Cucumis sativus*) [24]. Co-transformed transgenic roots developing on composite plants do not differ from wild-type adventitious roots and are suitable for studying the physiological and molecular mechanisms of root development. We have demonstrated that cucurbit composite plants are useful tools for such studies in root development, such as analyzing the expression patterns of specific genes [24, 25], elucidating the role of plant hormones (e.g., auxin) [26], performing gene overexpression combined with subsequent proteomics assays [27], and implementing CRISPR/Cas9-mediated gene knock-out [28].

Since only co-transformed transgenic roots should be used for experiments, special attention should be paid to their selection. During the transformation process, a tumor-like or callus-like structure develops at the inoculation site. This structure consists of both co-transformed and non-transformed cells. Both types of cells contribute to root regeneration, resulting in the development of at least three types of roots: co-transformed transgenic roots (Fig. 1a), non-transformed wild-type roots (Fig. 1a), and chimeric roots that consist of both cell types (Fig. 1b). For this reason, the T-DNA region of the vector for plant transformation should contain a marker for transgenicity.

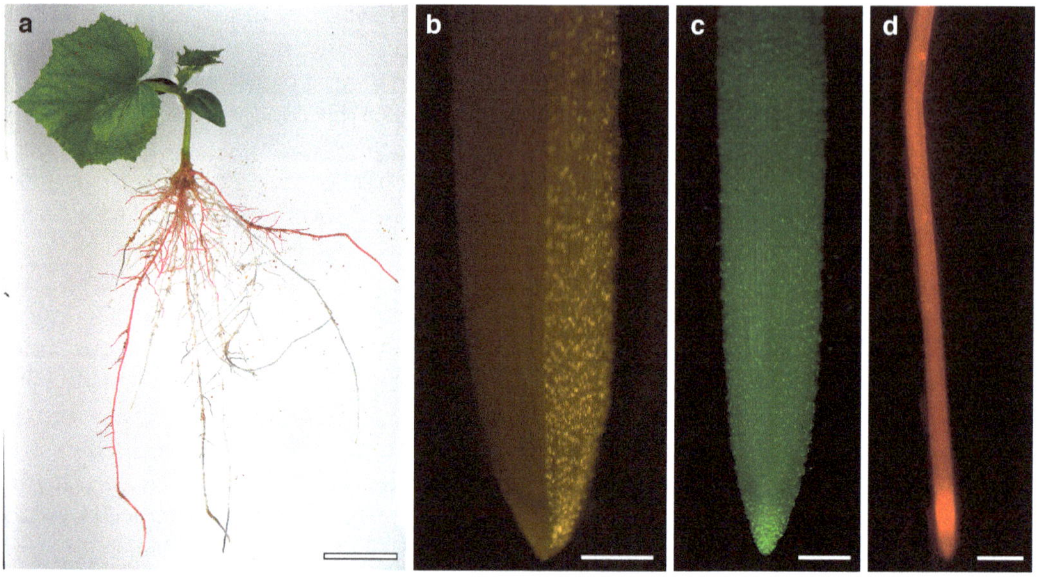

Fig. 1 Different types of screenable markers are used for the selection of co-transformed transgenic roots developed on cucurbit composite plants. (**a**) General view of cucumber composite plants. Co-transformed transgenic roots (purple) display the RUBY screenable marker; colorless (white) roots are either non-transgenic wild-type roots or roots with silenced T-DNA insertion, scale bar—2 cm. (**b**) Chimeric root of cucumber: half of the cells along the longitudinal axis express a *35S* CaMV*::mYongHong–H2B* fusion in non-fixed roots, which appears as a yellow protein. Red channel, scale bar—200 μm. (**c**) Co-transformed transgenic root of squash expressing the *pAtUBQ10::H2B–Venus* marker. Green channel, scale bar—200 μm. (**d**) Co-transformed transgenic root of cucumber expressing the *pAtUBQ10::DsRed1* marker. Red channel, scale bar—750 μm.

Selectable markers that confer tolerance to antibiotics or herbicides are not effective in tissue culture-free conditions, especially for roots. Screenable markers, which allow for the selection of transgenic roots by visual traits, can be divided into at least three groups.

The first group includes pigment-based markers that can be seen with the naked eye or with a simple stereomicroscope—for example, betalain [29, 30] or anthocyanin [31]. Synthesis of betalain by the RUBY marker was tested as a suitable screenable marker for cucumber composite plants (Fig. 1a). Nevertheless, such markers stain the roots and can interfere if examination of root sections is required. Betalain was previously demonstrated to posesess its own fluorescence with emission spectrum in the range 500–700 nm [32]. In addition, the application of pigment-based markers is inefficient if the hairy roots are naturally colored.

The second group of screenable markers includes genes encoding fluorescent proteins. Transgenic roots possessing such markers do not change color. However, a fluorescent stereomicroscope with a set of filters is needed for the visualization of fluorescent proteins. In addition, autofluorescence, normally displayed by plant tissues in the yellow–green part of the spectrum, can mask the fluorescent screening marker. The presence of compounds in plant tissues that cause autofluorescence is genotype-specific. For example, root tips of plants of the genus *Cucurbita* (pumpkin or squash) did not display autofluorescence in the green or red range of the spectrum (Fig. 2a–c), while root tips of several cultivars of cucumber showed strong green autofluorescence (Fig. 2d, f) and one of the tested cultivars showed red autofluorescence (Fig. 2e, g). Therefore, both green (e.g., eGFP, mNeonGreen) and red (e.g., tdTomato, mRuby3) fluorescent proteins can be used as screenable markers in plants of the genus *Cucurbita*. Moreover, fluorescent proteins fused with nuclear tags can mark nuclei instead of conventional dyes like DAPI while simultaneously serving as screenable markers (Fig. 1c) [33]. In contrast, only red (e.g., mRuby3) or orange (e.g., DsRed) fluorescent proteins can be used as screenable markers in roots of most cucumber cultivars (Fig. 1d). Green fluorescent proteins can be used in cucumber only as reporters for gene expression localization

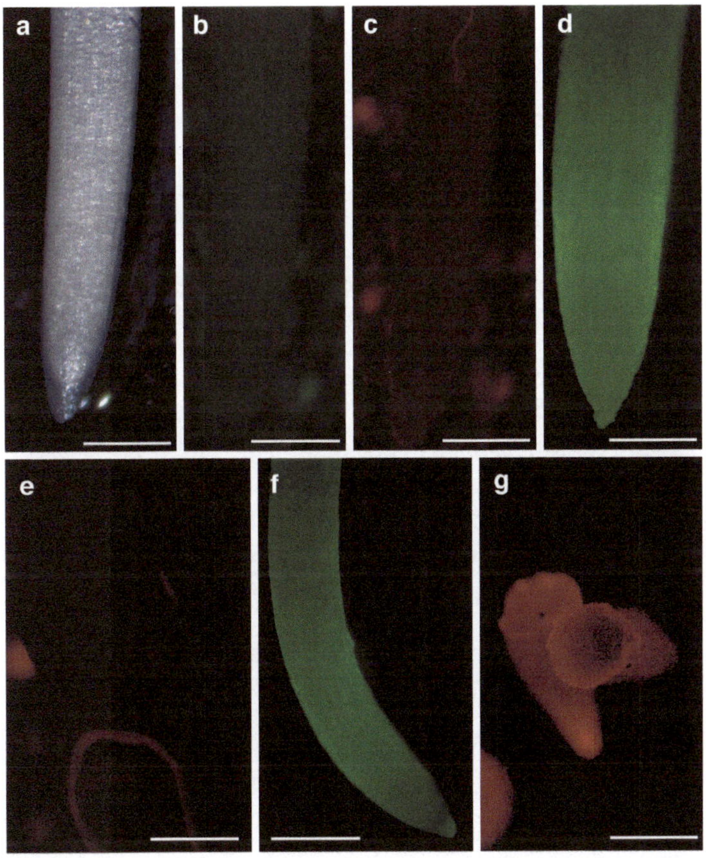

Fig. 2 Test of autofluorescence in cucurbit root tips. (**a–c**) Root tip of pumpkin (*Cucurbita maxima*), cultivar "Vitaminnaya." (**a**) Bright field. (**b**) Low autofluorescence, 38 filter set for GFP. (**c**) Low autofluorescence, 43 filter set for DsRed1. (**d, e**) Root tip of cucumber (*Cucumis sativus*), cultivar "Zasolochnyj." (**d**) Autofluorescence observed, 38 filter set. (**e**) Autofluorescence is low, 43 filter set. (**f–h**) Root tip of cucumber, cultivar "Parizhskij kornishon."(**f**) Autofluorescence observed, 38 filter set. (**h**) Autofluorescence observed, 43 filter set. Scale bar—500 μm.

experiments when using confocal laser scanning microscopy. A linear unmixing tool and an online fingerprinting algorithm allow for the separation of the fluorescent protein emission spectrum from the background autofluorescence [25,33].

The third group of screenable markers, based on enzyme activity, is mostly represented by the *gusA* (*uidA* gene, GUS marker) of *Escherichia coli*, which encodes β-glucuronidase [34]. Plants naturally lack β-glucuronidase, so any activity detected comes from the introduced *gusA* gene. β-Glucuronidase converts colorless X-Gluc to a bright blue insoluble substance easily detected by the naked eye. In some cases, staining for GUS-activity can be a limitation of the method, as it stops the living process in roots. In our studies, the synthetic auxin-sensitive promoter *DR5* [35] as a fusion with *gusA* (*DR5::gusA*) was used for analysis of auxin response maxima distribution in squash roots [26].

Special attention should be given to the matter of potential interaction between separate expression cassettes (i.e., promoter–ORF–terminator fusions) within a single T-DNA region of the vector for plant transformation. In our experience, the *35S* CaMV promoter is able to drive the expression of other reporter genes on the same T-DNA even if they lack any promoter of their own [17]. Therefore, it is recommended to carry out appropriate control experiments. Furthermore, the influence of the *DR5* promoter, which contains the minimal *35S* promoter [35], on the tissue-specific expression pattern of the cucumber orthologue of *SCARECROW* (*SCR*) gene of *Arabidopsis thaliana* (*CsSCR*) was examined (Fig. 3a–c); the expression of *CsSCR* in the quiescent center was suppressed when both constructs were present on the same T-DNA. In a similar vein, an impaired distribution of auxin response maxima was visualized by the *DR5::mRuby3–H2B* reporter fusion in roots simultaneously expressing the *pAtSCR::H2B–mNeonGreen* reporter fusion (Fig. 3d–f): *DR5* activity was detected in endodermal cells between lateral root primordia, while normally auxin response maxima are not present in this cell file [26]. In both cases, the two expression cassettes were adjacent to each other within the T-DNA of a binary vector.

The orientation of expression cassettes relative to each other (head-to-tail or tail-to-tail) on the same T-DNA is also relevant. For example, the activity of the *p35S::zCas9* cassette inhibited the expression of the adjacent screenable construct *pAtUBQ10::DsRed1*; specifically, fluorescence of DsRed1 was not detected if it was in tandem orientation relative to the *Cas9* cassette. In contrast, DsRed1 fluorescence was visible when the screenable cassette was in reverse orientation relative to the *Cas9* cassette [22].

In conclusion, composite plants are versatile and rapid tools for studying the hormonal and genetic mechanisms regulating various aspects of root development.

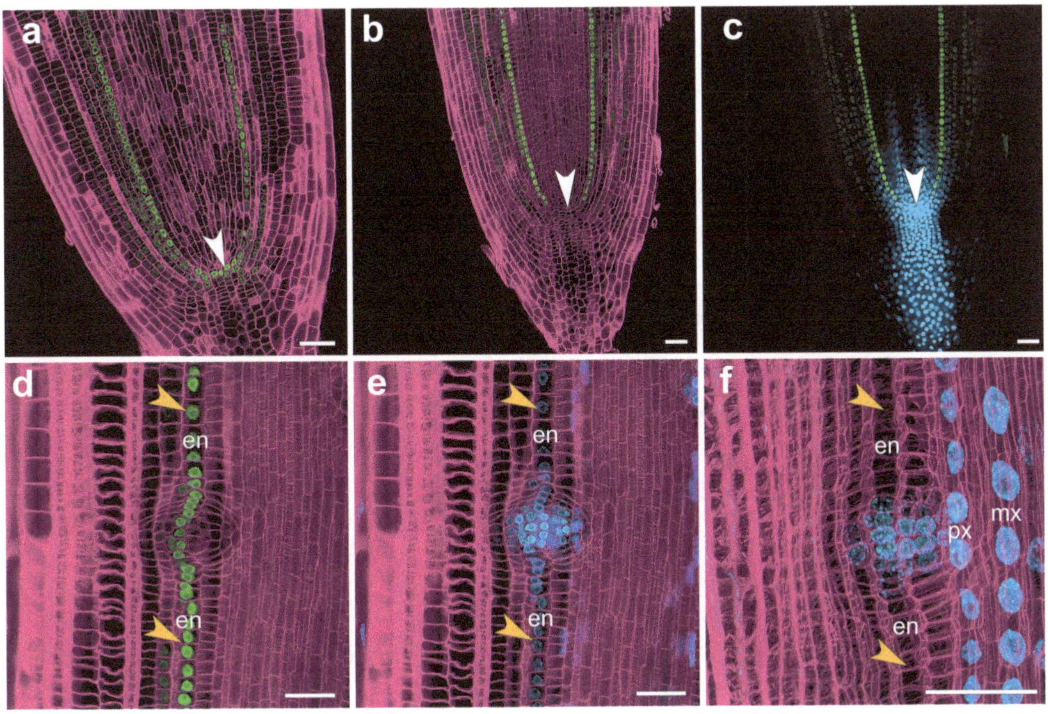

Fig. 3 Cross-activation between adjacent promoter–reporter fusions within the T-DNA of the binary vector. (**a–c**) Longitudinal section through the root tip. The white arrowhead indicates the quiescent center. (**a**) *CsSCR* expression pattern (*pCsSCR::mNeonGreen–H2B*) within the transgenic root tip: *CsSCR* is normally localized in the cells of the quiescent center, in the initials of the cortex and endodermis, in the cortex and endodermis cell files, and their derivatives. (**b**, **c**) *CsSCR* tissue expression pattern combined with the distribution of the cellular auxin response maxima (*pCsSCR::mNeonGreen–H2B*; *DR5::mRuby3–H2B*) within a transgenic root tip. Absence of *CsSCR* activity was observed in cells with a high level of cellular auxin response, for example, in cells of the quiescent center. In contrast, roots without *DR5::mRuby3–H2B* fusion exhibited *CsSCR* activity in quiescent center cells (**a**). (**d–f**) Endodermal cell files and lateral root primordia from the same area. Yellow arrowheads in (**d**) and (**e**) indicate endodermal cells exhibiting ectopic *DR5* promoter activity in roots co-expressing *pAtSCR::H2B–mNeonGreen* and *DR5::mRuby3–H2B* reporter fusions. (**f**) Transgenic root section expressing only *DR5::mRuby3–H2B* shows the normal auxin response pattern (*DR5* activity), i.e., absence of *DR5* activity in the endodermis (yellow arrowheads). Normally, maxima of cellular auxin response are observed in the root cap columella, the quiescent center, the metaxylem and protoxylem files, and in the lateral root primordia at all developmental stages. Green channel—fluorescence of the mNeonGreen protein in cell nuclei (*pCsSCR::mNeonGreen* and *pAtSCR::mNeonGreen*); cyan channel—auxin response maxima indicated by auxin-sensitive reporter fusion *DR5::mRuby3*; magenta channel—cell walls stained by SCRI Renaissance 2200. en endodermis, px protoxylem, mx metaxylem. Scale bars—50 μm.

2 Materials

All preparations are carried out at room temperature. Standard safety precautions must be followed when working with concentrated hydrogen peroxide and with a Bunsen burner during the sterilization of forceps and scalpels. Distilled water is sufficient for preparing all solutions and for rinsing the seeds after sterilization. Common consumables, such as Petri dishes, pipette tips, and other standard supplies, are not listed individually.

2.1 Surface Sterilization of Seeds

1. Cucumber or squash seeds of your choice (*see* **Note 1**).
2. Sterilization mixture: Mix one part of 33% hydrogen peroxide, two parts of 96% ethanol, and one part of water. For example, to prepare 40 mL of sterilization mixture, add 10 mL of hydrogen peroxide to 10 mL of water, mix, and then add 20 mL of ethanol (*see* **Notes 2 and 3**).
3. Laundry soap.
4. Plastic container.
5. Sterile Magenta™ vessel V8505 with sterile vermiculite for seed germination (*see* **Notes 4 and 5**). To prepare vermiculite for seed germination, pour dry, fine-particle vermiculite (0.3–1.5 mm) from the package into an appropriate plastic bowl and soak it in tap water at least overnight. Subsequently, remove excess moisture using a sieve and nonwoven material. Note that due to settling, the required dry volume of vermiculite is approximately 150% of the final wet volume needed. Fill Magenta™ vessels or other appropriate containers halfway with the moistened vermiculite, seal with lids, and autoclave for 30 min at 121°C (solid cycle).
6. Three sterile beakers for rinsing the seeds (*see* **Note 6**).
7. Sterile water (*see* **Note** 7).
8. A small sterile sieve to catch sterilized seeds.
9. Sterile paper towels for blot-drying of the sieve.
10. Sterile blotting paper for blot-drying of sterilized seeds.
11. Sterile spoon-like spatulas.
12. Sterile glass rods for stirring the seeds in beakers during rinsing.
13. Growth chamber: 26°C, darkness, 70% humidity.

2.2 Inoculation of Seedlings

1. Etiolated cucumber or squash seedlings of the required developmental stage, as detailed in Sect. 3.
2. YEB medium agar plates (*see* **Note** 7): weigh 5 g of beef extract, 1 g of yeast extract, 1 g of peptone, 5 g of sucrose, 1.2 g $MgSO_4 \times 7H_2O$, and dissolve in 900 mL of water. Adjust the pH of the solution to 7.2 ± 0.2 with KOH or HCl. Add 15 g of agar and fill up with water to 1 L. Autoclave for 20 min at 121°C (liquid cycle). After partial cooling, add filter-sterilized selective antibiotics and pour plates.
3. YEB plate with Agrobacteria of the chosen strain containing the appropriate binary vector (*see* **Notes 9 and 10**). Plate

Agrobacteria from glycerol stock on a YEB agar plate with the appropriate antibiotics and incubate at 28°C for approximately 24–48 h. The bacteria should cover the plate (*see* **Notes 11 and 12**).

4. Half-strength Murashige and Skoog medium (0.5× MS) [36] with agar: Weigh 2.2 g of MS salts and 10 g of sucrose and dissolve in 950 mL of water. Adjust pH to 5.6–5.8 with KOH or HCl. Add 8 g of agar and fill up with water to 1 L. Autoclave for 20 min at 121°C (liquid cycle) (*see* **Note 13**).
5. 0.5× MS medium slopes in square Petri dishes: in the laminar flow hood, put the lids of 10 × 10 cm or 12 × 12 cm Petri dishes on the sterile work surface (previously wiped with 70% ethanol) upside up and place on them one end of the bottoms of the Petri dishes, the other end on the work surface, so that they stand at an angle. Pour melted 0.5× MS medium with agar until the Petri dish is approximately half-filled (about 35–40 mL of MS medium per Petri dish) to generate the 0.5× MS medium slope, and allow to solidify (*see* **Note 14**).
6. Sterile blotting paper: cut blotting paper into rectangles matching half the size of square Petri dishes (length × 50% width) to cover the 0.5× MS medium slopes, and autoclave for 20 min at 121°C (solid cycle).
7. Growth chamber: 21°C, 16/8 light/dark at a light intensity of 600 μmol/m^2/s^1, 85% humidity.

2.3 Elimination of Agrobacteria

1. Three sterile beakers for every two Petri dishes with transformants.
2. Sterile water: 2 L for every two Petri dishes containing explants transformed with the same strain (vector).
3. Silver nitrate ($AgNO_3$): 5% (w/v) solution in water. Store at 4°C in a tube wrapped in aluminum foil (*see* **Note 15**).
4. Cefotaxime stock solution (250 mg/mL): dissolve 5 g of cefotaxime powder in 20 mL of water, filter-sterilize using a 0.22 μ m filter. Prepare aliquots and store at −20°C (*see* **Note 16**).
5. 0.5× MS medium with agar, supplemented with cefotaxime (500 μg/mL), in slopes in square Petri dishes.
6. Sterile blotting paper.
7. Sterile brown antiseptic paper from germination pouches: cut into rectangles matching the size of the agar slopes, and autoclave for 20 min at 121°C (solid cycle).
8. Sterile glass rods.
9. Growth chamber: 25°C, 16/8 light/dark, at a light intensity of 600 μmol/m^2/s^1, 70% humidity.

2.4 Growing of Composite Plants

1. Vermiculite of fine-particle size (0.3–1.5 mm).
2. Plastic vessels or nursery pots.
3. Macronutrient stock solutions: Prepare 1 M solutions of: KH_2PO_4; KNO_3; $Ca(NO_3)_2$; and $MgSO_4$. Autoclave for 20 min at 121°C (liquid cycle), store at room temperature (*see* **Note 17**).
4. Micronutrients stock solution: Weigh 2.86 g of H_3BO_3, 1.81 g of $MnCl_2 \times 4H_2O$, 0.22 g of $ZnSO_4 \times 7H_2O$, 0.08 g of $CuSO_4 \times 5H_2O$, and 0.025 g of $Na_2MoO_4 \times 2H_2O$. Dissolve in 900 mL of water and fill up with water to 1 L. Do not autoclave. Prepare 50 mL aliquots and keep one tube at 4°C for short-term storage and the other aliquots at −20°C for prolonged storage.
5. 0.1 M Iron stock solution: Weigh 5.56 g of $FeSO_4 \times 7H_2O$, and 7.45 g of Na_2EDTA and dissolve them separately in 450 mL of hot water (approximately 60°C). Pour both solutions together and fill up with water to 1 L. A yellow solution is obtained. Prepare 100 mL aliquots in amber bottles to protect from light. Autoclave for 20 min at 121°C (liquid cycle), store at room temperature.
6. 4× Hoagland's solution [37]: Add the following volumes of the different macronutrient stock solutions to 900 mL of distilled water: 4 mL of 1 M KH_2PO_4, 20 mL of 1 M KNO_3, 20 mL of 1 M $Ca(NO_3)_2$, 8 mL of 1 M $MgSO_4$. Add 4 mL of micronutrients stock solution and 4 mL of 0.1 M iron stock solution. Fill up with water to 1 L. Do not autoclave. Best to prepare the solution prior to use, but it is possible to store it at 4°C up to 3 weeks.

2.5 Other Tools

1. Forceps.
2. Cell spreaders or Drigalski spatula.
3. Scalpels.
4. Adhesive tape for sealing of Petri dishes (*see* **Note 18**).
5. Razor blades.

3 Methods

3.1 Presumptive Experimental Schedule

The process normally takes 12 days from seed sterilization to the transfer of transformants to non-sterile conditions, and 21 days until the first examination of transgenic roots. This schedule should be adjusted based on the chosen seed germination and seedling growth rate, and the cultivation time required for the *A. rhizogenes* strain used.

First week: Monday: surface sterilize seeds. Tuesday: Plate agrobacteria from glycerol stock (strain of choice containing the appropriate vector). Wednesday, Thursday, or Friday: Plate agrobacteria for inoculation. Thursday, Friday, or Saturday: Inoculate the seedlings.

Second week: Monday, Tuesday, or Wednesday: Eliminate agrobacteria. Friday: Transfer transformants to non-sterile conditions.

Third week: Assess co-transformed transgenic roots. Chimeric roots typically appear during this stage.

Fourth week: Harvesting of co-transformed transgenic roots: Begin harvesting and subsequent examination of transgenic roots. Up to seven to eight harvests of co-transformed transgenic roots may be obtained at approximately 7-day intervals.

3.2 Surface Sterilization of Seeds

1. Prepare the laminar flow hood according to the manufacturer's instructions. Sterilize forceps using a Bunsen burner, infrared micro-incinerator, or electric sterilizer.
2. Place the required number (*see* **Notes 19 and 20**) of squash or cucumber seeds into 50-mL Falcon polypropylene tubes, using no more than 20 mL of seeds per tube.
3. Transfer the seeds to a plastic container. Wash the seeds with laundry soap and tap water (*see* **Note 21**). Transfer seeds back to the polypropylene tubes (*see* **Note 22**).
4. Prepare a sterilization mixture: 30 mL for each tube with seeds.
5. Under the laminar flow hood, pour sterile water into each beaker.
6. Under the laminar flow hood, unpack the sieve and place three to four layers of sterile paper towels in a Petri dish or on sterile aluminum foil.
7. Pour the sterilization mixture into the tubes containing the seeds (*see* **Note 23**). Close the tubes and incubate on a rotary shaker for 8 min at 100 rpm. The seeds must be continuously agitated in the sterilization mixture. Alternatively, the tubes can be shaken by hand. This stage does not require a laminar flow hood.
8. Under the laminar flow hood, open the tubes with seeds and drain the sterilization mixture through a sieve. Touch the bottom of the sieve to the sterile paper towels to absorb the excess mixture, and transfer the seeds from the sieve into the first beaker of sterile water. Stir the seeds with a glass rod for 2–3 min.
9. Drain the water through the sieve and transfer the seeds from the sieve into the second beaker with sterile water. Rinse for 1–2 min, and then repeat procedure with the third beaker. Finally, transfer the seeds to a Petri dish lined with sterile blotting paper (*see* **Note 24**).

10. Sow the seeds into Magenta vessels with vermiculite (*see* **Note 25**). Use 30 cucumber seeds or 16 squash seeds per vessel.
11. Incubate at 26°C in the dark for germination (*see* **Note 26**). Cucumber seedlings with the required developmental stage will be ready in 3–4 days, depending on the cultivar, while squash seedlings will be ready in 4–5 days (*see* **Note 27**).

3.3 Inoculation of Seedlings by Agrobacteria

1. Prepare the laminar flow hood according to the manufacturer's instructions. Sterilize the forceps and scalpels (and Drigalski spatula, if you are using one) using a Bunsen burner, infrared micro-incinerator, or electric sterilizer.
2. Under laminar flow, wipe one to two vessels with seedlings with 70% ethanol and examine seedlings for the required developmental stage: hypocotyl length approximately 1.5–2 cm, any root length, and cotyledons that are folded or still under the seed coat (Figs. 4a and 5). Open the vessel lid and transfer 3 squash or 5 cucumber seedlings of the required developmental stage to a sterile round Petri dish; close the dish to prevent drying.
3. Take the YEB plate containing the agrobacteria (Fig. 4b) and disinfect the edges. Using a sterile cell spreader or Drigalski spatula, collect agrobacterial biomass from the agar surface into a bulk in the center of the plate.
4. Place one seedling on the lid of a new Petri dish using sterile forceps.
5. Cut off the root at the root–hypocotyl junction (Figs. 4c and 5), dip the cut site in agrobacterial paste.
6. Additionally, use a sterile scalpel dipped in the paste to make a 3–5 mm vertical incision at the base of the hypocotyl (*see* **Note 28**). Repeat this procedure with each seedling. In this protocol, the inoculated shoot parts of seedlings are referred to as apical explants (Fig. 4c).
7. Using sterile forceps, create vertical slots in the 0.5× MS medium slopes. Place the inoculated apical explant into these slots.
8. Repeat the inoculation procedure with the remaining seedlings. Each square Petri dish can accommodate up to six squash or ten cucumber apical explants.
9. Once all explants are positioned, verify that the agrobacterial paste on the cut of the hypocotyl is not smeared or removed during insertion.
10. Cover the apical explants mounted into 0.5× MS medium slopes with a sterile rectangle of blotting paper; this will secure the explants and prevent them from shifting (Fig. 4d).

Fig. 4 Different stages of cucurbit transformation. (**a**) Germination of cucumber seeds in sterile vermiculite, scale bar—1 cm. (**b**) Petri dish with agrobacterial lawn, scale bar—2 cm. (**c**) Apical explant prepared from a cucumber seedling, scale bar—0.5 cm. (**d**) Inoculation stage: apical explants of squash seedlings inoculated by agrobacteria on 0.5× MS medium slopes, scale bar—0.75 cm. (**e**) Elimination of agrobacteria: squash transformants on 0.5× MS medium slopes, scale bar—1 cm. (**f**) Transfer of rooted transformants to non-sterile conditions: cucumber plantlets in vermiculite, scale bar—2 cm. (**g**) Visual screening of co-transformed transgenic roots based on T-DNA marker (*pAtUBQ10::DsRed1*): cucumber roots display DsRed1 fluorescence, scale bar—0.5 cm. This stage is followed by further examination of co-transformed transgenic roots, for example, preparation of sections and slides, real-time qPCR, and genotyping

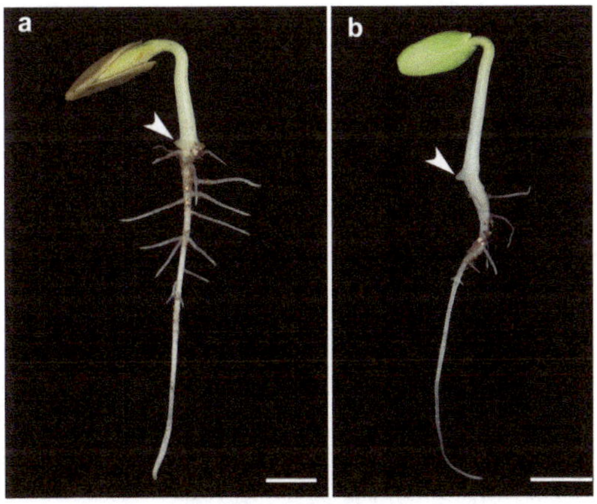

Fig. 5 General view of seedlings on the required developmental stage for inoculation. (**a**) Five-day-old squash seedling, scale bar—1 cm. (**b**) Four-day-old cucumber seedling, scale bar—1 cm. White arrowhead—peg, marking root–hypocotyl junction: the most suitable site for inoculation

11. Seal the Petri dishes with adhesive tape and place them in a vertical position (an appropriate holder may be used).
12. Cultivate the inoculated apical explants for 5 days at 20–21°C (*see* **Notes 29 and 30**) under a 16/8 h light/dark cycle at a light intensity of 600 μmol/m^2/s^1.

3.4 Eliminating Agrobacteria

1. Prepare the laminar flow hood according to the manufacturer's instructions. Sterilize the forceps using a Bunsen burner, infrared micro-incinerator, or electric sterilizer.
2. Under laminar flow, add 100 μL of a 5% silver nitrate ($AgNO_3$) solution per 1 L of sterile water, and distribute it (2 L of water) over three sterile beakers.
3. Remove the inoculated explants from the 0.5× MS medium slopes with sterile forceps, place them in the first beaker of water, and stir with a sterile glass rod. Rinse the explants for 1–2 min. Repeat the procedure with the two remaining beakers. Plates containing fungal contamination should be excluded from the experiment (*see* **Note 31**).
4. Blot-dry the transformants using sterile blotting paper.
5. Place the transformants onto 0.5× MS agar slopes supplemented with cefotaxime and cover them with rectangles of sterile brown antiseptic paper (Fig. 4e).
6. Seal all dishes with adhesive tape and place them in a vertical position (an appropriate holder may be used).
7. Cultivate transformants at 25°C under a 16/8 h light/dark, at a light intensity of 600 μmol/m^2/s^1, until roots develop (approximately 2–3 days) (*see* **Note 32**).

3.5 Cultivation of Composite Plants

1. Prepare the required number of sterile plastic vessels, filled with processed (Sect. 2.1.5) sterile vermiculite (*see* **Note 33**).
2. Moisten the vermiculite with 4× Hoagland's medium: it has to be dampish but not waterlogged.
3. Remove the transformed explants from the Petri dishes and rinse away any traces of agar with lukewarm water (about 25–30°C).
4. Using a razor blade, trim long or excessive roots from the transformants. Plant them into the vessels containing vermiculite and cover the vessels with their lids to maintain high humidity. After several days of cultivation and development of new roots, the lids can be removed. Fertilize plants with 4× Hoagland's solution as needed (Fig. 4f).
5. Cultivate the plants at 25°C under a light cycle of 16/8 h, at a light intensity of 600 μmol/m^2/s^1, and 85% humidity for 7–10 days.

6. Harvest the roots that have developed on the composite plants, using a razor blade (*see* **Note 34**). Collect them in an appropriate container filled with room temperature water, rinse off any vermiculite particles, and examine them according to the purpose of the experiment using the appropriate labware (Fig. 4g). A detailed description of root fixation, section preparation, and microscopy is given in a methodical paper [33] if further examination of root structure is needed.
7. Prepare the required number of fresh vessels with vermiculite, moistened with 4× Hoagland's solution, for each subsequent harvest, to transfer the explants to a fresh environment (*see* **Note 33**).
8. Harvest roots at approximately 7-day intervals (*see* **Note 35**). Up to eight harvests of co-transformed transgenic roots can be obtained.

4 Notes

1. The method presented here was tested on the squash cultivars "Beloplodniy" and "Konkurent," the cucumber cultivars "Kustovoy" and "Phoenix." Despite the fact that the host specificity of *A. rhizogenes* is broader than that of *A. tumefaciens* [18], the transformation efficiency can vary between different cultivars of cucumber or squash. A pilot experiment should be carried out for each new cultivar. We suppose that this method will be effective for other cucurbit species or cultivars.
2. Sterile water is not obligatory in this mixture; distilled water is sufficient. Wear gloves when working with hydrogen peroxide and the sterilization mixture.
3. Sterilization solutions based on sodium hypochlorite-containing bleach can also be used.
4. Magenta™ vessel is an autoclavable laboratory vessel consisting of a transparent polycarbonate body and polypropylene closure; dimensions (W×L×H) are as follows: 77 mm×77 mm×97 mm.
5. To estimate the required number of Magenta™ vessels: about 30 cucumber seeds or 16 squash seeds can be sown per vessel with vermiculite.
6. Seeds should be rinsed with an excess volume of water. It is convenient to use beakers with capacities of 500, 800, and 1000 mL. Beakers of this combination of volumes can be nested inside each other, which saves space in the autoclave.

7. The amount of water required depends on the quantity of seeds for sterilization. 1.5 L is sufficient to rinse up to 20 mL of seeds (measured using the calibration of the 50-mL Falcon tube). It is necessary to increase the volume of water proportionally if a larger volume of seeds is to be sterilized. In such cases, an additional set of beakers will be required as well.
8. YEB is a rich medium that supports robust growth of agrobacteria. However, other suitable media, such as YEP or LB, can be used if adjustments have to be made for the growth rate of the chosen strain.
9. The *A. rhizogenes* R1000 strain was shown to be the most virulent for squash plants: 85% of the resulting composite plants developed at least one transgenic root, with an average of six transgenic roots per plant [17]. The R1000 strain was also very virulent for cucumber seedlings, resulting in a similar percentage of composite plants [24]. The R1000 strain itself does not display any antibiotic resistance, making it compatible with any vector for plant transformation. R1000 is a transconjugant strain that was obtained by conjugation of the pRiA4b plasmid from *A. rhizogenes* A4T to *A. tumefaciens* C58 [38, 39].
   Another strain, MSU440, tested for transformation efficiency on squash seedlings, showed moderate results in comparison with R1000: only 33% of the squash seedlings transformed using this strain developed co-transformed transgenic roots, averaging three roots per plant [17]. MSU440 was also able to induce co-transformed transgenic roots in cucumber, but the efficiency was lower than in R1000-inoculation tests (unpublished data). MSU440 contains the pRiA4 plasmid and, like R1000, does not display any antibiotic resistance [40].
   Also, A4RS (an A4 derivative, carrying chromosomal markers for rifampicin and spectinomycin resistance) [41] and ARqua1 (a derivative of R1000 carrying a chromosomal marker for streptomycin resistance [42]) were shown to be virulent for squash cultivar “Kuand” in preliminary transformation experiments *in vivo* (data not published).
10. Each plate (9 cm in diameter) with a lawn of agrobacteria is sufficient to inoculate at least 12 squash or 20 cucumber seedlings.
11. It is very important to maintain the optimal temperature for the cultivation of agrobacteria: 28–30°C. At 37°C, agrobacteria cells lose their Ri plasmids [43]. This process, known as “plasmid curing,” is used to obtain new strains of agrobacteria.
12. The cultivation time for *A. rhizogenes* R1000 on YEB–agar plates is approximately 24 h. However, this can vary for other strains. It is recommended not to use over-aged bacterial

lawns, as the age of the bacterial culture has an impact on transformation efficiency [44]. The optimal stage is when the bacterial lawn has just formed and is sufficiently dense, but not overgrown (Fig. 4b).

13. 0.5× MS medium is preferred over full-strength MS medium formulation because high mineral content in the latter can slightly inhibit root induction. 0.5× MS medium formulation is sufficient to supply inoculated seedlings for 10 days with all necessary nutrients.
14. Each square Petri dish of such size contains up to six squash or nine cucumber apical explants (i.e., the shoot parts of seedlings without the primary root).
15. Since silver nitrate acts as an antiseptic [45], it can be prepared under non-sterile conditions.
16. Alternatively, carbenicillin or timentin can be used to inhibit the growth of agrobacteria.
17. As preparation of Hoagland's solution takes place in non-sterile conditions, always check solutions before use for bacterial or fungal contamination.
18. To ensure gas exchange, avoid sealing the dishes with Parafilm®; instead, use a more breathable material, such as 3 M Micropore™ surgical tape.
19. Estimating the required number of seeds: despite the high germination rate of cucumber and squash in vermiculite (about 95%), seedlings will be at different developmental stages. Usually, only five seedlings, suitable for inoculation, can be selected from one Magenta™ vessel containing cucumber seeds (30 seeds total) or two vessels with squash seeds (32 seeds total). For convenience, instead of counting, seeds can be weighed using the weight of 50 seeds as a reference.
20. To improve germination efficiency, it is recommended to warm up the dry seeds at 53–55°C for 2–4 h, prior to the sterilization procedure.
21. This stage is optional, but it is strongly recommended to remove surface contamination.
22. It is easier to estimate the volume of dry seeds than wet.
23. If there is more than one tube with seeds, it is better not to fill all tubes with the sterilization mixture at the same time, but rather at intervals of at least 1–2 min. This helps to avoid excessive sterilization of seeds, which would reduce the germination rate.
24. Certain seeds may be more severely contaminated than others by bacteria or fungi, which can be located under the seed coat. We do not peel seeds during sterilization because our

transformation method involves a large volume of seeds, and it is time-consuming and labor-intensive. If contamination is found during one of the *in vitro* stages, a fungicide or bactericide treatment should be included following the initial sterilization. Alternatively, surface-sterilized seeds can be germinated on slopes of water agar or MS agar supplemented with sucrose to detect contamination early.

25. For optimal results, push seeds into the vermiculite with the pointed end (the site of the micropyle) facing downwards. This orientation allows obtaining seedlings without curvature of the hypocotyl.

26. Germinate seedlings in the dark, as etiolated seedlings are more susceptible to inoculation by agrobacteria.

27. It is recommended to conduct a preliminary germination test to assess seed viability and determine the seed germination rate.

28. The choice of the correct inoculation site is the most crucial step in the entire experiment. We have demonstrated that proliferating cells capable of propagating T-DNA insertion are located at the base of the hypocotyl (hypocotyl–root junction) [17]. Specialized protuberances known as pegs, which develop on cucurbit seedlings, can serve as morphological markers of this region (Fig. 5). The biological function of the pegs is to assist the seedlings in emerging from the seed coat and anchor the young plant in the soil during root development. The primary root should be cut approximately at the level of the peg. Shifting the inoculation site (the cut site) to the middle of the hypocotyl will result in the development of non-transgenic adventitious roots and a drastic decrease in transformation efficiency.

29. A lower temperature (21°C) was shown to be preferable to 25°C for the transfer of T-DNA from agrobacteria to plant cells [23].

30. After the co-cultivation period, the base of the hypocotyl will swell slightly in squash explants due to intensive cell proliferation. The base of the hypocotyl in cucumber explants will remain unaffected, but it does not mean that the transformation has failed.

31. Bacterial overgrowth may occur. Most likely, it will not influence the further stages of the experiment, but a repetition of the procedure for the elimination of agrobacteria can be required. This includes rinsing the transformants with sterile water with silver nitrate as an antiseptic, cultivation on MS agar supplemented with the antibiotic cefotaxime to inhibit the growth of agrobacteria, and additional application of antiseptic paper to cover the MS agar slopes.

32. Once the first roots develop, the transformants are ready to be transferred to non-sterile conditions, for example, to vessels with vermiculite. Alternatively, aeroponics [46] or other systems [47] can be used for the further cultivation of transformants.
33. A dry-heat sterilization oven can be used for sterilizing vermiculite as an alternative to the autoclave. Vermiculite, which had been used for plant cultivation, can be washed properly with tap water, squeezed out using non-woven fabric, and dried in an oven for recycling.
34. It is important to cut off roots at the point of attachment to the plants. Cutting roots at other sites can prevent the development of new roots or lead to necrosis of the remaining tissue. Due to the robust regeneration of adventitious roots in cucurbits, it is optional to leave one to two roots in order to supply the plant with water and nutrients, as this can be necessary for cultivars/species with limited regenerative capacity.
35. It is important to maintain a strict schedule for root harvesting, as cucurbits develop robust root systems. If a harvest is missed, the roots can fill the entire vermiculite volume within the vessel, making it difficult to harvest them later.

Acknowledgments

This work was supported by an institutional research project of the Komarov Botanical Institute RAS (No. 124020100138-4), using the equipment of the Core Facilities Center "Cell and Molecular Technologies in Plant Science" at the Komarov Botanical Institute RAS (St. Petersburg, Russia).

Competing Interests The authors have no conflicts of interest to declare that are relevant to the content of this chapter.

References

1. Arahal DR, Bull CT, Busse H-J et al (2023) Judicial opinions 123–127. Int J Syst Evol Microbiol 72:005708. https://doi.org/10.1099/ijsem.0.005708
2. Young JM, Kuykendall LD, Martínez-Romero E et al (2001) A revision of Rhizobium Frank 1889, with an emended description of the genus, and the inclusion of all species of Agrobacterium Conn 1942 and *Allorhizobium undicola* de Lajudie et al 1998 as new combinations: *Rhizobium radiobacter*, *R. rhizogenes*, *R. rubi*, *R. undicola* and *R. vitis*. Int J Syst Evol Microbiol 51:89–103 https://doi.org/10.1099/00207713-51-1-89
3. Heeres P, Schippers-Rozenboom M, Jacobsen E et al (2002) Transformation of a large number of potato varieties: genotype-dependent variation in efficiency and somaclonal variability. Euphytica 124:13–22. https://doi.org/10.1023/A:1015689112703
4. Jung CS, Griffiths HM, De Jong DM et al (2005) The potato *P* locus codes for flavonoid 3′,5′-hydroxylase. Theor Appl Genet 110:269–275. https://doi.org/10.1007/s00122-004-1829-z

5. Nahirñak V, Almasia NI, González MN et al (2022) State of the art of genetic engineering in potato: from the first report to its future potential. Front Plant Sci 12:768233. https://doi.org/10.3389/fpls.2021.768233
6. Bartlett JG, Alves SC, Smedley M et al (2008) High-throughput *Agrobacterium*-mediated barley transformation. Plant Methods 4: 22. https://doi.org/10.1186/1746-4811-4-22
7. Han Y, Broughton S, Liu L et al (2020) Highly efficient and genotype-independent barley gene editing based on anther culture. Plant Commun 2:100082. https://doi.org/10.1016/j.xplc.2020.100082
8. Nalapalli S, Tunc-Ozdemir M, Sun Y et al (2021) Morphogenic regulators and their application in improving plant transformation. Methods Mol Biol 2238:37–61. https://doi.org/10.1007/978-1-0716-1068-8_3
9. Leva A, Rinaldi LMR (2017) Somaclonal variation. In: Thomas B, Murray BG, Murphy DJ (Eds.)Eds. Encyclopedia of applied plant sciences. Second. Academic Press, Oxford, 468–473pp
10. Cao X, Xie H, Song M et al (2023) Cut–dip–budding delivery system enables genetic modifications in plants without tissue culture. Innov 4:100345. https://doi.org/10.1016/j.xinn.2022.100345
11. Mei G, Chen A, Wang Y et al (2024) A simple and efficient in planta transformation method based on the active regeneration capacity of plants. Plant Commun 5:100822. https://doi.org/10.1016/j.xplc.2024.100822
12. Chilton M-D, Tepfer DA, Petit A et al (1982) *Agrobacterium rhizogenes* inserts T-DNA into the genomes of the host plant root cells. Nature 295:432–434. https://doi.org/10.1038/295432a0
13. Gutierrez-Valdes N, Häkkinen ST, Lemasson C et al (2020) Hairy root cultures—a versatile tool with multiple applications. Front Plant Sci 11:33. https://doi.org/10.3389/fpls.2020.00033
14. Awere CO, Rakkammal K, Mwaura MM et al (2024) Hairy-root technology: a metabolic engineering tool and specialized metabolite pathway elucidation and production of secondary metabolites. Results Eng 23:102697. https://doi.org/10.1016/j.rineng.2024.102697
15. Taylor CG, Fuchs B, Collier R et al (2006) Generation of composite plants using *Agrobacterium rhizogenes*. Methods Mol Biol 343:155–167. https://doi.org/10.1385/1-59745-130-4:155
16. Shanks JV, Morgan J (1999) Plant "hairy root" culture. Curr Opin Biotechnol 10:151–155. https://doi.org/10.1016/s0958-1669(99)80026-3
17. Ilina EL, Logachov AA, Laplaze L et al (2012) Composite *Cucurbita pepo* plants with transgenic roots as a tool to study root development. Ann Bot 110:479–489. https://doi.org/10.1093/aob/mcs086
18. Christey MC (2001) Use of Ri-mediated transformation for production of transgenic plants. In Vitro Cell Dev Biol Plant 37:687–700. https://doi.org/10.1007/s11627-001-0120-0
19. Leppyanen IV, Kirienko AN, Dolgikh EA (2019) *Agrobacterium rhizogenes*-mediated transformation of *Pisum sativum* L. roots as a tool for studying the mycorrhizal and root nodule symbioses. PeerJ 7:e6552. https://doi.org/10.7717/peerj.6552
20. Kaur R, Donoso T, Scheske C et al (2022) Highly efficient and reproducible genetic transformation in pea for targeted trait improvement. ACS Agric Sci Technol 2:780–787. https://doi.org/10.1021/acsagscitech.2c00084
21. Li G, Liu R, Xu R et al (2023) Development of an Agrobacterium-mediated CRISPR/Cas9 system in pea (*Pisum sativum* L.). Crop J 11:132–139. https://doi.org/10.1016/j.cj.2022.04.011
22. Kiryushkin AS, Ilina EL, Guseva ED et al (2022) Hairy CRISPR: genome editing in plants using hairy root transformation. Plants 11:51. https://doi.org/10.3390/plants11010051
23. Limpens E, Ramos J, Franken C et al (2004) RNA interference in *Agrobacterium rhizogenes*-transformed roots of *Arabidopsis* and *Medicago truncatula*. J Exp Bot 55:983–992. https://doi.org/10.1093/jxb/erh122
24. Kiryushkin AS, Ilina EL, Puchkova VA et al (2019) Lateral root initiation in the parental root meristem of cucurbits: old players in a new position. Front Plant Sci 10:365. https://doi.org/10.3389/fpls.2019.00365
25. Kiryushkin AS, Ilina EL, Guseva ED et al (2023) Lateral root initiation in cucumber (*Cucumis sativus*): what does the expression pattern of *Rapid Alkalinization Factor 34* (*RALF34*) tell us?. Int J Mol Sci 24:8440. https://doi.org/10.3390/ijms24098440
26. Ilina EL, Kiryushkin AS, Semenova VA et al (2018) Lateral root initiation and formation within the parental root meristem of *Cucurbita pepo*: is auxin a key player?. Ann Bot 122:873–888. https://doi.org/10.1093/aob/mcy052
27. Shumilina J, Kiryushkin AS, Frolova N et al (2023) Integrative proteomics and metabolomics analysis reveals the role of small signaling peptide Rapid Alkalinization Factor 34 (RALF34) in cucumber roots. Int J Mol Sci 24:7654. https://doi.org/10.3390/ijms24087654
28. Kiryushkin AS, Ilina EL, Kiikova TY et al (2024) Do DEEPER ROOTING 1 homologs

regulate the lateral root slope angle in Cucumber (*Cucumis sativus*)?. Int J Mol Sci 25:1975. https://doi.org/10.3390/ijms25041975
29. He Y, Zhang T, Sun H et al (2020) A reporter for noninvasively monitoring gene expression and plant transformation. Hortic Res 7:152. https://doi.org/10.1038/s41438-020-00390-1
30. Zhao Lab (2020) The RUBY reporter. https://zhaolab.biosci.ucsd.edu/ruby/
31. Fan Y, Wang X, Li H et al (2020) Anthocyanin, a novel and user-friendly reporter for convenient, non-destructive, low cost, directly visual selection of transgenic hairy roots in the study of rhizobia-legume symbiosis. Plant Methods 16: 94. https://doi.org/10.1186/s13007-020-00638-w
32. Scheuring D, Minina EA, Krueger F et al (2024) Light at the end of the tunnel: FRAP assays combined with super resolution microscopy confirm the presence of a tubular vacuole network in meristematic plant cells. Plant Cell 36:4683–4691. https://doi.org/10.1093/plcell/koae243
33. Ilina EL, Kiryushkin AS, Demchenko KN (2020) Features of fluorescent protein application to study the root system development of cucurbits (Cucurbitaceae). Agric Biol 55:1040–1055. https://doi.org/10.15389/agrobiology.2020.5.1040eng
34. Jefferson RA, Kavanagh TA, Bevan MW (1987) GUS fusions: beta-glucuronidase as a sensitive and versatile gene fusion marker in higher plants. EMBO J 6:3901–3907. https://doi.org/10.1002/j.1460-2075.1987.tb02730.x
35. Ulmasov T, Murfett J, Hagen G et al (1997) Aux/IAA proteins repress expression of reporter genes containing natural and highly active synthetic auxin response elements. Plant Cell 9:1963–1971. https://doi.org/10.1105/tpc.9.11.1963
36. Murashige T, Skoog F (1962) A revised medium for rapid growth and bioassays with tobacco tissue cultures. Physiol Plant 15:473–497. https://doi.org/10.1111/j.1399-3054.1962.tb08052.x
37. Hoagland DR, Arnon DT (1950) The water-culture method for growing plants without soil. In: California Agriculture Experiment Station. University of California, College of Agriculture, Berkeley, CA, USA, pp 4–32
38. Moore L, Warren G, Strobel G (1979) Involvement of a plasmid in the hairy root disease of plants caused by *Agrobacterium rhizogenes*. Plasmid 2:617–626. https://doi.org/10.1016/0147-619x(79)90059-3
39. White FF, Taylor BH, Huffman GA et al (1985) Molecular and genetic analysis of the transferred DNA regions of the root-inducing plasmid of *Agrobacterium rhizogenes*. J Bacteriol 164:33–44. https://doi.org/10.1128/jb.164.1.33-44.1985
40. Sonti RV, Chiurazzi M, Wong D et al (1995) Arabidopsis mutants deficient in T-DNA integration. Proc Natl Acad Sci USA 92:11786–11790. https://doi.org/10.1073/pnas.92.25.11786
41. Jouanin L, Tourneur J, Tourneur C et al (1986) Restriction maps and homologies of the three plasmids of *Agrobacterium rhizogenes* strain A4. Plasmid 16:124–134 https://doi.org/10.1016/0147-619x(86)90071-5
42. Quandt HJ, Puhler A, Broer I (1993) Transgenic root-nodules of *Vicia hirsuta*—a fast and efficient system for the study of gene expression in indeterminate-type nodules. Mol Plant-Microbe Interact 6:699–706
43. White FF, Nester EW (1980) Hairy root: plasmid encodes virulence traits in *Agrobacterium rhizogenes*. J Bacteriol 141:1134–1141. https://doi.org/10.1128/jb.141.3.1134-1141.1980
44. Sanità Di Toppi L, Pecchioni N, Durante M (1997) *Cucurbita pepo* L. can be transformed by *Agrobacterium rhizogenes*. Plant Cell Tissue Org Cult 51:89–93
45. Nacheva L, Ivanova V (2017) Silver nitrate and chlorhexidine gluconate – effective surface sterilization agents in disinfection procedures at the initiation of woody shoot tip and embryo culture. J BioSci Biotechnol 6:187–190
46. Peterson LA, Krueger AR (1988) An intermittent aeroponics system. Crop Sci 28:712–713. https://doi.org/10.2135/cropsci1988.0011183X002800040033x
47. Fernández-Trujillo, Pablo Juan, Sánchez C et al (2004) Quality of greenhouse cucumber fruit grown on perlite substrate or nutrient film technique. Acta Hortic 633:229–236. https://doi.org/10.17660/ActaHortic.2004.633.27

Chapter 2

Virus-Induced Gene Silencing for Nodulation Analysis in Soybean

Sylwia Struk, Kris Gevaert, Sofie Goormachtig, and Feng Qu

Abstract

Virus-induced gene silencing (VIGS) is a powerful reverse genetics tool that enables the transient knockdown of gene expression in plants by exploiting their natural antiviral defense mechanisms. This chapter presents a detailed protocol for applying a Cowpea severe mosaic virus (CPSMV)-based VIGS system to study gene function in soybean (*Glycine max*), with a focus on symbiotic root nodulation. VIGS allows systemic silencing of target genes, making it especially valuable to study pathways involving systemic signaling. The protocol includes Agrobacterium-mediated propagation of CPSMV variants in *Nicotiana benthamiana*, followed by mechanical inoculation of soybean plants and nodulation analysis. Detailed steps for vector construction, viral amplification, plant infection, and downstream molecular and phenotypic analyses are provided. This protocol is reproducible, accessible, and suitable for functional genomics studies in soybean, particularly for genes involved in nodulation but also in other developmental or stress-related processes.

Keywords Cowpea severe mosaic virus, VIGS, Soybean, Nodulation, Functional genomics

1 Introduction

Virus-induced gene silencing (VIGS) is a powerful reverse genetics tool that enables the transient downregulation of gene expression in plants by harnessing their natural antiviral defense mechanism [1]. Upon viral infection, plants detect double-stranded RNA intermediates of viral replication and initiate posttranscriptional gene silencing (PTGS), producing small interfering RNAs (siRNAs) that degrade viral RNA in a sequence-specific manner [2]. By engineering viral vectors to carry fragments of plant genes, this silencing response can be redirected to target endogenous transcripts, allowing rapid functional analysis without the need for stable transformation. This approach is particularly valuable in crop species like soybean (*Glycine max*), where stable transformation is often time-consuming and inefficient.

Steffen Vanneste and Inge Verstraeten (eds.), *Root Development*, Methods in Molecular Biology, vol. 3086,
https://doi.org/10.1007/978-1-0716-5603-7_2,

Soybean is a globally important legume crop, widely cultivated for its high-quality protein and oil used in food and feed [3]. As a legume, soybean has the unique ability to form nitrogen-fixing root nodules through symbiosis with rhizobia, reducing dependence on synthetic nitrogen fertilizers [4]. Understanding the genetic basis of nodulation and other developmental and stress-related processes in soybean is essential for improving crop performance and sustainability. While *Agrobacterium rhizogenes*-mediated hairy root transformation is widely used for studying root-specific gene functions in soybean, it has notable limitations [5, 6]. This method is unsuitable for investigating genes expressed in aerial tissues or those involved in systemic processes, such as autoregulation of nodulation, which relies on root–shoot–root signaling [7]. VIGS overcomes this limitation by enabling systemic silencing of endogenous genes throughout the plant, providing a versatile and efficient tool for functional genomics in soybean [8–12].

Among the available VIGS systems for soybean, the Cowpea severe mosaic virus (CPSMV)-based vector stands out for its effectiveness and technical simplicity, requiring no specialized equipment [9]. CPSMV is a bipartite, single-stranded RNA virus from the *Secoviridae* family, genus *Comovirus*, which naturally infects legumes. Its genome consists of two RNA segments, RNA1 and RNA2, each encoding a single open reading frame producing proteins involved in viral replication, translation, and virion assembly [13]. The CPSMV-based VIGS system utilizes two binary plasmids, QUIN and FZ, which contain cDNA copies of RNA1 and a modified RNA2, respectively. The RNA2 cDNA (FZ) includes a cloning site for insertion of a fragment from the plant gene of interest, enabling targeted silencing. Viral constructs are delivered via *Agrobacterium*-mediated infiltration of *Nicotiana benthamiana* leaves, where the modified virus is amplified. Crude extract from infected leaves is then used to mechanically inoculate soybean, resulting in systemic downregulation of the target gene [9]. CPSMV-based VIGS has already been successfully applied for functional analysis of nodulation in soybean, demonstrating its ability to silence key regulatory genes both in roots and shoots, resulting in significant changes in nodule formation [8].

This chapter provides a detailed protocol for the preparation and application of CPSMV-based VIGS vectors in soybean. It includes steps for vector construction, *Agrobacterium*-mediated delivery, plant inoculation, and monitoring of gene silencing. The protocol is optimized for reproducibility and efficiency, and is suitable for studying nodulation and other traits in soybean under controlled conditions.

2 Materials

Prepare all solutions using ultrapure water (resistivity of 18 MΩ. cm at 25°C), unless indicated otherwise. Prepare and store all reagents at room temperature. Autoclave all media and SOLi solution before use. Standard laboratory equipment and consumables such as pipettes, centrifuge, 12 mL tubes, 1.5 mL Eppendorf and PCR tubes, and 50 mL Falcon tubes are required throughout the protocol. Waste, including genetically modified bacteria and plants, must be disposed of in accordance with institutional biosafety regulations. When working with plant viruses, appropriate biosafety procedures should be followed as defined by institutional biosafety regulation guidelines.

2.1 VIGS Constructs Generation

2.1.1 Cloning into the FZ Vector

1. Vectors FZ-NbPDS and QUIN [9] (*see* **Note 1**).
2. Eco72I (PmlI) restriction enzyme and buffer.
3. NEBuilder® HiFi DNA Assembly (New England Biolabs).
4. DH5alpha thermo-competent *Escherichia coli* cells (1–3 × 10^9 CFU/μg of DNA), commercial or prepared as previously described [see 14].
5. Sterile LB medium: Mix 25 g of Luria broth, 1 g of glucose per 1 L of water, and autoclave. Add 15 g/L Select Agar prior to autoclaving to obtain solid LB medium
6. Sterile S.O.C. medium: Mix 20 g of tryptone, 5 g of yeast extract, 0.58 g of NaCl, 0.18 g of KCl, 2.03 g of $MgCl_2$, 2.46 g of $MgSO_4$, 3.96 g of glucose in 1 L of water.
7. Kanamycin stock solution: Dissolve 25 mg of Kanamycin per mL of water, filter sterilize, and store up to 1 month at 4°C.
8. Incubator and incubator shaker at 37°C.
9. Plasmid DNA and gel extraction kits.
10. PCR machine
11. PCR reagents: DNA polymerase (e.g., Taq); 10× PCR buffer; 10 mM dNTPs
12. Electrophoresis equipment and reagents: 1.5% agarose gel; 1×TAE buffer; electrophoresis equipment; DNA ladder (e.g., SmartLadder 10 kbp).

2.2 Propagation of CPSMV-Based VIGS Constructs in N. benthamiana

2.2.1 Nicotiana benthamiana Growth

1. *N. benthamiana* seeds.
2. Pots of 11 cm diameter.
3. Potting soil mix prepared at a 5:1 ratio with vermiculite.
4. Growth room (16 h light/8 h dark) at 22°C, 65% relative humidity, and light intensity of 250 μmol/s/m^2.

2.2.2 Transformation of A. tumefaciens by Electroporation

1. *A. tumefaciens* strain C58C1Rif^R (pMP90).
2. Sterile distilled (d)H_2O cooled to 4°C.
3. Antibiotics: 40 mg/mL gentamicin, 25 mg/mL kanamycin, 50 mg/mL rifampicin stock solutions, filter sterilized and stored at 4°C.
4. Electroporation cuvette of 0.2 cm and system.
5. Incubator and incubator shaker at 28°C.

2.2.3 N. benthamiana Infiltration

1. Infiltration buffer: 10 mM $MgCl_2$, 10 mM MES, pH 5.7, 100 μM acetosyringone in water.
2. Spectrophotometer for the determination of bacterial optical density at 600 nm (OD_{600}).
3. Plasmid carrying the P19 viral suppressor of gene silencing (see [15], *see* **Note 2**).
4. Incubator shaker at 28°C.
5. 1 mL syringe without a needle.

2.2.4 Confirmation of Viral Infection in N. benthamiana

1. Scalpel.
2. Plastic (poly) bag with zipper (3 × 5 in.).
3. Liquid nitrogen.
4. RNA extraction kit.
5. Microvolume UV–vis spectrophotometer.
6. cDNA synthesis kit.
7. PCR machine
8. PCR reagents: DNA polymerase (e.g., Taq); 10× PCR buffer; 10 mM dNTPs; and stock of each of the primers.
9. Primers (see Table 2).

2.3 Nodulation Analysis of VIGS-Infected Soybeans

2.3.1 Soybean Growth

1. Soybean seeds, e.g., Wiliams82 (Wm82, *see***Note 3**).
2. Pots of 13 cm diameter (1 L).
3. Vermiculite (e.g., Agra-Vermiculite RHP).
4. Prepare the following stock solutions separately in deionised water: A1: 20.2 g/L KNO_3, A2: 27.2 g/L KH_2PO_4, A3: 73.0 g/L $CaCl_2{\cdot}2H_2O$, A4: 24.6 g/L $MgSO_4{\cdot}7H_2O$, A5: 43.5 g/L K_2SO_4, B: 8.2 g/L EDTA·$FeNa_2$, C: Mix in this order 11 g of H_3BO_3, 6.2 g of $MnSO_4{\cdot}H_2O$, 10 g of KCl, 1 g of $ZnSO_4{\cdot}7H_2O$, 1 of g $(NH_4)_6Mo_7O_{24}{\cdot}4H_2O$, 0.5 g of $CuSO_4{\cdot}5H_2O$, 0.5 mL of H_2SO_4 in 1 L of water. Store at RT for max 5 months.
5. Nitrogen-poor solution "i" (SOLi) [see ref. 16]. Combine the following volumes of each stock: 25mL of A1, of A2, of A3, of A4, and of A5, 25 mL of B, and 1.35 mL of C. Adjust to 1 L with deionized water and set pH to 5.8 using 1 M KOH. Dilute 1:40 in deionized water and autoclave.

2.3.2 VIGS Infection of Soybean

1. Soybean infection buffer: 50 mM Na_3PO_4, pH 7.0, 1% Celite 545 in water.
2. Cotton swaps.
3. Mortar and pestle.

2.3.3 Detection of CPSMV RNA

1. RNA extraction kit, e.g., ReliaPrep RNA tissue Miniprep kit.
2. Microvolume UV–vis spectrophotometer.
3. Reverse transcriptase.
4. RNase inhibitor.
5. PCR machine.
6. Proofreading DNA polymerase.
7. 10 mM dNTPs, aqueous solution
8. RNase-free water

2.3.4 GOI Expression Analysis

1. cDNA synthesis mix.
2. 384-well plate and optically clear PCR sealing film.
3. SYBR® Green PCR Master mix.
4. Forward and reverse primers designed for real-time PCR of target genes and internal controls, e.g., *GmATP synthase* (Glyma20g25920) and *GmCons6* (Glyma12G051100) (see Table 2).
5. Quantitative real-time PCR machine.

2.3.5 Nodule Number Analysis

1. *Bradyrhizobium diazoefficiens* strain G49.
2. Sterile YMA medium: Mix 10 g of mannitol, 0.5 g of glutamate, 0.5 g of K_2HPO_4, 0.1 g of $MgSO_4 \cdot 7H_2O$, 0.05 g of NaCl, 0.04 g of $CaCl_2$, 0.004 g of $FeCl_3$, 1 g of yeast extract in 800 mL of distilled water and adjust pH to 6.8. Bring the final volume to 1 L with distilled water and autoclave. For solid YMA plates, add 15 g of Select Agar before bringing up to 1 L volume.
3. SOLi working solution (see Sect. 2.2)
4. Incubator shaker at 28°C.
5. Stereo light microscope with a magnification range of 6.3× to 30×.

3 Methods

3.1 Design and Cloning of the Target Gene Region into the FZ Vector

3.1.1 In Silico Design of the Target Region

1. Select a 399-nucleotide fragment of the gene of interest, ensuring it is in-frame and does not contain a stop codon (*see* **Note 4**).
2. Add the FZ vector-specific flanking sequences to both the 5′ and 3′ ends of the selected fragment (see Table 1).

3.1.2 Cloning into the FZ Vector

1. Synthesize the in silico-designed DNA fragment, including flanking sequences, using a commercial provider (sect. 3.1.1)
2. Digest the FZ-NbPDS vector with Eco72I restriction enzyme following the manufacturer's instructions.
3. Separate the digestion products on 1.5% (w/v) agarose gel and excise the FZ backbone band (~12 kb). Purify the DNA using a gel extraction kit.

Table 1
FZ flanking sequences

FZ 5′	CTGTTTAAGAGGAACCTAGACGATGTCTCGACTATCTCAGATACGATGCACGTG
FZ 3′	CACGTGACGTTCGATACCACACACGCAACCCTACAGTCTGGGGCTGATCTATTC

FZ flanking sequences were added to enable inserting the selected gene fragment into the CPSMV RNA1 vector using Gibson Assembly. Underlined: Eco72I restriction enzyme sites.

4. Assemble the synthesized gene fragment into the digested FZ vector using Gibson Assembly to obtain FZ-Gene-Of-Interest (FZ-GOI):
   - Mix 1 μL of synthesized fragment, 1 μL of gel-purified FZ vector, 2 μL of NEBuilder® HiFi DNA Assembly Master Mix, and 16 μL of water.
   - Incubate at 50°C for 15 min.
5. Per reaction, thaw 50 μL of DH5 alpha thermo competent *E. coli* cells for 10 min on ice. Add 2 μL of the assembly reaction, mix by pipetting, and leave it on ice for an additional 10 min. Transfer to an incubator or heat block at 37°C for 5 min. Add 1 mL of S.O.C. medium to cells immediately after heat shock and incubate cells in a 37°C shaker at 200 rpm for 45–60 min.
6. Plate all the transformed cells on LB agar containing 50 mg/L kanamycin using a sterile Drigalski spatula
7. Incubate plates overnight at 37°C.
8. Select a single colony for each construct using a toothpick or pipette tip and inoculate into 2 mL of liquid LB medium containing 50 mg/L kanamycin in 12 mL tubes. Grow the bacterial culture overnight in a shaker at 37°C at 200 rpm.
9. Extract the plasmid using a miniprep kit.
10. Confirm the sequence of the insert by Sanger sequencing using a primer in the vector backbone (e.g., CPSMR2-R primer, see Table 2).

Table 2
Primer sequences

Name	Sequence
CPSMR2-R	GAGAACTGTACCTGCAACTAATG
CPSMR2-F	GAGATCTGGCGATGATGCATGTGAAT
NbActin-F	TGGACTCTGGTGATGGTGTC
NbActin-R	CCTCCAATCCAAACACTGTA
GmATPs-F	GCGATTCTTAAGCCAGCCTTT
GmATPs-R	ACACACCCTGGAAACTGGTGA
GmCons6-F	AGATAGGGAAATGGTGCAGGT
GmCons6-R	CTAATGGCAATTGCAGCTCTC

3.2 Propagation of CPSMV-Based VIGS Constructs in *N. benthamiana*

3.2.1 *N. benthamiana* Growth

1. Sow *N. benthamiana* seeds in a potting soil–vermiculite mix (5:1) in an 11 cm diameter pot. Cover with transparent foil to maintain moisture until germination.
2. Once germinated, transplant individual seedlings into fresh 11 cm pots containing the same soil mix.
3. Grow plants under long-day conditions (16 h light/8 h dark) at 22–25°C for 3 weeks in a biosafety level 2 (BSL-2) growth room or growth chamber.

3.2.2 Transformation of *A. tumefaciens* by Electroporation (see Note 5)

1. Inoculate *A. tumefaciens* strain C58C1 (pMP90) in 2 ml of YEB medium containing rifampicin (100 μg/mL) and incubate overnight at 28°C with shaking at 200 rpm until OD_{600} of 0.6–0.8.
2. Harvest the culture by centrifugation at ≥12,000×*g* for 2 min at 4°C. Discard the supernatant and wash the pellet sequentially: once with 1 mL of cold sterile dH_2O, twice with 500 μ L of cold sterile dH_2O, and once with of 200 μL cold sterile dH_2O. Resuspend the final pellet in 200 μl of cold sterile dH_2O and keep on ice.
3. Add 5 μL of plasmid DNA to the obtained electrocompetent *A. tumefaciens* cells. Transfer the mixture to a prechilled 0.2 cm electroporation cuvette and electroporate using the following settings: Voltage: 2.5 kV, Capacitance: 25 μF, Resistance: 200 Ω , Pulse time: 5 ms.
4. Immediately add 800 μL of YEB medium to the cuvette and transfer the contents to a sterile 12 mL tube. Incubate at 28°C with shaking at 200 rpm for 2 h.
5. Plate the recovered cells on YEB agar supplemented with 100 μg/mL rifampicin, 40 μg/mL gentamicin, and 25 μg/mL kanamycin.
6. Incubate the plates at 28°C for 2 days until colonies appear.
7. Select a single colony and confirm the presence of the insert by a colony PCR. Transfer a small amount of the colony using a toothpick or pipette tip to the PCR reaction mix as the DNA template. Briefly lyse the cells by extending the first denaturation step to 10 min, and amplify the target region under standard PCR conditions and primers: CPSMR2-F and CPSMR2-R (see Table 2).

3.2.3 *N. benthamiana* Infiltration

1. Inoculate single colonies of *A. tumefaciens* harboring the constructs QUIN, as negative control FZ vector without an insert (FZ-NC), FZ vector with the Gene Of Interest (FZ-GOI), and P19 into 2 mL of liquid YEB medium supplemented with 100 μg/mL rifampicin, 40 μg/mL gentamicin, and 25 μg/mL kanamycin. Incubate overnight on a shaker at 28°C at 200 rpm.
2. Transfer 500 μL of each overnight culture into 9.5 mL of fresh YEB medium containing the same antibiotics, 10 mM MES (pH 5.7), and 100 μM acetosyringone. Grow in 50 mL Falcon tubes overnight at 28°C, 200 rpm.
3. Measure the OD_{600}.
4. Centrifuge cultures at 4000×*g* for 10 min and discard the supernatants
5. Resuspend bacterial pellets in the infiltration buffer to a final OD_{600} of 3.0.
6. Prepare the infiltration mixes:
   - Experimental mix: combine equal volumes of *Agrobacterium* suspensions carrying QUIN, FZ-GOI, and P19 (final OD_{600} of 1 for each construct).
   - Negative control: mix equal volumes of suspensions carrying QUIN, FZ-NC, and p19.
   - Nonviral control: use *Agrobacterium* carrying P19 only, diluted to a final OD_{600} of 1.0 in infiltration buffer.
7. Using a 1 mL needleless syringe, infiltrate the abaxial side of 2–3 leaves of a 3-week-old *N. benthamiana* plant with the prepared mixtures. Apply 1–4 infiltration spots per leaf. Infiltration is most effective during the light period when stomata are open (*see* **Note 6**).
8. Return plants to the growth room. Monitor for systemic viral spread, which typically manifests 5–8 days post-infiltration as newly emerging leaves are wrinkled, curled, and bumpy. Compare phenotypes to the nonviral P19 control (*see* **Note 7**, Fig. 1a).

3.2.4 Confirmation of Viral Infection in *N. benthamiana*

1. At 7–10 days post-infiltration, collect young, symptomatic leaves from the top of the plant. Store samples in zip-lock plastic bags at −70°C until further processing (*see* **Note 8**).
2. Isolate total RNA from symptomatic leaves and nonviral P19 control using a commercial RNA extraction kit, following the manufacturer's protocol.
3. Determine RNA concentration and quality using a microvolume UV–Vis Spectrophotometer.

Fig. 1 CPSMV-induced symptoms on leaves of infected plants. (**a**) *N. benthamiana* plant at 10 days post-infection with CPSMV, showing symptomatic leaves indicated by white arrows. (**b**) Soybean Wm82 cultivar at three weeks post-infection with P19 nonviral control, displaying no symptoms. (**c**) Soybean Wm82 cultivar at three weeks post-infection with CPSMV, exhibiting characteristic leaf symptoms

4. Use 1 μg of total RNA for cDNA synthesis with a commercial cDNA synthesis kit to the final volume of 20 μL, following the manufacturer's protocol.
5. Assess the presence of the transcript of the gene of interest fragment by RT-PCR using PCR reaction mix with primers specific to the fragment cloned into the VIGS vector (FZ) and 1 μL of cDNA as template. Perform an initial amplification with 28 cycles and increase the cycle number if the signal is weak. Include RT-PCR with *NbACTIN* as an internal control (Table 2). These RT-PCRs are used only to confirm expression of the gene-of-interest fragment, not for quantification (*see* **Note 9**).

3.3 Nodulation Analysis of VIGS-Infected Soybeans

3.3.1 Soybean Growth

1. Sow seeds of the soybean Wm82 variety in vermiculite soaked in SOLi solution in 13 cm round pots of 1 L (three seeds per pot). Cover with transparent foil to maintain moisture until germination.
2. Grow the soybean plants in a biosafety level 2 (BSL-2) growth room or growth chamber under long-day conditions (16 h light/8 h dark) at 22°C, at 65% relative humidity, and a light intensity of 250 μmol/s/m^2 for 7–10 days until the first leaves fully open.

3. Water the plants twice a week with SOLi solution (50 mL/pot) until plants are collected for the nodulation analysis.

3.3.2 VIGS Infection of Soybean

1. Grind two symptomatic *N. benthamiana* leaves per construct combination and nonviral P19 control in 3 mL of soybean infection buffer using a mortar and pestle at room temperature (*see* **Note 10**).
2. Rub the leaf extract onto fully expanded unifoliate soybean leaves using a cotton swab (*see* **Note 10**).
3. Return plants to the growth room and monitor for systemic viral symptoms, typically appearing 7–9 days post-infection as wrinkled, crinkled, and mottled newly emerging leaves. Compare phenotypes to the nonviral P19 control (Fig. 1b, c).

3.3.3 Detection of CPSMV RNA

1. Harvest leaf tissue in liquid nitrogen once the VIGS symptoms are clearly visible. Store at −70°C until use.
2. Extract total RNA using a commercial kit following the manufacturer's instructions.
3. For each sample, prepare an annealing mix with 1 μg of RNA, 1 μL of 10 μM CPSMR2-R primer (see Table 2), 2 μL of 10 mM dNTP mix, and 12 μL of RNase-free water. Incubate at 70°C for 5 min, then cool on ice.
4. Prepare two cDNA Synthesis Mixes per sample,
   - RT+: 4 μL of 5× RT buffer, 0.5 μL of reverse transcriptase, 0.5 μL of RNase inhibitor, 5 μL of RNase-free water.
   - RT-: 4 μL of 5× RT buffer, 0.5 μL of RNase inhibitor, 5.5 μ L of RNase-free water.
5. Add 10 μL of the annealing mix to RT+ and RT− tubes. Run cDNA synthesis at 42°C for 60 min, followed by 70°C for 15 min. Cool on ice.
6. Perform RT-PCR using PCR reaction mix with CPMV-specific primers (CPSMR2-F and CPSMR2-R, Table 2) and 5 μL of both RT+ and RT− cDNA as template to confirm CPSMV presence. The RT+ reaction should yield a clear amplicon of the expected size 1022 bp, whereas the RT− control (lacking reverse transcriptase) should show no amplification, confirming that the signal originates from viral RNA rather than contaminating DNA.

3.3.4 GOI Expression Analysis

1. Use 1 μg of total RNA for cDNA synthesis with a commercial kit following the manufacturer's protocol. Dilute the obtained cDNA 25-fold with ultrapure water. Use 0.5 μl of diluted cDNA as a template for qRT-PCR

2. Perform qRT-PCR reactions using a suitable instrument (e.g., Roche LightCycler 480-II Advanced PCR System).
3. Analyze relative gene expression using the 2−ΔΔCt method to determine silencing efficiency [see ref. 17]. For each sample, calculate the ΔCt value by subtracting the Ct of the housekeeping gene from the Ct of the GOI. Normalize expression levels using stable reference genes such as GmATP synthase (Glyma20g25920) and GmCons6 (Glyma12G051100) (Table 2). Determine ΔΔCt by comparing each VIGS-treated sample to the corresponding empty-vector (FZ-NC) or P19 control, and convert this value to relative expression using the 2−ΔΔCt formula. Silencing efficiency can then be expressed as the percentage reduction in GOI transcript levels relative to the control.

3.3.5 Nodule Number Analysis

1. Incubate *Bradyrhizobium diazoefficiens* strain G49 on solid YMA at 28°C for 2–5 days until colonies form.
2. Inoculate 2 mL of liquid YMA medium in a 50 mL Falcon tube with a single colony. Incubate overnight at 28°C with shaking at 200 rpm.
3. Add 8 mL of liquid YMA medium and incubate at 28°C with shaking at 200 rpm for 1–2 days, until OD_{600} reaches 0.8–1.2.
4. Measure the OD_{600}. Centrifuge cultures at 4000×*g* for 10 min and discard the supernatants. Resuspend bacterial pellets in the SOLi solution and dilute to an OD_{600} of 0.01.
5. Ten days post-viral infection, inoculate soybean plants by slowly applying 1 mL of bacterial suspension with a pipette to the root zone, just above the vermiculite (*see* **Note 11**).
6. Score nodules from 14 days post inoculation (dpi) in plants showing CPSMV symptoms using a stereolight microscope. Carefully remove the vermiculite from the root system and rinse the roots thoroughly with water to eliminate all substrate particles before counting (*see* **Note 12**).

4 Notes

1. The QUIN (Plasmid #184267) and FZ-NbPDS (Plasmid #184268) vectors can be ordered from www.addgene.org.
2. An *Agrobacterium* strain harboring a P19 viral suppressor of gene silencing is co-infiltrated to boost protein expression [see ref. 15].
3. The efficiency of VIGS can vary depending on the soybean variety. We recommend using Wm82, as it consistently

demonstrates robust silencing efficiency. If alternative soybean varieties are to be used, we advise conducting preliminary tests to assess if the selected variety is susceptible to CPSMV and evaluate silencing efficiency.

4. To design an effective VIGS construct, select a gene fragment predicted to generate multiple siRNAs using tools such as the RNAi Designer (https://rnaidesigner.thermofisher.com/rnaiexpress/). To ensure specificity, check the selected fragment for potential off-target effects by performing a BLAST search against the soybean genome. Because soybean is a tetraploid with many duplicated genes and highly similar paralogs, off-target silencing can be a concern. When paralogs share high sequence identity, it may be difficult to design a fragment that uniquely targets only one isoform. In such cases, either choose a region that is sufficiently divergent between isoforms or, if biologically appropriate, intentionally target both isoforms. It is recommended to test more than one fragment per target gene to confirm that observed phenotypes result from specific silencing rather than off-target effects.
5. For optimal propagation of CPSMV-based vectors in *N. benthamiana*, freshly transformed *A. tumefaciens* should be used. Glycerol stocks are not recommended.
6. Water *N. benthamiana* plants in the morning of the day that you want to VIGS, or the day before. Open stomata are important for the efficient bacteria inoculation into the leaves.
7. To avoid cross-contamination between the VIGS constructs, change gloves after infiltrating each combination. Grow *N. benthamiana* and soybean plants after infection on different trays. After harvesting symptomatic *N. benthamina* leaves, change the gloves and scalpel after each construct.
8. Any remaining symptomatic *N. bethamiana* leaves can be stored at −70°C until needed.
9. If non-specific bands or primer dimers appear, increase the annealing temperature by 2°C–4°C or reduce primer concentration. Failure to amplify both the GOI and *ACTIN* typically indicates degraded mRNA or an unsuccessful reverse transcription reaction.
10. When working with multiple viral constructs, it is essential to prevent any carry-over of viral material between samples. To avoid it, use a clean mortar and pestle for each construct, change gloves between samples, and thoroughly disinfect the working surface with 4% NaClO after processing each construct.

11. Do not add SOLi to the soybean plants on the day of inoculation with the *B. japonicum* G49 strain.
12. Use a small box or bucket to wash the roots. Discard the dirty wash water through a mesh or sieve to prevent vermiculite and plant debris from entering the sink. All plant material (soybean and tobacco), vermiculite, soil, and used pots must be disposed of as biological waste. Trays may be reused after thorough decontamination with 4% sodium hypochlorite. Clean the bench, growth-room shelves, stereomicroscope, and all reusable equipment with 4% sodium hypochlorite after finishing the experiment.

Acknowledgments

S.S. has received a research grant from Research Foundation Flanders (1280424N).

Competing Interests The authors have no conflicts of interest to declare that are relevant to the content of this chapter.

References

1. Waterhouse PM, Wang MB, Lough T (2001) Gene silencing as an adaptive defence against viruses. Nature 411:834–842. https://doi.org/10.1038/35081168
2. Guo Q, Liu Q, Smith NA et al (2016) RNA silencing in plants: mechanisms, technologies and applications in horticultural crops. Curr Genomics 17:476–489. https://doi.org/10.2174/1389202917666160520103117
3. Dilawari R, Kaur N, Priyadarshi N et al (2022) Soybean: a key player for global food security. In: Wani SH, Sofi NuR, Bhat MA, Lin F (eds) Soybean improvement: Physiological, Molecular and Genetic Perspectives. Springer International Publishing, Cham, pp 1–46. https://doi.org/10.1007/978-3-031-12232-3_1
4. Udvardi M, Poole PS (2013) Transport and metabolism in legume-rhizobia symbioses. Annu Rev Plant Biol 64:781–805. https://doi.org/10.1146/annurev-arplant-050312-120235
5. Kereszt A, Li D, Indrasumunar A et al (2007) *Agrobacterium rhizogenes*-mediated transformation of soybean to study root biology. Nat Protoc 2:948–952. https://doi.org/10.1038/nprot.2007.141
6. Huang P, Lu M, Li X et al (2022) An efficient *Agrobacterium rhizogenes*-mediated hairy root transformation method in a soybean root biology study. Int J Mol Sci 23:12261. https://doi.org/10.3390/ijms232012261
7. Ferguson BJ, Mens C, Hastwell AH t al (2019) Legume nodulation: the host controls the party. Plant Cell Environ 42:41–51. https://doi.org/10.1111/pce.13348
8. Struk S, Van Damme B, Yang SH et al (2025) Virus-induced gene silencing as a powerful tool for functional analysis of nodulation in soybean. Plant Cell Environ https://doi.org/10.1111/pce.15486
9. Zaulda FA, Yang SH, Han JP et al (2022) A cowpea severe mosaic virus-based vector simplifies virus-induced gene silencing and foreign protein expression in soybean. Plant Methods 18: 116. https://doi.org/10.1186/s13007-022-00950-7
10. Zhang C, Ghabrial SA (2006) Development of Bean pod mottle virus-based vectors for stable protein expression and sequence-specific virus-induced gene silencing in soybean. Virology 344:401–411. https://doi.org/10.1016/j.virol.2005.08.046
11. Shin SY, Park MR, Kim HS et al (2023) Virus-induced gene silencing shows that LATE FLOWERING plays a role in promoting flower development in soybean. Plant Growth Regul 99:229–239. https://doi.org/10.1007/s10725-022-00899-6
12. Ogata T, Nagatoshi Y, Yamagishi N et al (2017) Virus-induced down-regulation of GmERA1A and GmERA1B genes enhances the stomatal

response to abscisic acid and drought resistance in soybean. PLoS ONE 12:e0175650. https://doi.org/10.1371/journal.pone.0175650
13. Chen X, Bruening G (1992) Nucleotide sequence and genetic map of cowpea severe mosaic virus RNA 2 and comparisons with RNA 2 of other comoviruses. Virology 187:682–692. https://doi.org/10.1016/0042-6822(92)90471-z
14. Inoue H, Nojima H, Okayama H (1990) High efficiency transformation of Escherichia coli with plasmids. Gene 96:23–28. https://doi.org/10.1016/0378-1119(90)90336-p
15. Jay F, Brioudes F, Voinnet O (2023) A contemporary reassessment of the enhanced transient expression system based on the tombusviral silencing suppressor protein P19. Plant J 113:186–204. https://doi.org/10.1111/tpj.16032
16. Blondon F (1964) Contribution à l'étude du développement de graminées fourragères: ray-grass et dactyle. Rev Gén Bot 71:293–381
17. Livak KJ, Schmittgen TD (2001) Analysis of relative gene expression data using real-time quantitative PCR and the 2(T)(-Delta Delta C) method. Methods 25:402–408

Chapter 3

Protocols for Root Developmental Studies in White Lupin

Mélanie Gasser, Laurent Brottier, Laurence Marquès and Benjamin Péret

Abstract

White lupin (*Lupinus albus*) is a legume species that develops two remarkable root adaptations to nutrient limitation: nitrogen-fixing nodules and dense cluster roots specialized for phosphate acquisition. These traits, together with the species' agronomic relevance and strong developmental plasticity, make white lupin a powerful model for studying local and systemic regulation of root development under nutrient stress. This chapter provides a comprehensive collection of experimental procedures designed to investigate cluster root formation, nodulation, and associated physiological responses at both cellular and whole-plant levels. We describe optimized hydroponic culture systems that allow precise control of nitrogen and phosphate availability, together with reliable seed sterilization and germination procedures. We present a robust protocol for *Agrobacterium rhizogenes*-mediated hairy-root transformation to investigate gene function in root tissues. A controlled nodulation assay using *Bradyrhizobium lupini* enables reproducible symbiotic infection and nodule development. A grafting method is also provided to dissect local versus systemic contributions to root architectural responses. Finally, we detail three in situ physiological assays—phosphatase activity, ferric reductase activity, and proton efflux—that collectively allow the visualization of key functional traits associated with nutrient foraging. Together, these protocols provide an integrated toolkit for exploring the developmental and physiological mechanisms underpinning root plasticity and symbiotic regulation in white lupin.

Keywords White lupin, Cluster roots, Nodulation, Root plasticity, Hydroponic culture, Hairy-root transformation, *Bradyrhizobium lupini*, Grafting technique, Phosphate deficiency, Physiological assays

1 Introduction

The white lupin (*Lupinus albus*) is a Mediterranean legume valued for its high seed protein content and for its exceptional root developmental plasticity under nutrient limitation [1, 2]. In response to phosphate deficiency, white lupin forms dense cluster roots composed of numerous short third-order lateral roots specialized for the mobilization and acquisition of inorganic phosphate [3, 4]. Under nitrogen deficiency, it establishes a symbiotic interaction with *Bradyrhizobium lupini*, leading to the formation of nitrogen-fixing nodules. These two adaptive responses provide an experimentally tractable system

Steffen Vanneste and Inge Verstraeten (eds.), *Root Development*, Methods in Molecular Biology, vol. 3086,
https://doi.org/10.1007/978-1-0716-5603-7_3,

to study nutrient sensing, systemic regulation, and coordinated root architectural changes.

White lupin is therefore an increasingly used model for investigating the molecular and physiological mechanisms underlying root plasticity, nutrient foraging, and symbiotic development. However, the successful implementation of experimental approaches in this species requires tightly controlled growth conditions, reproducible transformation methods, and quantitative assays that capture key physiological activities associated with nutrient uptake.

This chapter presents a set of complementary methods enabling the functional and physiological analysis of cluster root development and nodulation. These protocols are designed to be accessible, robust, and adaptable, allowing researchers to combine developmental, physiological, and molecular approaches within a unified experimental framework.

This chapter describes five major procedures:

(1) hydroponic culture systems enabling precise control of nitrogen and phosphate supply;

(2) *Agrobacterium rhizogenes*–mediated hairy-root transformation for gene functional analysis;

(3) controlled nodulation assays using *B. lupini*;

(4) a stem grafting method to dissect local versus systemic regulatory pathways; and

(5) three in situ physiological assays quantifying phosphatase activity, ferric reductase activity, and proton efflux.

2 Materials

2.1 Plant Material

- White lupin plants: *Lupinus albus* cv. AMIGA (Florimond Desprez, France).

2.2 Bacteria Strain

1. *Bradyrhizobium lupini* MIAE428 (MIAE collection, AgroEcologie INRAE Dijon) (formerly LL13).
2. *Agrobacterium rhizogenes* ARqua1.

2.3 White Lupin Hydroponic Culture (see Note 1)

1. Phosphate-deficient nutrient solution: 400 μM $Ca(NO_3)_2$, 54 μM $MgSO_4$, 0.24 μM $MnSO_4$, 0.1 μM $ZnSO_4$, 0.018 μM $CuSO_4$, 2.4 μM H_3BO_3, 0.03 μM Na_2MoO_4, 10 μM Fe–EDTA, 200 μM K_2SO_4. Store at 4°C.
2. Phosphate-rich nutrient solution: 400 μM $Ca(NO_3)_2$, 54 μM $MgSO_4$, 0.24 μM $MnSO_4$, 0.1 μM $ZnSO_4$, 0.018 μM $CuSO_4$, 2.4 μM H_3BO_3, 0.03 μM Na_2MoO_4, 10 μM Fe–EDTA, 400 μM KH_2PO_4. Store at 4°C.

3. Nitrogen-free nutrient solution: 54 μM $MgSO_4$, 0.24 μM $MnSO_4$, 0.1 μM $ZnSO_4$, 0.018 μM $CuSO_4$, 2.4 μM H_3BO_3, 0.03 μM Na_2MoO_4, 10 μM Fe–EDTA, 400 μM KH_2PO_4. Store at 4°C.
4. Growth conditions: 16 h light/8 h dark, 25°C day/20°C night, 65% relative humidity, photon flux density 200 μmol/m^2/s.
5. 3-L pot or 200-L tank.
6. Air pump to ensure nutrient solution aeration.

2.4 Seed Sterilization and Germination on Plate

1. Sterilization solution: 20% solution prepared using Halonet (Proquimia, 7% sodium hypochlorite, 1% sodium hydroxide) and water.
2. 0.5× MS solid medium: 2.2 g/L Murashige and Skoog Basal medium, 10 g/L Saccharose, 0.5 g/L MES, 7 g/L Agar, pH 5.7.
3. Square Petri dishes (12 × 12 cm).
4. Sterile Milli-Q water.

2.5 Nodulation Assay

1. Substrate: 1:1 (v/v) mixture of zeolite 0.7–3 mm and zeolite 0.5–1.0 mm.
2. YM medium: 10 g/L Mannitol, 1 g/L yeast extract, 0.5 g/L K_2HPO_4, 50 mg/L NaCl, 100 mg/L $MgSO_4 \cdot 7H_2O$, 40 mg/L $CaCl_2$, 0.43 g/L glutamic acid, 4 mg/L $FeCl_3$. Sterilize using an autoclave.
3. Nalidixic acid stock: 20 mg/mL solution in sterile water.
4. Containers: Magenta GA-7 pots 77 × 77 × 97 mm (standard and perforated).
5. Cotton cord.

2.6 Hairy-Root Transformation

1. LB liquid and solid medium. 10 g/L Tryptone, 5 g/L Yeast extract, 5 g/L NaCl, add 15 g/L Agar for solid medium.
2. LB Glucose 1% solid medium. 10 g/L Tryptone, 5 g/L Yeast extract, 5 g/L NaCl, 15 g/L Agar, 10 g/L of D+ Glucose.
3. pK7m24GW-CR vector: pK7m24GW with constitutive pUbi:*mCherry*-ER:t35s cassette allowing identification of transformed roots by red fluorescence (*see* **Note 2**).
4. Antibiotics stock solution:
   Kanamycin 30 mg/mL solution in sterile water.
   Streptomycin 20 mg/mL solution in sterile water.
   Spectinomycin 100 mg/mL solution in sterile water.
5. Acetosyringone.

6. Tools and substrate: Scalpel blade; vermiculite; mini-greenhouse lid.
7. Hydroponic pots and aeration system.
8. FastGene Blue/Green LED Flashlight.

2.7 Physiological Activity Assay

2.7.1 Phosphatase Activity Assay

1. 5-Bromo-4-chloro-3-indolyl phosphate (BCIP) stock solution: 65 mg of BCIP dissolved in 5 mL of DMSO.
2. Phosphatase activity gel: 1.2 g of agar in 150 mL of 50 mM sodium acetate buffer (pH 5), supplemented with 1.5 mL of BCIP solution. Pour onto a 24 × 24 cm square Petri dish lid and allow to solidify (*see* **Note 3**).
3. Plastic wrap.

2.7.2 Ferric Reductase Activity Assay

1. Reagents: 100 mM Ferric Sodium EDTA (FeNaEDTA) solution; bathophenanthroline disulfonic acid disodium salt (BPDS).
2. Ferric reductase gel: 1.2 g of agar, 26 mg of K_2SO_4 in 150 mL of distilled water, supplemented with 150 μL of 100 mM FeNaEDTA and 24 mg of BPDS. Pour onto a 24 × 24 cm square Petri dish lid and allow to solidify (*see* **Notes 3–5**).
3. Plastic wrap.

2.7.3 Proton Exudation Assay

1. Bromocresol Purple solution (BCP) stock solution: 25 mg of BCP dissolved in 5 mL of 500 mM Tris-HCl buffer (pH 6).
2. Proton exudation gel: Dissolve 1.2 g of agar, 26 mg of $CaSO_4 \cdot 2H_2O$, 52 mg of K_2SO_4 in 150 mL of distilled water. Autoclave, let it cool, and add 1.5 mL of 100x BCP solution. Pour onto a 24 × 24 cm square Petri dish lid and allow to solidify (*see* **Note 3**).
3. Plastic wrap.

3 Methods

3.1 Seed Sterilization and Germination

1. Dispense up to 35 seeds into a 50 mL tube and add 20% Halonet solution. Incubate for 20 min on slow agitation (*see* **Note 6**).
2. Remove the sterilization solution and wash seeds 5–6 times with sterile Milli-Q water until no bubbles remain.
3. Immerse the seeds in sterile Milli-Q water for 2 h in a square Petri dish (*see* **Note 7**).

4. Remove the water and allow seeds to dry for 10 min in the laminar flow.
5. Place seeds on 0.5× MS solid medium in square Petri dishes with the hilum facing downward. Press seeds gently into the agar to ensure proper imbibition and prevent displacement.
6. Seal the plates with parafilm and cover them with paper to protect from light.
7. Incubate plates at an angle of 45° in a growth chamber for 3 days.

3.2 Hairy-Root Transformation

Hairy root transformation in *L. albus* is performed using *A. rhizogenes* strain ARqua1, generating composite plants with transformed roots and unaltered shoots. Transformed roots exhibit a characteristic hairy-root phenotype alongside non-transformed, unaltered roots. We used the pK7m24GW-CR vector, a Gateway-compatible destination vector, which was modified by introducing a constitutive pUbi:*mCherry*-ER:t35s cassette [5] *see* **Note 2**) into the pK7m24GW [6] vector (Fig. 1A), allowing identification of transformed roots by red fluorescence.

3.2.1 Preparation of Agrobacterium rhizogenes ARqua1

1. Streak *A. rhizogenes* ARqua1 from a glycerol stock onto LB agar containing 20 µg/mL streptomycin (bacterial selection) and 100 µg/mL spectinomycin (vector selection) if using a

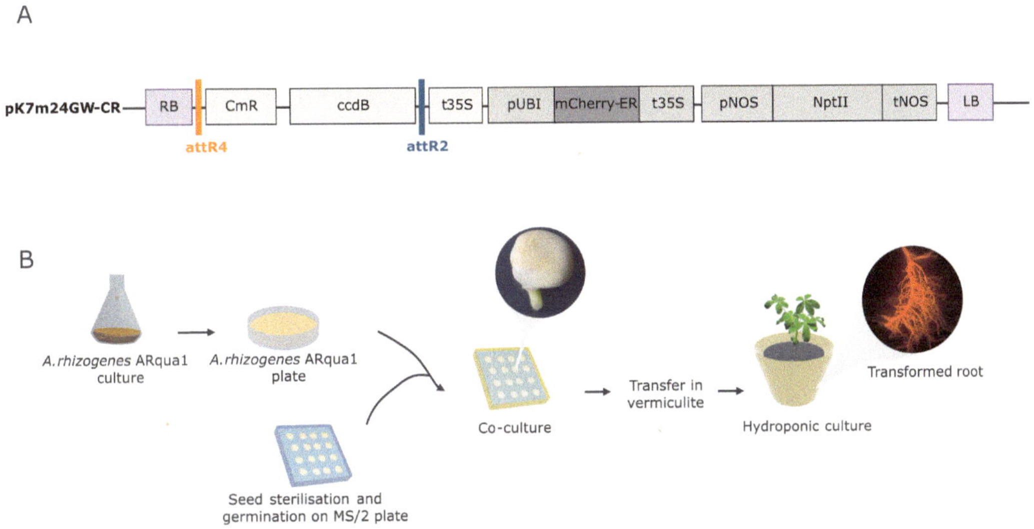

Fig. 1 Plasmid map of the Gateway backbone pK7m24GW-CR used for hairy-root transformation (**A**). Schematic overview of the major steps involved in *Lupinus albus* hairy-root transformation (**B**), including *Agrobacterium rhizogenes* culture, seed sterilization and germination, cocultivation with an image of a germinated seed with the primary root cut, transfer to vermiculite, and subsequent hydroponic culture. The image shows a transformed hairy root of *L. albus*. Schematics were designed using Blender software

pK7m24GW-CR Gateway-based vector. Incubate at 28°C for 3 days.

2. Inoculate a single colony into 2 mL of LB liquid medium with antibiotics and incubate in a shaker at 28°C, 150 rpm for 24 h.
3. The day before transformation, spread 200 μL of the overnight culture onto LB + 1% Glucose plates supplemented with 100 μM of acetosyringone and antibiotics. Incubate overnight at 28°C (*see* **Note 8**).

3.2.2 Transformation and Coculture (Fig. 1B)

1. Sterilize and germinate seeds as described in Sect. 3.1. Incubate seedlings for 43–45 h until primary roots reach 1–1.5 cm (*see* **Note 9**).
2. Cut each root at mid-length.
3. Gently scrape the cut root surface across the *A. rhizogenes* ARqua1lawn on the LB +1% glucose, 100 μM acetosyringone plate prepared the day before transformation (*see* **Note 10**).
4. Transfer each seedling onto 0.5× MS solid medium containing 25 μg/mL kanamycin, positioning the cut root surface in contact with the medium to prevent adventitious rooting (*see* **Note 11**).
5. Incubate plates at a 20° angle under long-day conditions (16 h light/8 h dark, 20°C).
6. After 6 days of coculture, transfer seedlings to moist vermiculite (*see* **Note 12**). Remove the seed coat when possible and place seedlings with the cut root surface facing downward. Ensure cotyledons rest on the vermiculite surface to prevent desiccation.
7. Cover with a mini-greenhouse lid for 10 days to aid acclimation (*see* **Note 13**).

3.2.3 Transfer to Hydroponics

1. Prepare hydroponic 3-L pots with nutrient solution and aeration.
2. Remove transformed seedlings from vermiculite after 10 days and rinse roots gently with water (*see* **Note 14**).
3. Transfer plants to hydroponic pots containing the appropriate nutrient solution.
4. Replace nutrient solution weekly (*see* **Note 15**).
5. For phosphate deficiency experiments, harvest cluster roots after 20–30 days of hydroponic culture (*see* **Notes 16 and 17**).
6. Chimeric roots containing non-transformed regions may be obtained. When using plasmid construction from pK7m24GW-CR, the transformation status of the analyzed root region should be verified by the presence of red fluorescence excited by a handheld FastGene Blue/Green LED flashlight.

3.3 Nodulation Assay

3.3.1 Bradyrhizobium Lupini MIAE428 Culture

1. Three days before inoculation, inoculate *B. lupini* MIAE428 into 5 mL of YM medium containing 20 µg/mL nalidixic acid
2. Incubate at 28°C, 150 rpm for 3 days until OD_{600} reaches the appropriate value (*see* **Note 18**).

3.3.2 Plant Inoculation and Nodulation

1. Prepare a 1:1 (v/v) mixture of zeolite 0.7–3 mm and zeolite 0.5–1.0 mm.
2. Wash the substrate with Milli-Q water until rinse water runs clear (*see* **Note 19**).
3. Thread a cotton wick through the perforated Magenta GA-7 pot so that it reaches the bottom of the second, non-perforated pot.
4. Fill the upper pot completely with the substrate mixture, ensuring the cotton wick is fully embedded. Pack down well.
5. Cover with two sheets of aluminum foil and sterilize the entire assembly (*see* **Note 20**).
6. Sterilize and dispense seeds on solid 0.5× MS medium as described in Sect. 3.1 and incubate the plates at an angle of 45°C in a growth chamber.
7. Transfer 3-day-old seedlings to the sterilized pots and keep the foil cover to maintain humidity (*see* **Note 9**).
8. Inoculate 7-day-old seedlings with 500 µL of *B. lupini* culture applied close to the root system.
9. Refill the above tank with medium as soon as necessary.
10. Observe nodules 3–4 weeks after inoculation (Fig. 2).

3.4 V-Shape Grafting

1. Germinate seeds for 4 days in vermiculite.
2. Transfer seedlings to aerated hydroponic nutrient solution (*see* **Note 14**).
3. At day 6, cut the stem horizontally to separate the scion and rootstock using a scalpel with a new blade.
4. Cut the top of the rootstock vertically and prepare the scion with a thin bevel (V-shape graft) (Fig. 3).
5. Insert the scion bevel into the slit of the rootstock and wrap tightly with parafilm to avoid any gaps (*see* **Note 21**; Fig. 3).
6. Place grafted seedlings in a large (200 L) tank of nutrient solution (*see* **Notes 22 and 23**).
7. Cover plants with a greenhouse dome for 4 days to maintain high humidity.
8. Over the next 21 days, inspect grafts regularly. Remove root growth emerging through the parafilm and rewrap with fresh parafilm as needed (*see* **Note 24**).

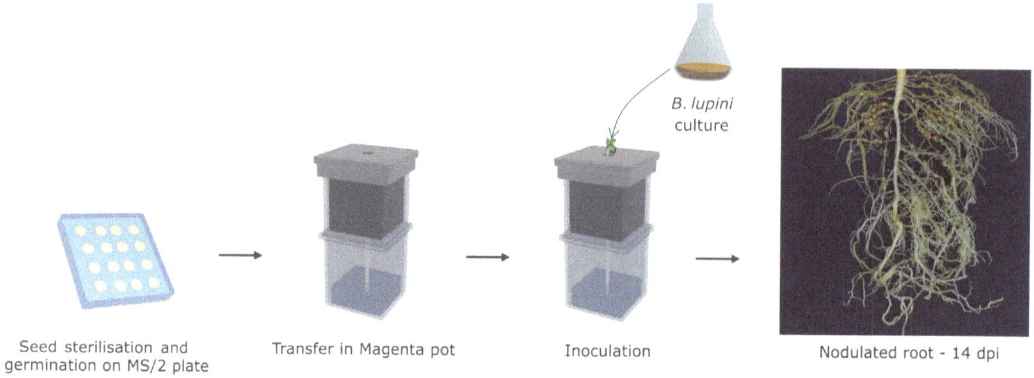

Fig. 2 Overview of the major steps of the nodulation assay in *Lupinus albus* with *Bradyrhizobium lupini*: Seeds were sterilized and germinated, then transferred to Magenta GA-7 pots containing a zeolite mixture and inoculated with the symbiotic bacterium *B. lupini* MIAE428. Nodules can be visible at 14 days post inoculation (dpi), as shown in the photograph of a lupine root grown under nitrogen-deficient conditions at 14 dpi. Schematics were designed using Blender software

Fig. 3 V-shaped grafting of lupin seedlings: the first panel shows the horizontal cut of the rootstock and the V-shaped bevel prepared on the scion. The second panel shows the scion inserted into the rootstock, and the final panel shows the assembled graft, tight with parafilm

3.5 Physiological Activity Assay

3.5.1 Phosphatase Activity Assay

1. Rinse the root system with Milli-Q water and spread it flat onto the phosphatase activity gel.
2. Cover roots with plastic wrap, removing air bubbles (*see* **Notes 25 and 26**).
3. Incubate for the required time until blue signal develops (*see* **Notes 27–30**; Fig. 4A).

3.5.2 Ferric Reductase Activity Assay

1. Rinse the root system with Milli-Q water and spread it flat onto the ferric reductase gel.
2. Cover with plastic wrap and remove air bubbles (*see* **Notes 25 and 26**).
3. Protect from light and incubate 4–5 h until a red/pink coloration develops (*see* Notes **28–30**; Fig. 4B).

3.5.3 Proton Exudation Assay

1. Rinse the root system with Milli-Q water and spread it flat onto the proton exudation gel.
2. Cover with plastic wrap and remove air bubbles (*see* **Note 26**).
3. Protect from light and incubate 4–5 h until purple-to-yellow color changes reveal acidification zones (*see* **Notes 28–31**; Fig. 4C).

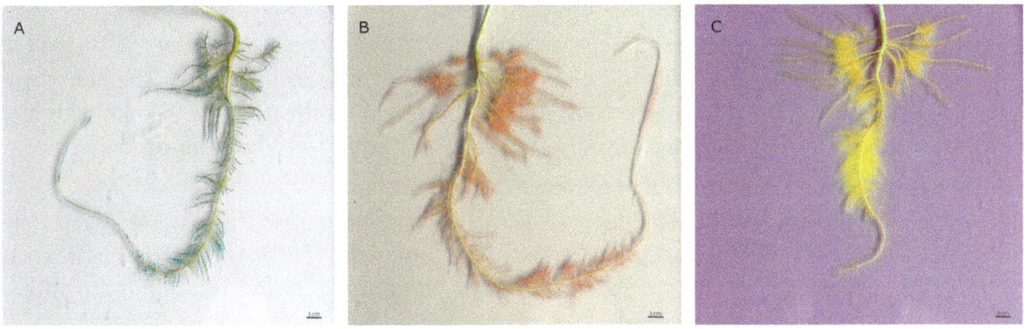

Fig. 4 Physiological activity assays. Phosphatase activity: image of a plant root system grown in a phosphate-rich, iron-free nutrient solution and after 4 h of incubation on gel (**A**); ferric reductase activity assay: image of a plant root system grown under phosphate-deficient conditions and after 2 h of incubation on gel (**B**); proton exudation assay: Image of a plant root system grown in phosphate-deficient conditions and after 4.5 h of incubation on gel (**C**)

4 Notes

1. Adequate aeration using an air pump in the growth medium is a critical parameter. In addition, culture volume affects lupin root system architecture: the developmental kinetics of cluster roots differ between plants grown in 3-L pots and those grown in 200-L tanks.
2. Briefly, the pUbi:*mCherry*-ER:t35s cassette from Fournier et al. 2018 [5] was PCR-amplified using specifically designed primers (5′-CGGCACTAGTCTAAAACCAAAATC-3′ and 5′-GCGCACTAGTAGGTCACTGGATTT-3′), each containing a *SpeI* restriction site. The resulting PCR product was ligated into the pk7m24GW vector, which had been previously digested with *XbaI* enzyme.
3. When preparing agar gels, repeatedly stop the microwave and swirl gently to avoid local overheating. Overheating can caramelize sugars in buffer components, introducing color backgrounds that interfere with colorimetric detection.
4. The ferric reductase assay must always be carried out in the dark. Even brief exposure to light during incubation will affect BPDS chemistry and generate spurious red signals.
5. FeNaEDTA solutions should be prepared fresh whenever possible. Aged solutions show partial iron oxidation and diminish the sensitivity of the ferric reductase assay.
6. Seeds that turn strongly yellow or orange during sterilization should be discarded. They generally reflect compromised seed coats or internal damage and germinate poorly.
7. Ensure seeds remain fully immersed during the 2 h water imbibition step. Uneven hydration leads to heterogeneous germination and variable root lengths at the time of transformation.
8. LB Glucose 1% plates supplemented with acetosyringone should be prepared on the same day as the bacterial streaking step. Acetosyringone degrades rapidly, reducing the induction of virulence genes and transformation efficiency.
9. For hairy-root transformation, seedling age and root length uniformity are critical. Transforming seedlings with roots >1.5 cm drastically reduces the frequency of transformed roots and increases callus formation.
10. When scraping seedlings on *A. rhizogenes*, avoid collecting excessive bacterial mass. Overloading the wound increases contamination risks and reduces plant survival. A thin, even smear of bacteria is sufficient.

11. Ensure the cut root end of the seedling remains in tight contact with kanamycin-containing medium during coculture. Poor contact allows escape of wild-type adventitious roots, reducing the proportion of transgenic roots.
12. Removing the seed coat before transplanting to vermiculite improves survival but should be done carefully. Removing it too early or too aggressively can detach cotyledons, which severely reduces plant vigor.
13. Maintaining stable humidity under the mini-greenhouse lid is essential during the first 10 days post-transformation. Drying episodes commonly result in the collapse of transformed roots or the cessation of growth.
14. For hydroponic transfer, rinse vermiculite completely to avoid the transfer of fine particles that may accumulate around roots and create local hypoxic zones in the nutrient solution.
15. Weekly replacement of the nutrient solution is essential for consistent phosphorus or nitrogen treatments. White lupin roots modify their rhizosphere chemistry intensely; old solutions quickly become chemically heterogeneous.
16. When analyzing cluster root responses, ensure plants are grown at comparable developmental stages. Cluster root induction is highly stage-dependent and can vary several-fold between plants differing by only a few days.
17. For long experiments (20–30 days of hydroponics), algae growth can occur in the nutrient solution. Use opaque covers or frequent cleaning to avoid light penetration, as algal biofilms alter nutrient availability and local pH. This problem is much less prevalent in phosphate deficiency conditions.
18. *Bradyrhizobium lupini* grows slowly. Ensure cultures are incubated for a full 72 h for optimal OD_{600} values. Under-grown cultures reduce nodulation efficiency, while over-grown cultures may lead to clumping and uneven inoculation.
19. Rinse the zeolite mix substrate thoroughly until the wash water runs completely clear. Residual dust significantly reduces aeration and may retain inhibitory ions, interfering with early nodulation.
20. During the first days after transfer into the Magenta pots, maintaining high humidity with aluminum foil is essential. Seedlings dessicate rapidly when the substrate dries even slightly.
21. During grafting, the scion bevel must match the width of the rootstock slit precisely. Any lateral gap greatly promotes oxidative events, resulting in graft failure within 48 h.

22. Grafting is very difficult to achieve in phosphate-deficient conditions.
23. After transferring grafted plants to hydroponics, slight movement of the nutrient solution can disturb fragile graft unions. Reduce aeration strength during the first days until graft healing stabilizes.
24. When rewrapping parafilm around graft junctions, avoid constricting the stem. Restrictive wrapping inhibits vascular reconnection and may lead to swelling or cracking of the junction.
25. Roots must be fully flattened on the gels for all physiological assays. Air pockets between root tissue and agar lead to incomplete contact and false-negative zones in the reaction patterns.
26. When spreading roots for physiological assays, avoid mechanical damage. Damaged tissues can release oxidative compounds that react nonspecifically with BCIP or BPDS, causing artifacts.
27. Phosphatase activity development time varies with plant age and phosphate status. Strongly induced cluster roots may develop color within 30–60 min, whereas control roots may require several hours. Avoid over-incubation, which can cause background staining.
28. For physiological assays involving color reactions, always include a negative control root (e.g., from nutrient-replete conditions) to validate specificity. Background coloration occasionally develops in gels stored too long.
29. When imaging roots after assays, keep gels hydrated and avoid direct exposure to strong lights. Both BCIP and bromocresol purple signals can fade under intense illumination.
30. If roots detach from gels before imaging, gently reposition them with forceps under plastic wrap. Reexposure to air will dry roots rapidly and distort signals.
31. Proton exudation patterns depend on both genotype and nutrient history. Roots previously exposed to high-phosphate conditions tend to acidify more slowly than those grown under phosphate deficiency. Interpret timings accordingly.

Competing Interests

The authors have no conflicts of interest to declare that are relevant to the content of this chapter.

References

1. Neumann G, Martinoia E (2002) Cluster roots—an underground adaptation for survival in extreme environments. Trends Plant Sci 7:162–167. https://doi.org/10.1016/S1360-1385(02)02241-0
2. Cheng L, Bucciarelli B, Liu J et al (2011) White lupin cluster root acclimation to phosphorus deficiency and root hair development involve unique glycerophosphodiester phosphodiesterases. Plant Physiol 156:1131–1148. https://doi.org/10.1104/pp.111.173724
3. Le Thanh T, Hufnagel B, Soriano A et al (2021) Dynamic development of white lupin rootlets along a cluster root. Front Plant Sci 12 https://doi.org/10.3389/fpls.2021.738172
4. Massonneau A, Langlade N, Léon S et al (2001) Metabolic changes associated with cluster root development in white lupin (*Lupinus albus* L.): relationship between organic acid excretion, sucrose metabolism and energy status. Planta 213:534–542. https://doi.org/10.1007/s004250100529
5. Fournier J, Imanishi L, Chabaud M et al (2018) Cell remodeling and subtilase gene expression in the actinorhizal plant *Discaria trinervis* highlight host orchestration of intercellular Frankia colonization. New Phytol 219:1018–1030. https://doi.org/10.1111/nph.15216
6. Karimi M, Bleys A, Vanderhaeghen R, Hilson P (2007) Building blocks for plant gene assembly. Plant Physiol 145:1183–1191. https://doi.org/10.1104/pp.107.110411

Chapter 4

Imaging of *Ceratopteris richardii* Root Anatomy

Annie Espinal-Centeno, Alejandro Aragón-Raygoza, Domingo Méndez-Alvarez and Alfredo Cruz-Ramírez

Abstract

Here, we present detailed protocols for histological staining and imaging of root cross sections of the fern *Ceratopteris richardii*, providing high-resolution visualization of cellular and tissue structures essential for understanding root development, nutrient and water uptake, and responses to environmental cues. This chapter describes the preparation of root samples, fixation, embedding, sectioning, and the application of specific histological stains that highlight key anatomical features, including the epidermis, cortex, endodermis, vascular tissues, and root cap. The protocol also covers imaging techniques using confocal microscopy to obtain clear, reproducible images suitable for quantitative and qualitative analyses. This standardized approach facilitates comparative studies of root anatomy and development, offering a valuable tool for plant developmental biology research while ensuring the preservation of tissue morphology.

Keywords Root development, Cellular organization, *Ceratopteris richardii*, Vascular tissue, Fluorescence microscopy, Anatomy, Fern

1 Introduction

Roots are fundamental organs in vascular plants, playing essential roles in water and nutrient uptake, anchorage, and interaction with soil microorganisms. Understanding root anatomy and development provides insights into how plants adapt to diverse environmental conditions and regulate internal growth patterns. Histological analysis of root tissues allows detailed visualization of cellular organization, including the epidermis, endodermis, cortex, vascular tissues, and the root cap. Visualizing cellular organization is critical for dissecting processes such as cell differentiation, tissue patterning, and vascular development in roots [1].

Ferns, particularly *Ceratopteris richardii*, also known as C-fern, serve as valuable model systems for studying root development due to their relatively simple root organization, short life cycle, and amenability to genetic and cellular studies [2]. Recent work has shown the

Steffen Vanneste and Inge Verstraeten (eds.), *Root Development*, Methods in Molecular Biology, vol. 3086,
https://doi.org/10.1007/978-1-0716-5603-7_4,

mitotic activity of the root apical cell (RAC) in *C. richardii*, helping to define features of meristem organization in ferns [2].

In an evolutionary context, ferns such as *C. richardii* (Monilophytes) are positioned between bryophytes and seed plants, providing a key perspective on the innovations in root architecture that occurred during vascular plant evolution. Meanwhile, the early-diverging group of vascular plants known as Lycophytes occupies a basal position in the Tracheophyte tree and exhibits distinct root structures compared to ferns and seed plants. Fossil and developmental evidence support that roots originated at least twice: once within Lycophytes and independently in the Euphyllophytes (ferns + seed plants) [3,4]. Comparative studies between Lycophytes and ferns, such as *C. richardii*, can illuminate both conserved and derived features of root anatomy, vascular organization, and tissue differentiation.

Histological staining and imaging of root cross sections provide a precise and reproducible approach to examining these features. By applying these techniques, researchers can map cell types, track developmental stages, and quantify anatomical traits across species, enhancing our understanding of how roots evolved and diversified. In this chapter, we describe protocols to visualize the cell walls of all root cell layers (Fig. 1A), along with staining approaches that reveal specific biochemical compounds characteristic of each tissue's differentiation state and function. These methods allow us to detect markers enriched in specific layers, for example, lignin, which typically accumulates in the xylem vessels and xylem parenchyma, and in the Casparian strip of the endodermis of mature roots (Figs. 1 and 2). By combining these stains with confocal microscopy, we provide a detailed view of how structural polymers and cell-specific features contribute to the organization and functional specialization of vascular plant roots. This study not only advances developmental biology in ferns but also contributes to broader evolutionary studies of plant root systems, bridging insights between early vascular plants and more derived lineages [2].

2 Materials

All solutions should be prepared using analytical-grade reagents and ultrapure water (18 MΩ cm at 25°C, purified and deionized), unless otherwise stated. Reagents should be stored at room temperature unless specific storage conditions are indicated. Dispose of all chemical and biological waste according to local institutional and governmental regulations.

1. Plant material: *Ceratopteris richardii* Hn-n first shoot-born root of 30 daf (days after fertilization). Plant growth and culture conditions are described in "Methods."

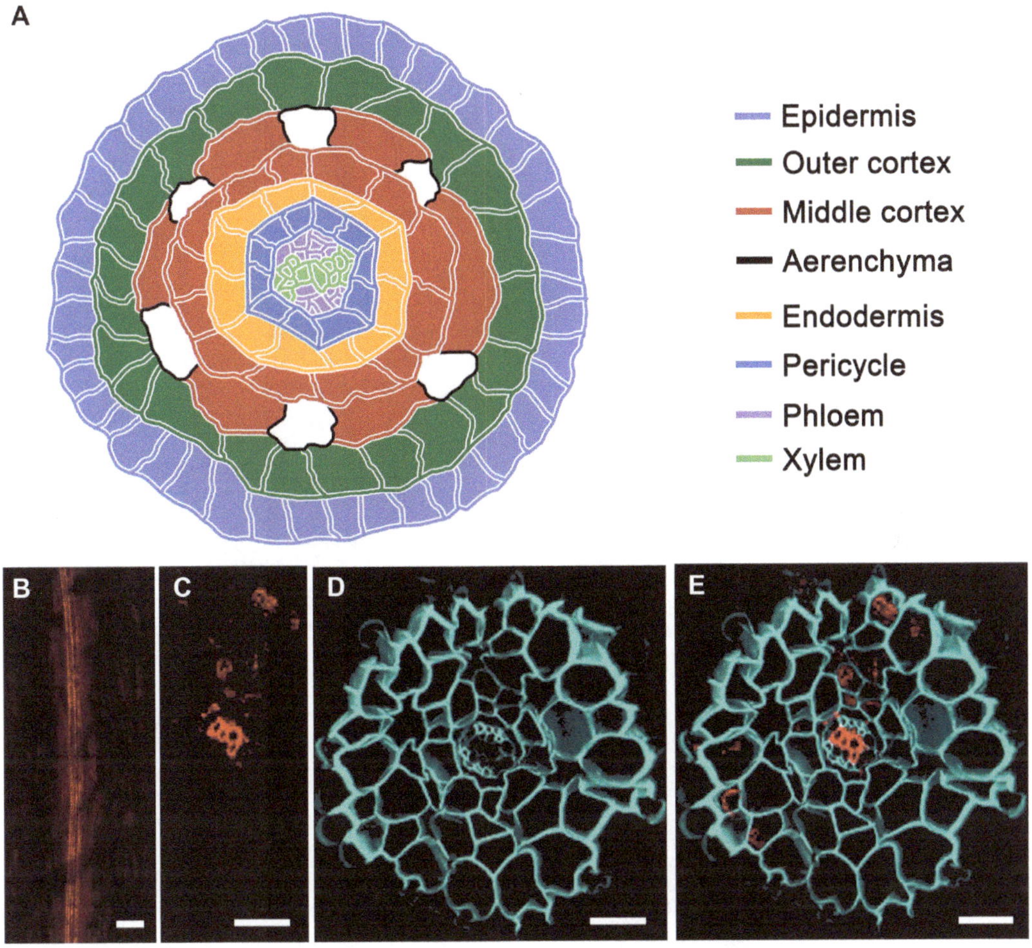

Fig. 1 Graphic representation of the concentric organization of the differentiation zone of a *C. richardii* root, and CLSM images of root sections stained with Basic Fuchsin (BF) and SCRI-Renaissance. A graphic representation of all layers present and observed in a transverse section of the differentiation zone of *C. richardii* (**A**). BF staining was used to observe vascular tissue in whole-mount-stained roots in a longitudinal section of the root (**B**) as well as in a transverse section (**C**). SCRI-Renaissance 2200 was used to stain the cell walls of the same transverse section (**D**). The merge of both previously described staining is observed in (**E**). Scale bars in all panels: 50 μm

2. 10× Macronutrients Stock Solution: Dissolve 1.25 g of NH_4NO_3, 5 g of KH_2PO_4, 1.20 g of $MgSO_4.7H_2O$ and 0.26 g of $CaCl_2{\cdot}2H_2O$ in 900 mL of water. Top up to 1 L with water and store at 4°C.
3. 100× Chelated Iron (Fe–EDTA) Stock Solution: Dissolve 2.78 g of $FeSO_4{\cdot}7H_2O$ and 3.73 g of Disodium EDTA·$2H_2O$ separately in about 450 mL of water each. Raise the pH of the EDTA solution using 1 M NaOH to 7.0 or higher to fully dissolve the powder, then readjust the pH to 7.0 using 2 M HCl once dissolved. Bring the EDTA solution to a boil

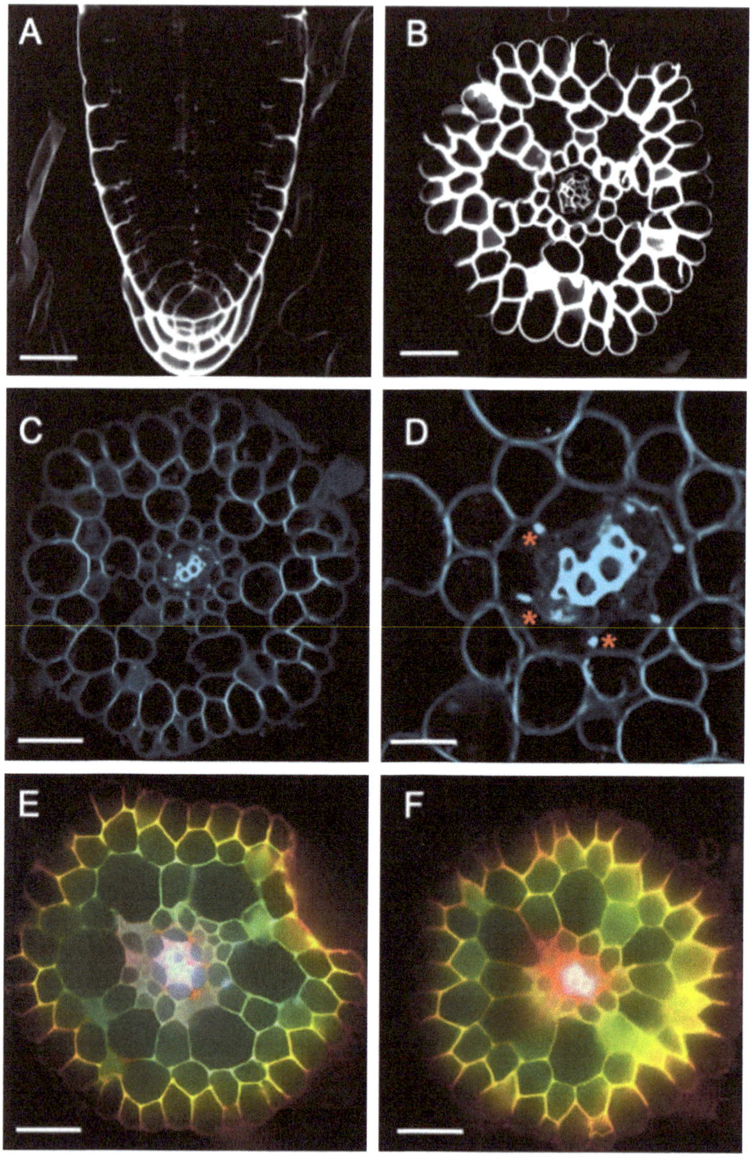

Fig. 2 Microscopy of *C. richardii* roots stained with Calcofluor white (CW), Berberine (Be), Aniline Blue (Ab), and Safranin (S). CW staining was used to observe the cell walls in agar sections of primary root meristem (**A**) and the diverse layers of the differentiated zone in a root section (**B**). Calcofluor white staining revealed the organization of the Ceratopteris root, showing six concentric cell layers surrounding a central cylinder (**A**). Based on cell morphology and previous reports, the central cylinder contains vascular tissues, while the adjacent layers correspond to the pericycle and endodermis (**A**, **B**). Transverse section of a root stained with Be and Ab (**C**). Zoom in on the same image showing the vasculature, endodermis, and xylem cell walls and the Casparian strip (**D**), respectively (red asterisks). Images in **A**–**D** were recorded using CLSM. Transverse sections of roots stained with Be and S, observed in an epifluorescence microscope, show the diverse root layers in the differentiation zone (**E**, **F**). Scale bars in all panels: 50 μm

using a hot plate and then combine it with the $FeSO_4$ solution. Cover and boil the mixture for one hour, then let it cool completely. Top up to 1 L with water.

4. 200× Micronutrients stock solution: Dissolve 0.05 g of $MnSO_4 \cdot 1H_2O$, 0.074 g of $CuSO_4 \cdot 5H2O$, 0.104 g of $ZnSO_4 \cdot 7H_2O$, 0.37 g of H_3BO_3 and 0.0074 g of $(NH4)_6Mo_6O_{24} \cdot 4H_2O$ in 900 mL of water. Adjust the final volume to 1 L with water and store at 4°C.
5. C-fern medium (CFM): Mix 100 mL of 10× Macronutrients Stock Solution, 10 mL of 100× Fe–EDTA Stock Solution, 5 mL of 200× micronutrients stock solution, and 0.5 g of MES in 900 mL of water and adjust the pH to 6.0 using 1 M NaOH solution. Adjust the final volume to 1 L with water, add 8 g of plant agar, and autoclave the medium at 121°C for 20 min. Pour into 100 ×15 mm Petri dishes. Let the medium solidify, store at 4°C until use.
6. 10× Phosphate buffered saline (PBS): Weigh 80 g of NaCl, 2 g KCl, 14.4 g of Na_2HPO_4, and 2.4 g of KH_2PO_4, transfer to a 1 L glass beaker containing about 800 mL of water. Mix and adjust pH to 7.2 using HCl or NaOH. Adjust final volume to 1 L using water. Store at 4°C. To make 1 L of 1× PBS, add 100 mL of 10× PBS to 900 mL of water.
7. Paraformaldehyde (PFA) solution: Weigh 0.4 g of PFA and transfer it to a heat-resistant container in a fume hood. Add 10 mL of 1× PBS to obtain a final concentration of 4% (w/v) PFA. Place the mixture on a temperature-controlled magnetic stirrer and heat it to approximately 65°C, ensuring the solution does not reach boiling. Gradually increase the pH by dropwise addition of 1 M NaOH until the solution becomes clear. Once the PFA has fully dissolved, adjust the pH to 6.9 by carefully adding small amounts of HCl. Allow the solution to cool to room temperature before use (*see* **Notes 1 and 2**).
8. ClearSee solution: Add 10 g of xylitol, 25 g of sodium deoxycholate, and 25 g of urea to a glass beaker containing about 60 mL of water. Place the beaker on a magnetic stirrer and stir vigorously until all solids are completely dissolved and the solution is clear (approximately 30 min). Add water to a final volume of 100 mL (*see* **Note 3**).
9. Basic fuchsin (BF) staining solution: Weigh 20 mg of basic fuchsin and dissolve in 10 mL of ClearSee solution, filter by using a 0.22 μm membrane filter, and store protected from light at 4°C no longer than one week (*see* **Note 4**).
10. SCRI Renaissance 2200 staining solution: Weigh 5 mg of SCRI Renaissance 2200 (Renaissance Chemicals, UK) and dissolve in 10 mL of ClearSee solution, filter by using a 0.22 μ

m membrane filter, and store protected from light at 4°C (*see* **Note 5**).

11. Calcofluor white (CW) staining solution: Weigh 10 mg of Calcofluor White and dissolve in 10 mL of ClearSee solution, filter by using a 0.22 µm membrane filter, and store protected from light at 4°C no longer than one week (*see* **Note 6**).
12. Aniline blue staining solution: Weigh 50 mg of Aniline blue and dissolve in 10 mL water, filter by using a 0.22 µm membrane filter, and store protected from light at −20°C for 1 month and for 6 months at −80°C.
13. Berberine hemisulfate staining solution: Weigh 10 mg of Berberine hemisulfate salt and dissolve in 10 mL of lactic acid, filter by using a 0.22 µm membrane filter, and store protected from light at 4°C no longer than one week (*see* **Note 4**).
14. Safranin staining solution: Weigh 50 mg of Safranin O and dissolve in 10 mL of 50% ethanol, filter by using a 0.22 µm membrane filter, and store protected from light at 4°C (*see* **Note 4**).
15. Agarose solution: Dissolve 4 g of low gelling temperature agarose powder in 10 mL of distilled water and heat until fully melted. Allow the agarose to cool to about 25°C–30°C. Keep the solution in a water bath to prevent it from solidifying.
16. Glycerol 50% (v/v)
17. Growth cabinet set to 25°C with a 16-h light/8-h dark photoperiod and a relative humidity (RH) of 60%–70%.
18. 100 ×15 mm Petri dishes
19. Microscopy slides
20. Cover slips
21. Razor blades
22. Sterilization solution: Dilute commercial bleach (6% sodium hypochlorite) 5-fold in deionized water
23. Sterile water
24. 0.22 µm membrane filters
25. Pipette or plastic inoculation loop
26. Parafilm
27. Small molds ~10 ×5 × 5 mm. Can be made using aluminum foil

3 Methods

3.1 Plant Growth and Culture Conditions

1. Collect spores from mature *Ceratopteris richardii* Hn-n plants grown in the growth cabinet. Surface sterilize spores by immersing them in sterilization solution for 1 min. Rinse

spores thoroughly with sterile water five times to remove any residual sterilant. Store the spores in darkness for 3 days at room temperature before plating.

2. Under sterile conditions, plate the sterilized spores evenly onto the solidified C-Fern Medium using a sterile pipette or plastic inoculation loop. Seal the Petri dishes with parafilm or plastic wrap to prevent contamination.
3. Incubate the Petri dishes at 25°C with a photoperiod of 16 h/8 h light/dark (*see* **Note 7**).

3.2 Tissue Fixation and Clearing

1. Fix roots from *Ceratopteris richardii* first shoot-born root of 30 daf (days after fertilization) immediately after collection by submerging them in Paraformaldehyde solution for 12 h (*see* **Note 8**).
2. Wash the tissues three times for 10 min with 1× PBS.
3. Immerse tissues in ClearSee solution for at least 7 days. Refresh the ClearSee solution every 24 h.

3.3 Agarose-Embedded Root Cross-Sections

1. Keep the melted agarose solution at 25°C–30°C in a water bath.
2. Carefully cut 1 cm of fixed *Ceratopteris richardii* tissue.
3. Place the root samples in small molds, then pour the melted agarose over the roots, ensuring they are fully embedded. Allow the agarose to solidify at room temperature.
4. Using a sharp razor blade to manually cut transverse or cross sections of the agarose-embedded roots at the desired thickness.

3.4 Tissue Staining

3.4.1 Basic Fuchsin for Lignin Staining

1. Transfer root tissues to Basic Fuchsin Solution and incubate for 1 h at room temperature, protected from light. For staining of root transverse sections, reduce incubation time to 15 min.
2. Wash three times with ClearSee solution for 15 min each, followed by one wash with water for 10 min (*see* **Note 9**).
3. Transfer the samples to 50% glycerol and store at 4°C in the dark until visualization.
4. Place the stained roots on a microscopy slide and cover with a coverslip.
5. Visualize the samples using a confocal laser scanning microscopy: use a 561 nm excitation laser and detect it at 600–650 nm (*see* **Note 10**).

3.4.2 SCRI Renaissance 2200 for Cell Wall Staining

1. Transfer root tissues to SCRI Renaissance 2200 staining solution for 1 h at room temperature, protected from light. For staining of root transverse sections, reduce incubation time to 15 min [see 5, 6].
2. Wash with 50% ethanol for 10 min (*see* **Note 9**).
3. Perform two additional washes with ClearSee solution for 10 min each (*see* **Note 9**).
4. Transfer the samples to 50% glycerol and store at 4°C in the dark until visualization (*see* **Note 10**).
5. Place the stained roots on a microscopy slide and cover with a coverslip.
6. Visualize the samples using a confocal laser scanning microscope with a 405 nm excitation laser and detect it at 460–465 nm.

3.4.3 Basic Fuchsin and Renaissance Double Staining

1. Follow Steps 1 and 2 described in Sect. 3.4.1.
2. Continue with Steps 1–5 described in Sect. 3.4.2.
3. Visualize the samples using a confocal laser scanning microscopy: to image Basic Fuchsin use 561 nm excitation laser and detect it at 600–650 nm (Fig. 1B, C). To visualize the Renaissance, we used a 405 nm excitation laser and detected it at 460–465 nm (Fig. 1D, E). Use sequential scanning to avoid bleed-through.

3.4.4 Calcofluor White for Cell Wall Staining

1. Incubate roots with Calcofluor White Staining solution for 1 h at room temperature, protected from light. For staining of root transverse sections, incubate briefly for 15 min (see ref. 6).
2. Rinse the stained roots gently with ClearSee solution to remove excess stain (see **Note 9**).
3. Transfer the samples to 50% glycerol and store at 4°C in the dark until visualization (see **Note 10**).
4. Place the stained samples on a microscope slide and cover with a coverslip.
5. Visualize the tissues using a confocal laser scanning microscopy: Use a 405 nm excitation laser and detect it at 425–475 nm (Fig. 2A, B).

3.4.5 Aniline Blue for Casparian strip (CS) and Cell Walls Staining

1. Immerse the root cross sections in Aniline Blue staining solution for 30 min at room temperature, protected from light. For staining, the roots' transverse sections are incubated briefly for 15 min.
2. Rinse the stained roots with water for 15 min (*see* **Note 9**).
3. Transfer the samples to 50% glycerol and store at 4°C in the dark until visualization (*see* **Note 10**).
4. Place the stained tissues on a microscope slide and cover with a coverslip.

5. Visualize the root cross sections using a confocal laser scanning microscopy: To image Aniline Blue using a 488 nm excitation laser and detect it at 475–525 nm (Fig. 2C, D).

3.4.6 Berberine for Suberin and Callose Staining

1. Incubate root transverse sections in lactic acid for 1 h.
2. Transfer the tissues to Berberine hemisulfate staining solution for 1.5 h at room temperature, protected from light.
3. Wash three times with water for 15 min (*see* **Note 9**).
4. Transfer the samples to 50% glycerol and store at 4°C in the dark until visualization (*see* **Note 10**).
5. Place the stained sections on a microscope slide and cover with a coverslip.
6. Visualize using a confocal laser scanning microscopy: using 405 nm excitation laser and detect it at 540–560 nm.

3.4.7 Safranin for Lignin Staining

1. Incubate root transverse sections in 0.5% Safranin O (S) in ethanol for 1 min.
2. Wash three times with water for 15 min (*see* **Note 9**).
3. Transfer the samples to 50% glycerol and store at 4°C in the dark until visualization (*see* **Note 10**).
4. Place the stained sections on a microscope slide and cover with a coverslip.
5. Visualize using a confocal laser scanning microscopy: to visualize Safranin used 488 nm excitation laser and detected it at 590–595 nm.

3.4.8 Berberine and Safranin Double Staining

1. To perform the double staining, follow Steps 1–3 described in Sect. 3.4.6.
2. Continue with Steps 1–4 described in Sect. 3.4.7.
3. Visualize using a confocal laser scanning microscopy: to image Berberine using 405 nm excitation laser and detect it at 540–560 nm, to visualize Safranin used 488 nm excitation laser and detected it at 590–595 nm (Fig. 2E, F).

4 Notes

1. Whenever possible, use freshly prepared PFA solution and store it at −20°C for no more than one week.
2. Please note that PFA is a hazardous substance. Review the MSDS prior to handling this chemical. Always wear gloves, use safety goggles and respiratory protection, and prepare solutions in a fume hood.

3. ClearSee Solution can be stored at room temperature up to 6 months.
4. The use of freshly prepared dye solutions, one week or less prior to the experiments, yields better results.
5. SCRI Renaissance 2200 staining solution can be stored long-term (6 months), aliquoted into 1 mL portions, and kept at −20°C. Always use fresh aliquots for staining. Avoid multiple free–thaw cycles.
6. Calcofluor White and Safranin staining solutions can be stored at −20°C up to 1 year.
7. The vertical growth of plants helps prevent the roots from penetrating the growing medium.
8. Do not apply vacuum during fixation to prevent structural damage.
9. Washing procedures should be done gently to prevent any damage to the roots.
10. It is recommended to image the stained tissues preferably during the first week to achieve optimal results.

Acknowledgments

AC-R acknowledges the Secretariat of Science, Humanities, Technology, and Innovation of Mexico (SECIHTI) for National Research System Fellowship (SNII-2), AE-C acknowledges SECIHTI for National Research System Fellowship (SNII-C), DM-A acknowledges the SECIHTI for PhD Fellowship No. 960257.

Conflict of Interest The authors have no conflicts of interest to declare that are relevant to the content of this chapter.

References

1. Jones VA, Dolan L (2012) The evolution of root hairs and rhizoids. Ann Bot 110:205–212. https://doi.org/10.1093/aob/mcs136
2. Aragón-Raygoza A, Vasco A, Blilou I et al (2020) Development and cell cycle activity of the root apical meristem in the fern *Ceratopteris richardii*. Genes 11:1455. https://doi.org/10.3390/genes11121455
3. Hetherington AJ, Dolan L (2018) Stepwise and independent origins of roots among land plants. Nature 561:235–238. https://doi.org/10.1038/s41586-018-0445-z
4. Fang T, Motte H, Parizot B et al (2021) Early "Rootprints" of plant terrestrialization: *Selaginella* root development sheds light on root evolution in vascular plants. Front Plant Sci 12:735514. https://doi.org/10.3389/fpls.2021.735514
5. Maceda-López LF, Espinal-Centeno A, Ordaz-Ortiz JJ et al (2025) Multi-omics analyses on maya-land tomatoes shed light on plant strategies to thrive in high temperatures. Physiol Plant 177:e70429. https://doi.org/10.1111/ppl.70429
6. Ursache R, Andersen TG, Marhavý P et al (2018) A protocol for combining fluorescent proteins with histological stains for diverse cell wall components. Plant J 93:399–412. https://doi.org/10.1111/tpj.13784

Chapter 5

A Microscopy Toolkit for Root Studies in *B. distachyon*

Cristovão De Jesus Vieira Teixeira, Kevin Bellande and Joop E.M. Vermeer

Abstract

The study of plant organogenesis presents significant challenges due to the small number of cells involved in its initial stages. Observing the initial cell divisions becomes increasingly challenging when transitioning from *Arabidopsis thaliana* (Arabidopsis) to other species. Lateral root (LR) initiation is an essential process for enhancing a plant's ability to access water and nutrients. In most cases, LR formation starts in the pericycle of the parent root, giving rise to a new meristem. Understanding the mechanisms coordinating LR development is important for improving plant resilience to biotic and abiotic stresses. The prediction of the precise timing and location of LR initiation along the root axis remains challenging, even in Arabidopsis. This is magnified when attempting to observe LR development in crops. *Brachypodium distachyon* (Brachypodium) has emerged as a versatile model for cereal crops. However, studying Brachypodium LR development requires a revision of protocols and methodologies to be applied from seedling growth to root imaging. Here, we present protocols for seed preparation, in vitro growth, and tissue clearing for Brachypodium. Whereas ClearSee appeared to be unsuitable for rendering Brachypodium root tissues transparent after a reasonable incubation period, our modified DEEP-Clear method resulted in improved efficiency in tissue clearing and is compatible with major fluorescent proteins and dyes. Finally, we introduce a simple and straightforward approach to locally synchronize LR development. These tools and methodologies are crucial for advancing our knowledge of plant root system architecture, transitioning from a model plant to agronomically important species.

Keywords *Brachypodium distachyon*, Lateral root development, Tissue clearing, DEEP-Clear, ClearSee, Lateral root synchronization, Microscopy, Plant organogenesis, Root system architecture, Monocot

1 Introduction

Plant roots are composed of different cell types arranged in concentric layers around the vascular tissue [1, 2]. In Arabidopsis, this vasculature is diarchic, with two xylem and phloem poles [3, 4]. In contrast, plants such as maize and rice have a polyarchic vascular structure, with multiple alternating xylem and phloem elements [5]. The vascular tissues are surrounded by the pericycle, which in turn is surrounded

These authors "Cristovão De Jesus Vieira Teixeira and Kevin Bellande" contributed equally to this work.

Steffen Vanneste and Inge Verstraeten (eds.), *Root Development*, Methods in Molecular Biology, vol. 3086,
https://doi.org/10.1007/978-1-0716-5603-7_5,

by the endodermal, cortical, and epidermal cell layers. Lateral root (LR) formation in angiosperms often originates from a distinct set of pericycle cells, known as founder cells [6]. In Arabidopsis, as well as in onion, radish, and sunflower, these founder cells are located at the xylem poles, whereas in maize, rice, and carrot, pericycle cells at the phloem poles are particularly prone to initiate LRs [7]. The initial step in LR formation is an asymmetric anticlinal division that occurs simultaneously in three or more adjacent pericycle founder cell files [4]. Further periclinal and anticlinal divisions lead to the formation of a dome-shaped primordium that eventually emerges from the parent root [8, 9]. However, the timing and location of these initial divisions are difficult to predict. The limited number of cells involved makes the study of LR initiation particularly challenging [10]. This challenge increases when moving from Arabidopsis to larger plant species.

Brachypodium (*Brachypodium distachyon*) was proposed as a plant model more than 20 years ago [11–13]. Its versatility makes it an excellent choice for a wide range of research areas in plant biology. It provides a robust basis for studying aspects such as plant diversity, genome organization at the molecular level, plant responses to environmental cues, and cell wall structures and their modification [13]. These characteristics are particularly relevant for gaining insights into key developmental processes that can be applied to agricultural species. However, working with Brachypodium presents significant challenges compared to Arabidopsis, particularly in the context of LR formation. In this chapter, we present a detailed set of tools and updated protocols aimed at accelerating and optimizing the visualization of LR development in Brachypodium and likely other monocot species. Approaches such as improved growth conditions, seed preparation, LR synchronization and histological methods are presented as tools to improve our understanding of LR development in Brachypodium. Recent studies have successfully used this protocol [14, 15].

2 Materials

All solutions should be prepared using analytical-grade reagents and ultrapure water (18 MΩ cm at 25°C, purified and deionized), unless otherwise stated. Reagents should be stored at room temperature unless specific storage conditions are indicated. Dispose of all chemical and biological waste according to local institutional and governmental regulations.

2.1 Biological Materials

Brachypodium distachyon seeds (Accessions Bd21-3 and Bd21)

2.2 Reagents and Chemicals

1. *In vitro* growth medium: Dissolve 4.802 g/L Murashige & Skoog Medium Incl. MES buffer (Duchefa, M0254) in water, set pH to 5.7, add 0.8% Plant Agar (Gel strength $\geq$ 1100 g/ cm^2, Ash < 3%, Acid-insoluble ash < 0.5%) and autoclave.
2. Seed sterilization solution: 6% sodium hypochlorite (6%–14% active chlorine basis), 0.1% Tween in water.
3. DEEP-Clear solution: 25% Urea (w/v), 3% Triton X-100 (v/v), 6% N,N,N′,N′-Tetrakis(2-hydroxypropyl)ethylenediamine (Theed) (v/v), 50 mM Na_2SO_3 in water. Mix the solution for 30 min to 1 h on the magnetic stirrer until complete dissolution. Can be stored at room temperature for a week.
4. 10× Phosphate-buffered saline (PBS): Mix 1.37 M NaCl, 27 mM KCl, 100 mM Na_2HPO_4, and 18 mM KH_2PO_4 in water. Dilute tenfold in water to achieve 1× PBS.
5. Fixation buffer: Weigh 4 g of PFA powder, add to 90 mL of 1× PBS. Heat to ~60°C under a fume hood with stirring (do not exceed 70°C). Caution: PFA is a known carcinogen; wear gloves! Add 1 M KOH dropwise until clear. Adjust pH to ~6.9 with HCl and adjust the volume using water to reach 100 mL. Cool before use. Use fresh or store at +4°C (1 week) or −20°C (2 weeks).
6. Optional fluorescent stains:

   - Basic Fuchsin: 0.1% in DEEP-Clear;
   - Calcofluor White M2R (Fluorescent Brightener 28): 0.1% in DEEP-Clear;
   - Renaissance SR2200: 0.1% in DEEP-Clear;
   - Fluorol Yellow 088 (FY): 0.01% in DEEP-Clear;
   - Aniline Blue: (Ink Blue R; 4-[4-[[4-(4-aminooxysulfonylanilino)phenyl]-[4-(4-aminooxysulfonylphenyl) iminocyclohexa-2,5-dien-1-ylidene]methyl]anilino]benzenesulfonic acid): 0.5% in water.

2.3 Equipment

1. Fine forceps.
2. Growth chambers (capable of continuous light or long-day cycles at 22°C).
3. Vacuum chamber and pump.
4. Magnetic stirrer with heating function.
5. pH meter or indicator strips.
6. Hand microtome.
7. Shaking incubator (optional, for clearing protocols).

8. Confocal laser scanning microscope (Leica TCS SP8-MP or equivalent), equipped with lasers and water immersion objectives.
9. Image analysis software (e.g., Fiji/ImageJ, Adobe Illustrator).

3 Methods

Carry out all procedures at room temperature unless otherwise specified.

3.1 Brachypodium from Seed to Seed

1. Carefully remove the seed husks using forceps, avoiding damage to the embryo. Preselect viable seeds (darker, stiffer) from nonviable ones (lighter, softer) (Fig. 1).
2. Place selected seeds in a 15 mL Falcon tube (up to 50% capacity). Add seed sterilization solution to fully submerge the seeds. Shake vigorously for at least 1 min. Remove seed sterilization solution and wash seeds at least five times with sterile distilled water. Ensure complete bleach removal (*see* **Note 1**).
3. Leave sterilized seeds in sterile water in the dark at 4°C for 2–4 days. Seeds should enlarge, indicating readiness for planting.
4. Rinse stratified seeds. Prepare 12 × 12 cm plates with *in vitro* growth medium. Place seeds in a row, 2 cm from the plate top, while ensuring the embryo faces upward and does not touch the medium (*see* **Note 2**). Do not place more than 15 seeds in a row on a 12×12 cm plate
5. Transfer plates to growth chambers (continuous light, 22°C). Angle plates at ~20° (or ~15° for Bd21-3) to keep roots on the medium surface. Use larger plates for growth beyond 7 days.
6. For propagation, transfer sterilized seeds directly to soil (4:1 vermiculite mix recommended). Grow at 22°C with a 20-h photoperiod (or 16-h to prolong vegetative phase) (*see* **Note 3**).
7. Water the soil-grown plants until they appear dry. Harvest seeds ~2 weeks after stopping watering. Remove seeds carefully (wear gloves). Place seeds in paper bags at 37°C for at least 2 days. Store dried seeds at room temperature.

3.2 Tissue Clearing (DEEP-Clear) Protocol for Plant Tissues

1. Harvest Brachypodium roots (e.g., from 5 to 7 day old seedlings) and submerge samples in fixation buffer. Use a soft brush to keep samples submerged if needed (*see* **Note 4**). Incubate maximally 2 h at room temperature, or at 4°C for samples with fluorescent proteins, with gentle agitation. Apply vacuum 3–5 times, each time slowly releasing the vacuum, to infiltrate the fixation buffer. Wash fixed tissues at least 3 times (1 min

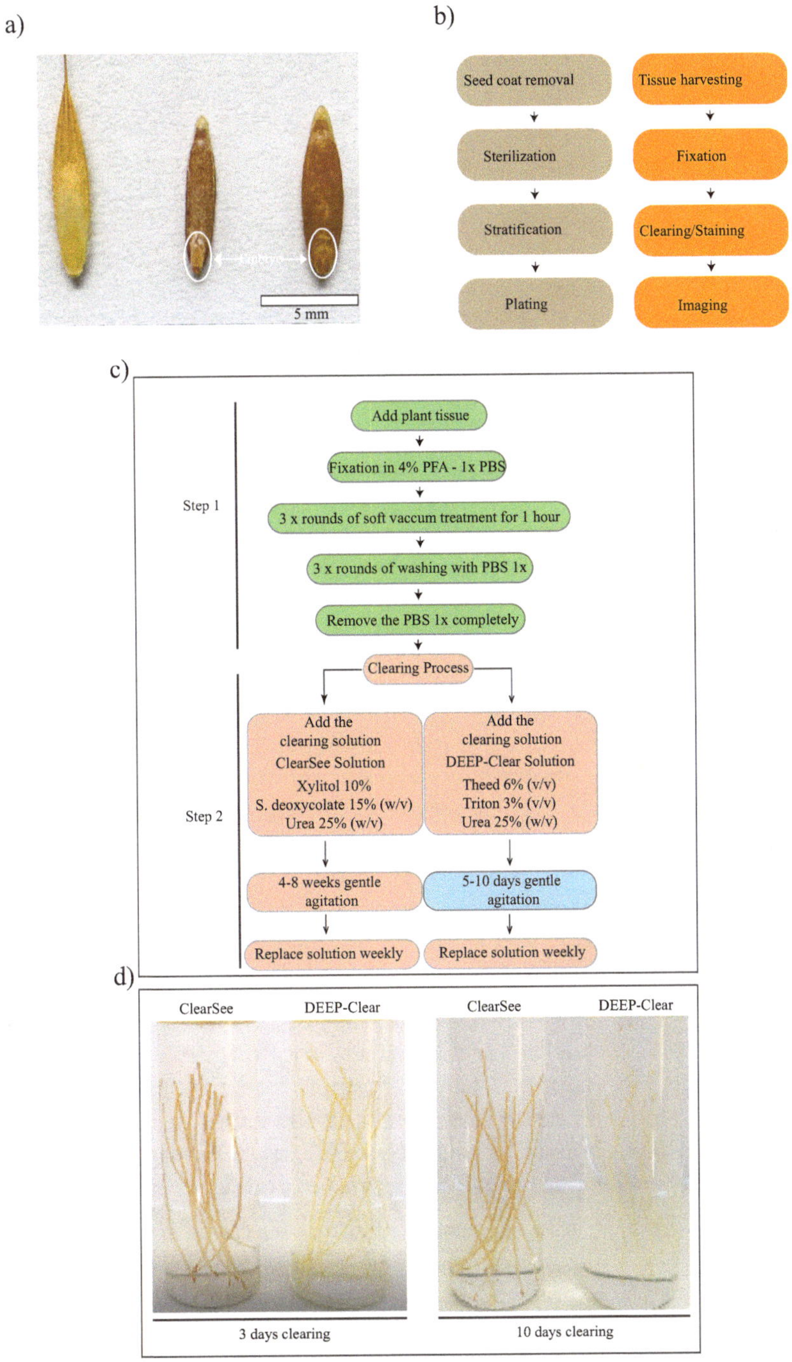

Fig. 1 Workflow from seed to imaging of Brachypodium LR development: (**a**) Seed preparation is essential for growing Brachypodium in vitro: untreated seed (left), seed with husk removed (middle), and seed after 3 days at 4°C, ready for plating (right). (**b**) Overview of the major steps in the protocol for *Brachypodium* root tissue preparation. (**c**) Comparison of the ClearSee and DEEP-Clear protocols: root tissue transparency can be achieved in 5–10 days (highlighted in blue). (**d**) Brachypodium root tissue treated with ClearSee and DEEP-Clear after 3 and 10 days of incubation: DEEP-Clear achieves quicker tissue transparency compared to ClearSee

each) in 1× PBS. Apply one round of vacuum infiltration with 1× PBS during the last wash. Complete removal of fixative is crucial.

2. Transfer samples to DEEP-Clear solution. Apply three rounds of gentle vacuum infiltration (slow release). Incubate at room temperature with gentle agitation for samples with fluorescent proteins. Alternatively, incubate at 28°C, 100 RPM in a shaking incubator. Incubate for 5–10 days, replacing the DEEP-Clear solution weekly or if it turns yellow/brown. Roots should show reduced pigmentation within 3 days and significant transparency by 10 days (*see* **Notes 5 and 6**).
3. For cell wall staining (Optional): Prepare 0.1% Calcofluor White or Renaissance solution in DEEP-Clear. Incubate samples in staining solution for 24 h (apply 1 round of vacuum for thick roots). Rinse once in DEEP-Clear. Wash samples overnight in DEEP-Clear (this method is also compatible with 15 µ M Propidium Iodide, 0.1% Basic Fuchsin (if sodium-sulfite is omitted), and 0.01% Fluorol Yellow) (*see* **Note 7**).
4. Mount samples in DEEP-Clear on chambered cover glasses or using slide spacers. Image using appropriate microscopy (e.g., two-photon, confocal, or epifluorescence). Image Calcofluor White with 405 nm excitation, detected at 425–475 nm. For identifying lateral root primordia, it can be helpful to start with orthogonal views (*see* **Note 8**) (Fig. 2).

3.3 Lateral Root Induction via Root Tip Excision (RTE-LRIS)

1. Grow Brachypodium seedlings (Bd21-3, Bd21) *in vitro* for at least 6–7 days. Ensure roots have not reached the plate bottom. Prepare more seedlings than needed to account for variability (*see* **Note 9**).
2. In a sterile flow hood, open the plate. Using a sterile scalpel, quickly cut the primary root tip just above the root hair zone (approx. removal point). Use gentle back-and-forth movements to avoid pushing the root into the agar. Seal the plate immediately (*see* **Note 10**). Return plates to the growth chamber for the desired time interval (e.g., 1 h to 32 h post-excision).
3. Carefully harvest ~5 mm to 1 cm of root tissue above the cut site at each time point. To speed up, align multiple seedlings and make a single cut across them.
4. Immediately transfer harvested root segments to Fixation buffer. Proceed with the DEEP-Clear protocol (Sect. 3.2, Steps 3–5). Stain with 15 µM Propidium Iodide (or other suitable stains) if desired.
5. Mount cleared samples on microscopy slides (*see* **Note 8**).
6. Examine using fluorescence or confocal microscopy. Start examining later time points (e.g., 32 h) to identify emerged

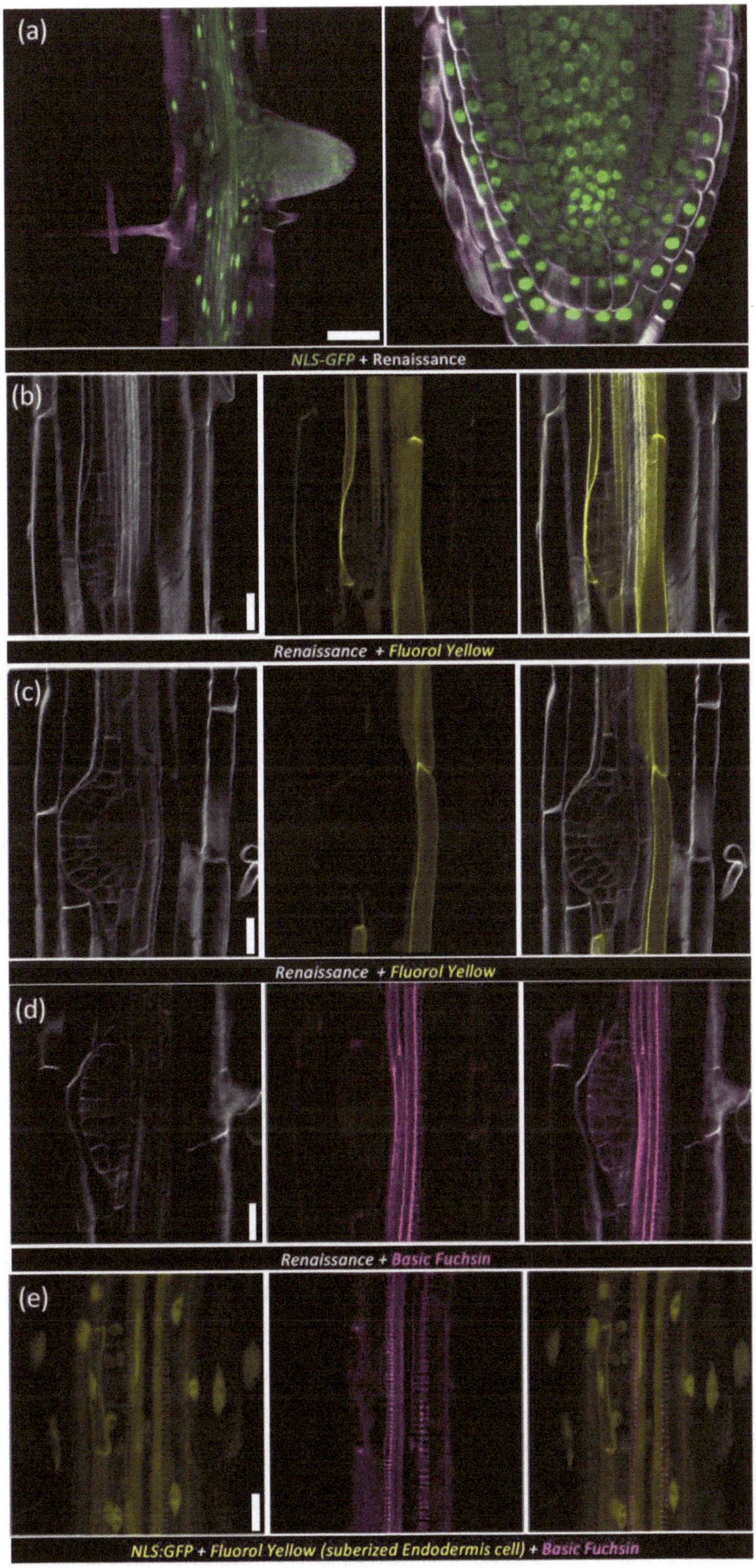

(a)
NLS-GFP + Renaissance
(b)
Renaissance + Fluorol Yellow
(c)
Renaissance + Fluorol Yellow
(d)
Renaissance + Basic Fuchsin
(e)
NLS:GFP + Fluorol Yellow (suberized Endodermis cell) + Basic Fuchsin

LRs, then work backward to observe earlier stages. Initial pericycle divisions can be observed around 1 h, endodermal divisions around 2 h, cortex divisions around 4 h, root cap formation between 12 and 16 h, and emergence around 32 h post-excision. LRPs typically develop from phloem-associated pericycle cells facing the medium. Staging can be performed within the first 3 mm from the cut (Fig. 3).

4 Notes

1. Gas sterilization is not recommended for Brachypodium as it reduces germination. Ethanol sterilization was less effective than 6% bleach + 0.1% Tween. Ensure tubes are not overfilled with seeds (<50%) and shaken vigorously for efficient sterilization. Complete removal of bleach post-washing is crucial. Falcon tubes can be reused.
2. 0.3% Sucrose can be added, but is not essential for satisfactory growth. Ensure correct embryo orientation (upward, not touching medium). The plate angle is critical; 20° is generally needed, but Bd21-3 grows well at 15°, while Bd21 may require >20° to prevent roots growing out of the medium.
3. For faster seed setting: Store plates with *in vitro* grown 7-day-old seedlings in the dark at 4°C for ~10 days before transferring to soil.
4. Brachypodium roots (esp. Bd21-3) brown quickly upon air exposure; immerse immediately in fixative. Use vacuum infiltration for fixative penetration (3–5 rounds, max 2 h total). Fix at 4°C for preservation of fluorescent proteins (if relevant). Handle samples gently, using soft brushes if needed.
5. Standard ClearSee is unsuitable for Brachypodium roots due to inefficiency in clearing later developmental stages, causing tissue browning/shrinkage. Preconditioning with diluted ClearSee partially mitigates shrinkage but does not improve transparency. DEEP-Clear provides superior transparency for Brachypodium roots within 5–10 days, reduces browning within 3 days, and is compatible with fluorescent proteins, Renaissance, Calcofluor White, Propidium Iodide,

Fig. 2 The Deep-Clear method is compatible with fluorescent proteins and dyes. (**a**) Arabidopsis fully emerged LRP (left) and root tip expressing *NLS::GFP* with cell walls stained by Renaissance for cellulose (gray). (**b**) LRP at stage 4 and (**c**) stage 7, both stained sequentially with Renaissance (gray) and Fluorol Yellow (yellow) to highlight endodermal suberization. (**d**) LRP region stained sequentially with Basic Fuchsin for lignin (magenta) and Renaissance (gray). (**e**) *NLS::GFP* combined with Basic Fuchsin and Fluorol Yellow, illustrating Casparian strips as dot-like structures near a suberized endodermal cell covering a young LRP (arrowheads). Scale bars: 20 μm

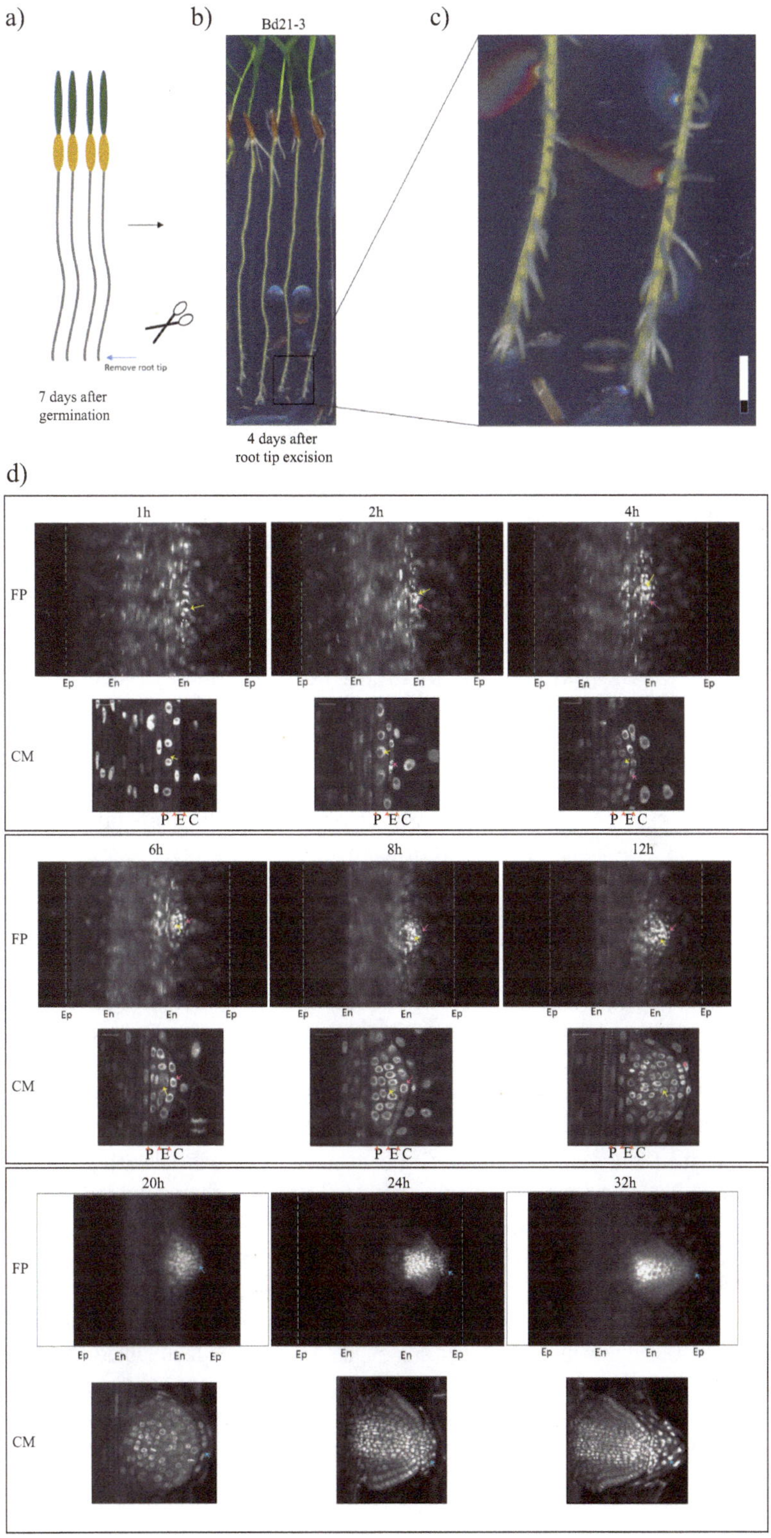

a)
b)
c)
Bd21-3
Remove root tip
7 days after germination
4 days after root tip excision
d)
1h
2h
4h
6h
8h
12h
20h
24h
32h
FP
CM
Ep
En
P E C

and Fluorol Yellow. For Basic Fuschin, prepare DEEP-Clear without the Sodium Sulfite as it is not compatible with this staining.

6. Ensure complete fixative removal before clearing. Use vacuum infiltration (three rounds) when starting clearing. Incubate at 28°C (shaking, 100 RPM) in screw-cap glass tubes for faster clearing. Change the solution if it discolors.
7. Apply vacuum (one round) for stain penetration in thicker tissues. Overnight washing after staining (Calcofluor/ Renaissance) reduces background. DEEP-Clear allows quick visualization of PI-stained LRPs with a standard fluorescence microscope. DEEP-Clear works well with many other stainings; for more information on multiple staining, please refer to the following protocol [15].
8. Chambered cover glasses (Nunc™ Lab-Tek™) are helpful to ensure tissue integrity during the mounting (only for inverted microscopes). Gently place a block of plant agar on top of the cleared roots to avoid movement or dehydration of the sample. Use slide spacers (2–3 strips of parafilm) in case these chambers are not available or when using an upright microscope. This protocol can be used for *A. thaliana* as well (Fig. 2).
9. Seedlings must grow for at least 5–7 days before RTE for optimal LR induction; shorter periods favor adventitious root growth. LRs near the excision site (~5 mm) show more synchronized development. Handle roots carefully during cutting. Collecting multiple samples at once reduces air exposure.
10. Properly dried and stored seeds can remain viable for at least 5 years. Harvesting green seeds reduces viability and increases contamination risk.

Fig. 3 LR synchronization in Brachypodium. (**a**) Compensatory LR growth following root tip excision as an LR Induction System (LRIs). Representative plants were grown on ½ MS medium for 7 days before the root tip was excised. (**b**, **c**) After 4 days, LRs near the excision site showed visually homogeneous emergence. This observation prompted us to explore this method as a potential system for inducing lateral root formation. Scale bar: 3 mm. (**d**) Root tip excision triggers cell divisions in the pericycle within 1 h and LR emergence after approximately 32 h. Root tip excision synchronizes LRP development at predefined time intervals following the removal of the root tip. Clearing with DEEP-Clear and staining with propidium iodide enables visualization of LRPs from the initial cell divisions in the pericycle, employing both fluorescence microscopy (FP) and confocal microscopy (CM). Yellow arrowheads indicate cell divisions in the pericycle; purple arrowheads denote cell divisions in the endodermis; blue arrowheads highlight the formation of the columella and root cap. Ep epidermis, En, E endodermis, P pericycle. Scale bars: 100 μm or FP and 100 μm for CM

Acknowledgments

Many thanks to the Vermeer lab members for their continuous feedback during this project. Work in the Vermeer lab was supported by the Swiss National Science Foundation (Project 157524 and 197568 awarded to JEMV) and the University of Neuchâtel. KB was supported by Marie Curie Global Fellowship (PLANT-ID, grant agreement: 101106663).

Author Contributions CDJVT and JEM designed the protocols and the figures. KB and CDJVT performed the experiments and analyses. KB wrote the manuscript with input from all coauthors.

References

1. Kenrick P, Strullu-Derrien C (2014) The origin and early evolution of roots. Plant Physiol 166:570–580
2. Dolan L, Roberts K (1995) The development of cell pattern in the root epidermis. Phil Trans: Biol Sci 350:95–99
3. Malamy JE, Benfey PN (1997) Organization and cell differentiation in lateral roots of *Arabidopsis thaliana*. Development 124:33–44
4. Yu P, Hochholdinger F, Li C (2019) Plasticity of lateral root branching in maize. Front Plant Sci 10:439375
5. Beeckman T, Eshel A (2024) Plant roots: the hidden half, 5th edn. CRC Press, Boca Raton
6. Casero PJ, Casimiro I, Lloret PG (1995) Lateral root initiation by asymmetrical transverse divisions of pericycle cells in four plant species: *Raphanus sativus*, *Helianthus annuus*, *Zea mays*, and *Daucus carota*. Protoplasma 1995 188:49–58
7. Stoeckle D, Thellmann M, Vermeer JEM (2018) Breakout—lateral root emergence in *Arabidopsis thaliana*. Curr Opin Plant Biol 41:67–72
8. Amtmann A, Bennett MJ, Henry A (2022) Root phenotypes for the future. Plant Cell Environ 45:595–601
9. Péret B, Larrieu A, Bennett MJ (2009) Lateral root emergence: a difficult birth. J Exp Bot 60:3637–3643
10. Garvin DF (2007) Brachypodium: a new monocot model plant system emerges. J Sci Food Agric 87:1177–1179
11. Vogel J (2008) Unique aspects of the grass cell wall. Curr Opin Plant Biol 2008 11:301–307
12. Raissig MT, Woods DP (2022) The wild grass *Brachypodium distachyon* as a developmental model system. Curr Top Dev Biol 147:33–71
13. De Jesus Vieira Teixeira C, Bellande K, van der Schuren A et al (2024) An atlas of *Brachypodium distachyon* lateral root development. Biol Open 13:bio060531
14. Bellande K, Vieira Teixeira CDJ, Malai M et al (2026) Root tip excision-induced exodermis lignification impacts lateral root emergence in Brachypodium distachyon. New Phytol, 249: 2894–2908
15. Ursache R, Fujita S, Dénervaud Tendon V, Geldner N (2021) Combined fluorescent seed selection and multiplex CRISPR/Cas9 assembly for fast generation of multiple Arabidopsis mutants. Plant Methods 17:111

Chapter 6

Live Imaging of Grass Roots Growth Dynamics Using Microfluidics

Lev Levenets, Dominik Kralík, Vasyl Brykov and Matyáš Fendrych

Abstract

Microfluidic tools enable manipulation of microliter-scale volumes within a stable microenvironment. Microfluidics has been widely adopted in various biological fields for, among other applications, controlled delivery of sensory, chemical, and mechanical stimuli to the observed organisms. In this chapter, we describe the design and usage of a custom-made double-layered PDMS microfluidic device, named BUGchip, developed for the purpose of studying plant root growth and metabolic responses in rapidly changing conditions with a high time-scale resolution. We demonstrate the utility of this device by a showcase experiment with a hyperosmotic treatment of *Brachypodium distachyon* roots, in which changes in growth rate and relative pH across the root surface in response to the treatment can be observed within minutes.

Keywords Microfluidics, Microscopy, Live imaging, Root growth, Noninvasive treatment, *Brachypodium distachyon*

1 Introduction

Typically, microscopical observations of live plants utilize mounting of the sample either between a glass slide and a coverslip (if the sample is small enough) or between a block of transparent gelled medium and a coverslip. While easy in execution, this common technique has several drawbacks. For example, there is inhomogeneity in the environment created at the border between the glass coverslip and the block of gelled growth medium; some plant organs (e.g., root hairs) are intrinsically sensitive and do not survive mounting manipulations well, and, most importantly, a controlled application of treatments while live imaging is a complicated and hard-to-replicate procedure. These shortcomings can be overcome by microfluidics, a technology widely used in the study of cell and tissue cultures that has found applications in animal and plant biology as well [1]. A microfluidic chip is an array of tiny channels etched in or molded on a block of translucent material, with capillary tubes connecting its inlets to a

Steffen Vanneste and Inge Verstraeten (eds.), *Root Development*, Methods in Molecular Biology, vol. 3086,
https://doi.org/10.1007/978-1-0716-5603-7_6,

pressure-based controller, allows for the manipulation of small volumes (μL) in a hydrodynamically stable microenvironment, with high precision in time and space [2], therefore making it a superb solution for studying rapid reactions such as calcium transients, membrane depolarization, pH changes and growth inhibition in plants [3–5]. These microfluidic chips are usually fabricated from Polydimethylsiloxane (PDMS), a type of silicone polymer, valued for its properties such as optical transparency, biocompatibility, chemical inertness, hydrophobicity, and gas permeability, combined with relative cheapness and plasticity as a material [6].

In the following chapter, we provide a detailed description of the design and use of a closeable double-layered PDMS microfluidic device (henceforth referred to as BUGchip, owing to the pattern formed by the device channels) manufactured via photolithography and PDMS casting. The BUGchip (Fig. 1A) is composed of two

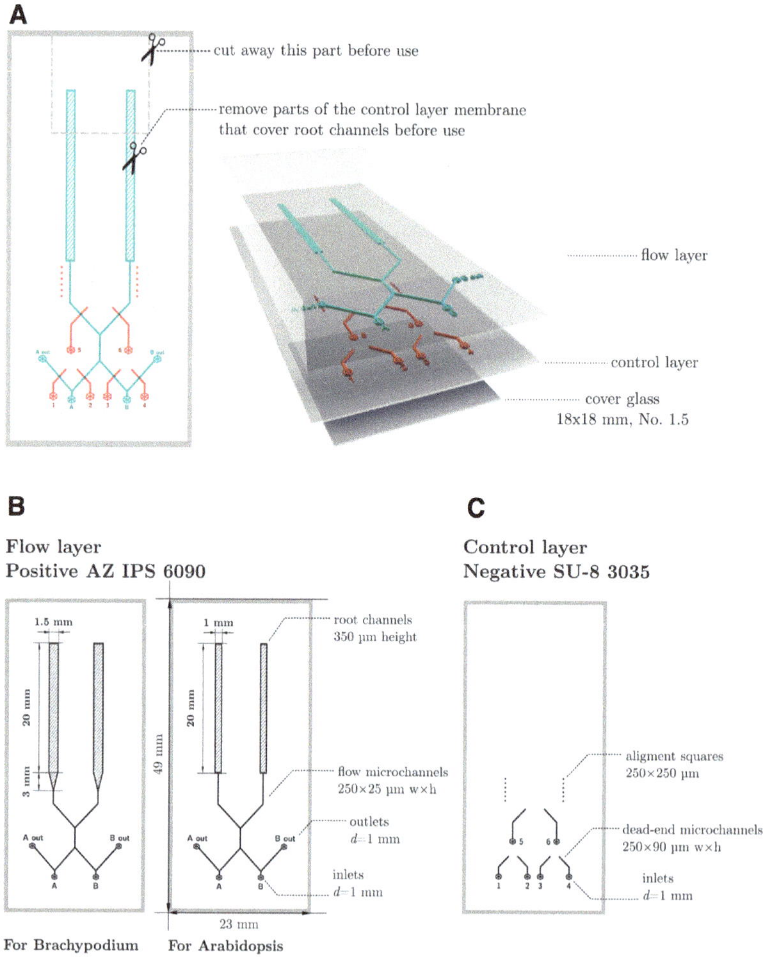

Fig. 1 Design of the BUGChip. (**A**) Overall chip in the assembled state, top view (left) and side view (right). Scheme and dimensions for the fabrication of (**B**) flow layer and (**C**) control layer

layers of channels: (1) a flow layer, containing inlet/outlet flow ports for growth media and channels meant to accommodate plants, and (2) a control layer with pressurized microvalves located below the flow layer microchannels, allowing for rapid treatment switches. The BUGchip is designed to allow a constant medium flow of both mock and treatment sources. The control over the type of solution that is currently reaching the plants is achieved by Quake pneumatic microvalves [7] grounded in the chip's double-layered design. These valves apply selective pressure to the chip control layer, which leads to the PDMS membranes bulging inwardly, thereby occluding the fluid flow in the flow layer channel below. Unlike many common microfluidic devices employed in plant biology, such as the Root-Chip [8], the BUGchip's root channels have an open construction [9], so plants can be sampled from conventional agar plates and mounted into the chip directly before imaging, minimizing sample contamination risks and increasing the throughput of the system. In addition, thanks to the open construction, the BUGchip can be reused countless times. Minor amendments in the chip design during the fabrication process allow for the creation of devices able to accommodate plants of sizes several times bigger than *Arabidopsis thaliana* seedlings (Fig. 2B). Therefore, in addition to the chip and microfluidic set-up description, we demonstrate its use with seedlings of the common grass model *Brachypodium distachyon*. We showcase an example experiment where the back-and-forth exchange between mock medium and hyperosmotic 150 mM D-Mannitol solution causes rapid and reversible root growth inhibition, cell shrinkage, and changes in the pH gradient along the root surface. We monitor root

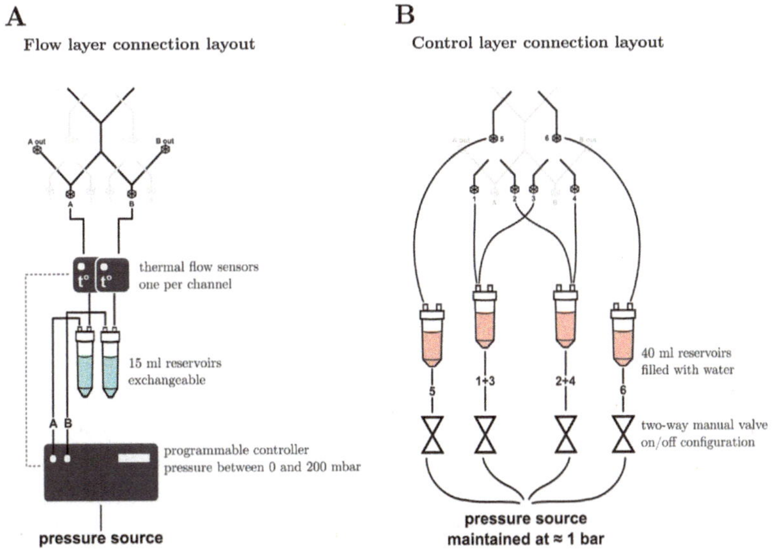

Fig. 2 Connection layout scheme for the BUGchip. (**A**) Flow layer connection layout. (**B**) Control layer connection layout

growth dynamics using a vertical-staged microscope; nevertheless, the microfluidic chips can be mounted on any microscopy system of choice, given that the spatiotemporal resolution is sufficient to analyze the desired imaging readout.

2 Materials

2.1 BUGchip and Microfluidic Hardware

The BUGchip is a closable double-layer polydimethylsiloxane (PDMS) microfluidic device for studying seedling root growth. It currently comes in two versions, the main difference being the size and shape of the root channels. One is optimized to accommodate *Arabidopsis* seedlings and the other for *Brachypodium*, or similarly sized plants (Fig. 1B). The fabrication of the BUGchip itself requires specialized equipment; we therefore recommend collaborating with a team experienced with and equipped for photolithography and PDMS devices fabrication or to fabricate the chip using commercially available services.

2.1.1 BUGchip Design and Fabrication

The BUGchip is a rectangular PDMS block of approximately 49 × 23 × 6 mm, consisting of two layers, conventionally referred to as "flow layer" and "control layer." Its flow layer (Fig. 1B), residing on a thick PDMS block, contains two channels meant to accommodate plant roots (root channels), two inlet ports (labeled "A" and "B" on the scheme), each branching into two microchannels, and two outlet ports (labeled "out A" and "out B" on the scheme). The diameters of channels and ports are 250 μm and 1 mm, respectively. Microchannel structures of this layer have a height of around 20 μ m and are rounded on one side, forming a D-shaped cross-section. This roundness is achieved by using a positive photoresist along with higher temperatures during the fabrication process. The chip control layer (Fig. 1C) is made from a thin PDMS membrane adhered to the flow layer PDMS block and consists of six dead-end microchannels (labeled 1–6 on the scheme in Fig. 2) located directly below the flow layer microchannels. Control layer microchannels are 250 μm wide and 90 μm in height. From the lower side they are protected by a 18 × 18 mm coverslip glass covalently linked to the PDMS block. Pressure application to the microchannels of this layer leads to membrane bulging inward, which allows for rapid occlusion of flow layer channels above them, thereby allowing for flow direction control.

The BUGchip's fabrication process consists of two steps: photolithography and PDMS casting. Photolithography is done using a 6.35 cm phosphor bronze wafer pretreated with aqueous nitric acid to remove surface imperfections and to create a mat finish. A thin layer of photoresist is applied to the wafer using a spin coater. The chip's flow layer is fabricated using positive photoresist AZ IPS 6090, while the control layer is fabricated with negative photoresist SU-8 3035. Lithography is performed using a maskless optical recorder, MicroWriter ML3, which allows for direct design on the photoresist using focused UV light.

PDMS casting is done separately for the flow and control layers with RTV 615 as the PDMS material for both. The flow layer is cast out of 17 g of RTV 615 premix (5:1 base to curing agent ratio) by pouring it into a 3D printed mold used to accommodate a 2.5-in. wafer with the flow layer's structure, followed by desiccating and baking it. The control layer is cast using around 4 g of deaired RTV 615 premix (20:1 base to curing agent ratio) by spin coating and then baking the coated wafer. Once baked, the flow layer mold is cut into rectangles of the appropriate size, and four holes for the inlets and outlets are punched using a 0.5 mm inner diameter (ID) biopsy punch needle (ref). The flow layer rectangles are aligned to the control layer wafer coated with the thin control PDMS layer and baked once again. Double-layered rectangles are carefully separated from the wafer, the excess thin layer is cut away using a sharp blade, and the remaining inlets for the control layer are made using the same 0.5 mm ID biopsy punch needle. A cleaned and degreased 1.5 18 × 18 mm coverglass is then bonded to the PDMS chip using a corona plasma treater (ref).

2.1.2 BUGChip Installation

All the following steps can be performed in a standard plant biology laboratory.

The flow control scheme is depicted in Fig. 2. The flow layer's hardware (Fig. 2A) consists of a pressure controller (OB1, 0–200 mbar; ElveFlow) independently connected to two reservoirs corresponding to inlets A and B, and two thermal flow sensors (MFS-2+, 0–7 μL/min; ElveFlow), one per microchannel. This configuration ensures a constant and controlled flow of medium into the chip flow layer, which is a prerequisite for controlled experimental conditions. The flow layer's hardware is operated via the ESI software (ElveFlow), which enables communication between the pressure controller and the flow sensors, allowing regulation of either a preset flow rate or pressure.

The control layer's hardware (Fig. 2B) comprises four water-filled reservoirs connected to a compressed air source via a manual pressure regulator (Fig. 3A), which sets the input pressure to approximately 1 bar. Two of the reservoirs (1 + 3 and 2 + 4) supply pressurized liquid to the control layer of the chip, enabling redirection of the medium from inlets A or B either into the root channels or toward the waste outlet ("A out" or "B out"). The remaining two reservoirs (5 and 6) independently switch off medium flow to the root channels. This arrangement allows roots in two root channels to be treated either sequentially or simultaneously.

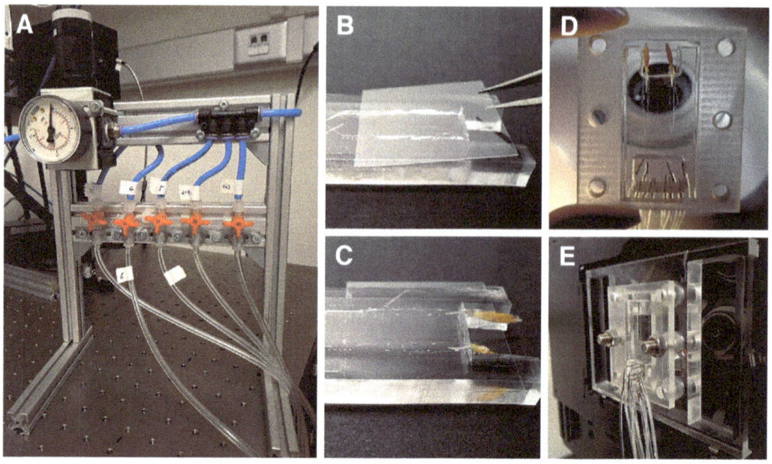

Fig. 3 (**A**) Rack with manual valve switches used to control the BUGchip's control layer. (**B**) The BUGchip is being sealed with a coverslip. (**C**) *B. distachyon* seedlings mounted in the root channels. (**D**) Sealed and connected the BUGchip with seedlings inside. (**E**) The BUGchip is mounted in its custom-made chip holder on a vertical microscope stage

Immediate connection of the pressure source to the liquid reservoirs for each layer is done via Tygon LMT-55 tubes with an outer diameter (OD) of 5.5 mm. The liquid reservoir to flow layer connection is made by PTFE Tubing (1.5 mm OD, 0.5 mm ID); the liquid reservoirs to control layer connection is made by Tygon Tubing (1.5 mm OD, 0.5 mm ID). Connecting the tubes to the chip's inlets and outlets is done through stainless steel connection pins (0.64 mm OD, 0.33 mm ID; New England Small Tube code 23RW).

2.2 Seed Sterilization and Germination

1. ½ Murashige and Skoog (MS) agar plates: 2.25 g/L MS basal salt mix, 0.5 g/L 2-morpholinoethanesulfonic acid monohydrate (MES), 8–10 g/L agar, and optionally, 10 g/L sucrose. pH was adjusted to 5.7 using 0.1 KOH solution. The medium was autoclaved and poured into plates (approximately 50 mL per plate). These plates can be stored at 4°C for up to 1 month.
2. Bleach solution: 5.25% NaOCl.
3. Tween 20 stock solution: 10% in water.
4. Any type of small sterile closable container—microcentrifuge tubes, falcons, small flasks.
5. Sterile filter paper.
6. Laminar flow hood.
7. Growth chamber or thermostat suitable for germination of *B. distachyon* seeds, set to 23°C–28°C. Light conditions are irrelevant as seedlings are used 24–36 h after germination.

2.3 Experimental Treatments

We recommend filtering all solutions that will come in contact with microfluidic hardware using 0.22 μm syringe filters. This is to avoid blocking the microchannels by introducing unwanted particles.

1. ½ liquid MS medium: 2.25 g/L MS salts, optionally add 10 g/L sucrose. Adjust pH to 5.7 using 0.1 M KOH. Considering that this solution contains no buffering reagent, perform the pH adjustment in small steps (*see* **Note 1**).
2. D-Mannitol: 98+%, powder.
3. Fluorescein-5-(and-6)-sulfonic acid (FS) Trisodium salt stock solution: 50 mM, in water—used for assessment of relative pH. Store aliquots at −20°C, protect from light.
4. Tetramethylrhodamine dextran (TMR-Dextran), 10,000 MW, stock solution: 10 mg/mL in water—used as tracer dye to monitor medium switch during the experiment (*see* **Note 2**). Store aliquots at −20°C, protect from light.
5. Centrifuge tubes of 15 mL.

2.4 Other Solutions, Reagents, and Supplies

1. Confocal fluorescent microscope equipped with 405, 488, and 561 nm lasers.
2. Dissecting microscope or eye loupe.
3. Any computer suitable for running the ImageJ software.
4. Fine metal tweezers.
5. Coverslip glasses, 22 × 22 or 24 × 24 mm.
6. Microcentrifuge tubes of 1.5 or 2 mL.
7. Tissue wipes, preferably lint-free.
8. Adhesive tape.
9. Rubber dust-blower.
10. Ethanol, 70%.
11. Sharp blade or modeling knife.

3 Methods

3.1 Seed Sterilization and Plant Germination

1. Select dry, ripe *Brachypodium* seeds. Dehusk each seed by grabbing the seed awn either with fine tweezers or your fingers and peeling the seed lemma away. Collect the dehusked seeds in a microcentrifuge tube (up to 50–60 seeds fit in one 2 mL tube) or any other container. Flood the seeds with 20% bleach solution and add a drop of 10% Tween 20 solution. Invert the tube a few times to break the surface tension and allow the seeds to settle in between. Afterward, place the tubes on a shaker for 15 min.

2. Transfer the tubes with bleach-soaked seeds to a sterile laminar flow. Remove the bleach solution either by decanting or by pipetting, fill the tubes with sterile water, and invert them a few times. Decant this washing solution, and replace it with fresh water. Repeat this washing step at least three times. Once bleach residues are removed, fill the tubes with a last volume of sterile water, close them, and keep them soaking overnight at 4°C.
3. Prepare ½ MS agar plates. Move the imbibed seeds to a laminar flow cabinet, dry the seeds on a piece of sterile filter paper, then transfer them to the prepared plates using sterilized tweezers. Seeds can be comfortably planted in one or two rows of 15–25 seeds per row. The seeds should be placed with their palea side (flat side) facing the agar and the embryo pole facing down. Seal the plates with micropore tape and move them to a growth chamber or thermostat. Plates should be placed under a slight incline to ensure germination in contact with the agar. Plate seeds approximately 24–36 h before a planned microscopy experiment (*for Arabidopsis, see* **Note 4**).

3.2 Microfluidic System Start, Finish, and General Handling

1. When using it for the first time, place the BUGchip backward on a hard surface. At the chip's end, opposite the flow inlets, cut out a small block of PDMS (approximately 0.5 × 0.5 cm), including a small part of the root channels. You have now freed a zone meant to accommodate the seedlings. Cutting is most convenient with a chiseling blade. Then, under a dissecting microscope using a razor blade, cut away part of the PDMS membrane layer covering the root channels (Fig. 1A, top view). During this and all subsequent handlings, take extreme care not to break the cover glass on the control layer.
2. Ensure that all microfluidic reservoirs (for both flow and control layer) are filled with ultrapure water and connect the chip. Place the chip facing down on a hard surface, take a pin connected to capillary tubing and one of the control layer bottles (inlets labeled 1–6), pressurize the selected bottle, let the tubing fill with water, and allow free flow for a few seconds. Stop the flow and gently insert the pin in the respective inlet. Repeat this for all other control layer inputs. Once all connections are inserted, pressurize all the control layers' inlets and let water fill the channels, pushing away the air. Monitor water flow by checking the chip under a dissecting microscope. When done for the first time (or after disconnecting chips for a long time), this step can take up to 30 min. Once the control layer is prepared, connect the flow layer channels (inlets labeled "A" and "B") and outlets ("out A" and "out B") similarly. Wash them with water for several minutes (*see* **Note 3**) or until the air bubbles are gone from the microchannels. Afterward, test the

tightness of the pneumatic valves by setting the pressure in the flow layer to 100 mbar while increasing the pressure in the control layer to completely stop all medium flow. Usually, this happens around 1 bar pressure, depending on the thickness of the membrane of the control layer and the shape of the flow layer's channels. Now the system is ready, so swap the water-filled reservoirs for the flow layer to those containing your media and proceed with the experiment.

3. Once the experiment is finished, perform the microfluidics clean-up steps to reuse the chip (see below).

3.3 Treatment Preparation, Sample Mounting

1. Mix ½ liquid MS medium with FS dye: use 1 μL of stock solution per mL of medium. Separate this into two aliquots: one for the mock and one for the treatment. To the treatment solution, add D-Mannitol to achieve a final concentration of 150 mM and adjust the pH to be as close as possible to that of the mock solution. Filter both solutions into 15 mL falcons using a nylon membrane filter.
2. Add 0.25 μL of tracer dye stock solution per mL of medium to the mock solution and 0.5 μL per mL to the treatment solution tubes, mix by inverting.
3. Replace the water-filled falcons for the flow layer with the falcons containing the experimental solutions. Connect the mock to inlet "A" and the treatment to inlet "B." Pressurize both inlets to replace the water. Then pressurize control channels 1 and 3, thereby diverging the treatment medium into "out B" and allowing only mock medium into root channels.
4. Fill a microtube with ½ liquid MS medium. Collect several *Brachypodium* seedlings from the agar plate and put them vertically (roots down) into the microtube. When moving seedlings, hold them with fine tweezers by the seed that is still attached to the plant at that stage. Try to carefully stick the seedlings to the wall of the microtube to prevent them from sinking or freely floating around. Leave for 5–15 min. This step is necessary to remove mucilage and border cells shed from the root cap (skip this step if working with Arabidopsis).
5. Meanwhile, prepare the microfluidic chip. Depressurize the microchannels in the flow layer and wipe away medium droplets collected on the chip surface during Step 3. Degrease the chip surface with ethanol and let it dry for a minute. Remove any dirt and dust particles from the chip surface using a rubber dust blower and adhesive tape. Take a 22 × 22 or 24 × 24 mm coverslip and dedust it the same way, place the coverslip over the root channels, and carefully press it to ensure it is sealed along its perimeter (Fig. 3B). Fill the now sealed root channels with mock medium.

6. With fine tweezers, take a *Brachypodium* seedling out of the microtube. Holding it by the seed, gently press the sides of its root tip against a piece of filter paper to remove shed border cells. With one hand, hold the sealed chip vertically at eye level, with the other insert the *Brachypodium* seedling from the top directly into the channel opening with the flat side of the seed facing the coverslip (Fig. 3C, D). If necessary, adjust the seedling's position to center the root tip in the channel (*for mounting of Arabidopsis seedlings, see* **Note 4**).
7. Make sure that after sample mounting, the chip is properly sealed and there are no air bubbles in any channels (*see* **Note 5**). Set the medium flow rate to 3–5 μL/min. Make sure that the liquid level is maintained and the liquid flow is constant when holding the chip vertically above the level of the liquid reservoirs. In case of improper sealing or any other unwanted result, remove the plants, clean the surface of the chip, and repeat the sealing and mounting procedure. Once the desired result is achieved, put the chip into its holder and mount it on the microscopy stage (Fig. 3E). Before starting imaging, let the mounted seedlings acclimatize in the chip with a running flow of mock solution for at least 20 min.

3.4 Timelapse Acquisition

1. In your microscope software, set up either simultaneous or sequential imaging with 488/520 nm Ex/Em (emission channel for FS dye), 405/520 nm Ex/Em (reference channel for FS dye) [10], and 555/580 nm Ex/Em wavelength (red tracer tracking channel) and, optionally, the brightfield channel. To capture the whole root tip, use a 5× or 10× objective.
2. Depending on your experimental goal, set up the time interval and desired number of captures. For experiments with rapid medium replacement, intervals between 30 s and 2 min usually work best.
3. Keep treating the samples with the mock solution (control layer valves 1 and 3 are pressurized).
4. Without stopping the image acquisition, depressurize the control layer microchannels 1 and 3 and simultaneously pressurize microchannels 2 and 4, thereby diverting mock solution to "out A" and allowing the treatment solution to flow into the root channels. Verify successful medium replacement by monitoring the tracer dye fluorescence in the 555/580 nm channel (*see* **Note 6**).
5. If desired, several back-and-forth switches between mock and treatment solutions can be performed in series to assess reversibility of the observed reactions.
6. Finish the experiment and clean the system. Swap the flow layer reservoirs back to those containing ultrapure water,

depressurize all the control layer channels, and flush the system with water long enough to get rid of all solutions used during the experiment. Then, swap the flow layer reservoirs to reservoirs containing 70% EtOH and flush the system sufficiently to replace all water with ethanol to prevent any kind of microbial contamination. Unless you want to replace the currently used microfluidic chip with another one, it is neither necessary nor advisable to disconnect it from the system. It is safe to keep it connected to the system.

3.5 Simple Semi-Manual Image Processing Using Fiji (ImageJ)

3.5.1 Estimate the Time of the Treatment Arrival by Tracking Changes in Tracer Dye Fluorescence

1. Either split the obtained stack into separate channels by going to Image > Color > Split Channels tool and select the one corresponding to tracer dye emission, or switch to this channel in the image stack.
2. On the first image of the time series, select a sizable area directly below the root tip using the Rectangle selection tool from the ImageJ toolbar.
3. Add this selection to the ROI manager either by going to Analyze > Tools > ROI Manager and clicking "Add" or by using the hotkey "t."
4. In the ROI manager, go to More > Multi Measure, check the box for "Measure all N slices."
5. In the output results table, make sure that one of the columns corresponds to the mean gray value (column header usually labeled as "Mean"). If no such column is presented, go to Results > Set Measurements and check that "Mean gray value" is ticked, then repeat the measurement.
6. Export the Results table by going to File > Save as... in the Results window.
7. Plot the Mean gray values of the tracer dye against time (depending on the selected interval in Sect. 3.4) to follow the exchange of mock and treatment media over time (Fig. 4A, B).

3.5.2 Estimate Root Growth Rate

1. Using the rectangular selection, in the first timeframe, select a small area solely containing the very end of the root tip.
2. Go to Plugins > Registration > Correct 3D Drift.
3. Check boxes for multi-time scale computation, sub-pixel drift correction, and edge enhancement. For registration, choose the channel with the highest background-to-foreground contrast (likely brightfield), and proceed by clicking "Ok."

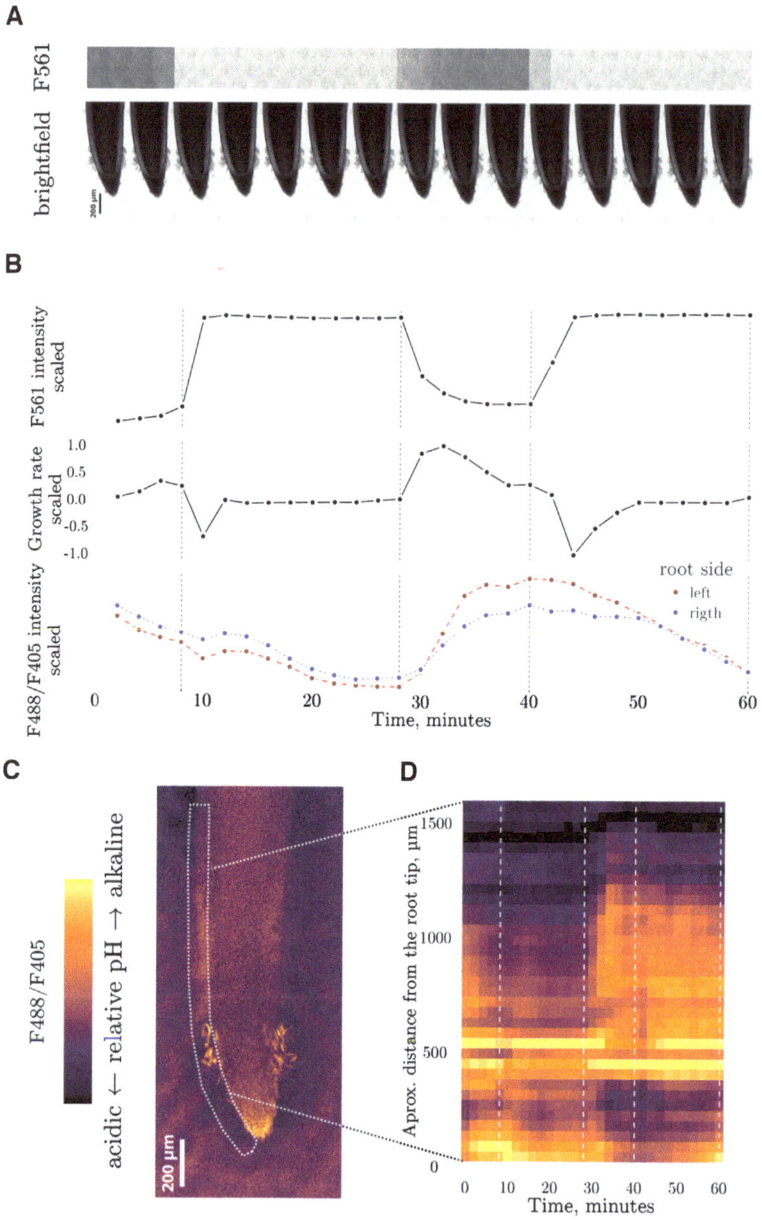

Fig. 4 Example of the acquired data. (**A**) Frame-by-frame time series of the tracer dye fluorescence in the 561 nm channel (top), together with the root tip growth captured in the brightfield channel (bottom). (**B**) Scaled tracer dye fluorescence intensity (F561), growth rate, and relative root surface pH (F488/F405 ratio) plotted together over time. Dashed vertical lines indicate medium replacements. (**C**) Ratiometric image demonstrating the relative pH gradient along the root tip surface. (**D**) Kymograph-like plot of the region outlined in subfigure C, demonstrating pH fluctuations along the root surface over time. The white vertical dashed lines indicate medium replacements

4. Assess the quality of the registered stack. If satisfactory, perform the same procedure once again, but this time with the "*Only compute drift vectors*" option selected.
5. The output text file will contain cumulative estimated displacements in XYZ, in pixels, for the region of interest for each time point. Calculate vector magnitude for each time point to obtain an estimation of the overall increment in root length for the time elapsed, and calculate the difference in displacement between two time points to obtain an estimation of the growth rate.
6. Plot the obtained growth measurements together with the information about the tracer dye fluorescence to correlate the observed changes in the growth rate with the specific treatment (Fig. 4A, B).
7. The registered and cropped image stacks can be used in some of the following examples to compensate for the root movement and simplify further measurements across the time series.

3.5.3 Create a Ratiometric Image Reflecting Relative pH Values along the Root Surface

1. Go to Image > Color > Split Channels.
2. Go to Process > Image Calculator.
3. Select the FS dye emission channel as the first image, the FS dye reference channel as the second image, and division as the operation type, check the box to create a 32-bit (float) result.
4. On the resulting ratio image (Fig. 4C), brighter areas will correspond to zones with relatively higher pH and vice versa.

3.5.4 Create a Mask Outline and Measure Relative pH Changes along the Root Surface Across Timepoints

1. Select the channel with the highest background-to-foreground contrast (likely brightfield).
2. Optionally, apply a smoothing filter to reduce noise and enhance the segmentation results. Go to Process > Filters and select either Gaussian, Mean or Median filter, setting the filter kernel value between 2 and 15 will usually produce good results.
3. Press "Ctrl + Shift + T" to evoke the Threshold menu. Select the desired threshold algorithm (Default or Otsu usually produce good results) and click "Auto" to automatically calculate the threshold value for segmentation.
4. Click "Apply" to create a segmentation mask. In the pop-up menu, select the "Dark" option for the background, select the option to calculate the threshold for each image, and, optionally, select the option to create a new stack. If necessary, you can manually clean up the segmentation mask to remove undesired high-contrast elements (e.g., air bubbles forming on the channel wall, floating border cells around the root tip).

5. Duplicate the newly created masked stack by pressing "Ctrl+Shift+D." Make sure that in the pop-up menu, the option to "Duplicate stack" is inactive.
6. Select either one of the masked stacks, go to Process > Find edges, and apply it to the whole stack.
7. Select the stack with edge outlines, go to Process > Binary > Options, in the pop-up menu select "Do: Dilate" and set the iteration count between 5 and 15.
8. Go to Process > Image Calculator... and apply the "AND" operation between the initial masked stack and the stack containing the dilated edge outlines. The obtained mask will contain an n-pixel-thick outline of the root, but it excludes the root surface itself.
9. Convert background zero values to NaNs. To do so, select the outline masked stack, evoke the Threshold menu again (Ctrl+Shift+T) and click "Auto," do not apply a calculated threshold value. Go to Process > Math > NaN Background and apply it to the full stack. Remove threshold highlighting by pressing "Reset" in the threshold menu.
10. Optionally, convert the segmented parts of the mask to the value 1 to preserve the values of the original image after mask application. Go to Process > Math > Divide... and divide the masked stack by 255.
11. Apply the mask to the ratiometric image of the FS dye by Process > Image Calculator... and multiplying the ratiometric image by the outline mask, make sure that the result image is in 32-bit format.
12. To measure the mean ratio value in the outlined region over time, either select the whole image or a specific region of the outline (e.g., only the left or right side of the root), add it to the ROI manager by pressing the "t" hotkey, and in the ROI manager, go to More > Multi Measure and measure all N slices.
13. Export the Results table by going to File > Save as... in the "Results" window.
14. Plot the obtained ratio values over time together with the information about tracer dye fluorescence and the growth rate to correlate observed changes with the specific treatment (Fig. 4B).

3.5.5 Measure the Relative pH Gradient Along the Root Surface

1. Select the FS output ratio image.
2. In the ImageJ toolbar, right-click the line icon and select the "Segmented Line" option.

3. Double-click on the line icon and specify the line width parameter between 10 and 25 pixels.
4. Starting from the root tip, draw a segmented line following the root outline on one side.
5. Either go to Analyze > Plot Profile or evoke it by using the "Ctrl + k" shortcut. The profile plot shows the mean (unless the line width was set to 1 px) gray value for each pixel along the line.
6. Repeat this measurement for the opposite side of the root to compare the pH gradient symmetry between the left and right root sides.
7. The values can be exported by going to Data > Save Data... in the plot window.
8. Given that the image stack was previously registered to negate root movement in time, it is possible to obtain the profile plot values for each slice in the stack using the loop call. In the ImageJ menu bar, go to File > New > Script..., select ImageJ Macro as the scripting language. Paste the next command into the scripting window.

```
"'
n = nSlices;
for (i = 1; i <= n; i++) {
            setSlice(i);
            roiManager("Add");
};
run("Select All");
roiManager("Multi Plot");
"'
```

9. Firstly, select the first timeframe of the FS ratio output stack and draw a segmented line on one side of the root as described above. Then, running a loop in the script window will propagate the defined ROI to all other slices in the stack and then call the function to plot the profiles.
10. Combined profile plots of each timeframe will appear. Export the plot values as before.
11. For multiple time frames, a more convenient way to represent fluctuations in pH gradient over time is to create a kymograph-like plot (Fig. 4D) where (usually binned) distance and time values are mapped to the XY axes and pH ratio values are mapped to the color.

4 Notes

1. Unbuffered media are only used for pH assessment using the FS dye. Otherwise, use ½ MS liquid medium buffered with 2-morpholinoethanesulfonic acid monohydrate.
2. A wide range of dextran conjugates with various Ex/Em spectra is available on the market. TMR-Dextran is used in this showcase because its emission spectrum does not overlap with that of the FS dye.
3. The time needed to properly wash the system for the first time and/or replace water for the experimental media depends on the flow rate and, consequently, the pressure maintained in the system. For washing the system or full medium replacement, keep the pressure above 200 mbar.
4. To mount Arabidopsis seedlings, clean the chip surface, place the chip horizontally and prefill root channels with mock medium, use a 10 μL pipette to evenly spread the medium droplet along the channel cavity. With fine tweezers, carefully take a 3- to 4-day-old Arabidopsis seedling from the agar plate, holding it by the hypocotyl, and place it into the channel. Try to place the root tip in the middle of the channel and then gently drag the seedling up the channel. Seal the chip using a 22 × 22 mm coverslip. Work without gloves to prevent the formation of static electricity.
5. Air bubbles formed in the root channels can sometimes be removed by increasing the medium flow rate and tapping on the coverslip glass that seals the root channels.
6. If both mock and treatment medium have already replaced the water in the system, the speed of switching between media in the chip itself depends on the flow rate and the position of the chip with regard to the liquid reservoirs. For a vertically mounted chip, staged approximately 20 cm above the reservoirs and with a flow rate maintained at 3 μL/min, medium replacement takes around 30–45 s for the chip adapted to Arabidopsis and about twice that for the chip adapted for Brachypodium.

Acknowledgments

This project was supported by Charles University Grant Agency (Grant No. 515726).

References

1. Frey N, Sönmez UM, Minden J, LeDuc P (2022) Microfluidics for understanding model organisms. Nat Commun 13: 3195. https://doi.org/10.1038/s41467-022-30814-6
2. Merrin J (2019) Frontiers in microfluidics, a teaching resource review. Bioengineering 6:109. https://doi.org/10.3390/bioengineering6040109
3. Fendrych M, Akhmanova M, Merrin J et al (2018) Rapid and reversible root growth inhibition by TIR1 auxin signalling. Nat Plants 4:453–459. https://doi.org/10.1038/s41477-018-0190-1
4. Dubey SM, Han S, Stutzman N et al (2023) The AFB1 auxin receptor controls the cytoplasmic auxin response pathway in *Arabidopsis thaliana*. Mol Plant 16:1120–1130. https://doi.org/10.1016/j.molp.2023.06.008
5. Li L, Verstraeten I, Roosjen M et al (2021) Cell surface and intracellular auxin signalling for H+ fluxes in root growth. Nature 599:273–277. https://doi.org/10.1038/s41586-021-04037-6
6. Nielsen JB, Hanson RL, Almughamsi HM et al (2020) Microfluidics: innovations in materials and their fabrication and functionalization. Anal Chem 92:150–168. https://doi.org/10.1021/acs.analchem.9b04986
7. Unger MA, Chou H-P, Thorsen T et al (2000) Monolithic microfabricated valves and pumps by multilayer soft lithography. Science 288:113–116. https://doi.org/10.1126/science.288.5463.113
8. Grossmann G, Guo W-J, Ehrhardt DW et al (2011) The RootChip: an integrated microfluidic chip for plant science. Plant Cell 23:4234–4240. https://doi.org/10.1105/tpc.111.092577
9. Serre NB, Kralík D, Yun P et al (2021) AFB1 controls rapid auxin signalling through membrane depolarization in *Arabidopsis thaliana* root. Nat Plants 7:1229–1238. https://doi.org/10.1038/s41477-021-00969-z
10. Serre NB, Wernerová D, Vittal P et al (2022) The AUX1-AFB1-CNGC14 module establishes a longitudinal root surface pH profile. eLife 12:e85193. https://doi.org/10.7554/eLife.85193

Check for updates

Chapter 7

High-Throughput Root Phenotyping of Maize Seedlings

Jodi Bianca Callwood, Clarice Felix Gonzales and Dior Rose Kelley

Abstract

Imaging has revolutionized phenotyping analysis by providing accurate documentation and records of observed phenotypes. Analyzing imaging data remains a major bottleneck in root phenotyping analysis. Advances in technology offer a plethora of analysis options at varying automation levels and price points, providing a reliable open-source pipeline for image analysis that delivers stable, reproducible results. Here, we discuss a maize seedling image analysis pipeline powered by open-source tools that enables rapid analysis of seedling root phenotypes. This image analysis pipeline uses Root Painter, an optional AI cleanup step, and RhizoVision Explorer to extract phenotypic data from maize seedlings grown on rolled towel assays, which are commonly used, enabling rapid analysis of these images.

Keywords Machine learning, Phenomics, Roots, Maize, Seedlings

1 Introduction

Root phenotyping is an essential component of plant growth research, but quantifying root traits across developmental time poses challenges due to the nature of rolled towel assays, which remain a bottleneck in phenotyping [1–3]. Advances in technology enable the capture of assay images for record-keeping and later quantification, either via smartphones, cameras, flatbed scanners, or low-cost image platforms. The bottleneck for phenotyping, especially for roots, is the analysis of the images, as this requires significant time for annotation and quantification. Automating this process can reduce the time and staff required to process images of seedlings (roots). One common method for growing maize seedlings to analyze root traits is the so-called "rolled towel assay" [2, 3]. Rolled towel assays allow synchronous and controlled germination and growth of maize seedlings and enable the capture of root traits free of soil in a uniform method. Rolled towels containing seedlings are easy to photograph with a smartphone attached to a ring light or a camera.

Steffen Vanneste and Inge Verstraeten (eds.), *Root Development*, Methods in Molecular Biology, vol. 3086,
https://doi.org/10.1007/978-1-0716-5603-7_7,

Here, we present an updated high-throughput phenotyping pipeline for quantifying maize seedling root phenotypes from rolled-towel assays. This method improves root trait measurement accuracy and increases the number of assays that can be processed in a short window of time. This protocol details a pipeline that combines minimal image processing, image segmentation, and phenotype extraction using publicly available software [4–10] to enable high-throughput root phenotyping from rolled towel assays based on our previously published study [11]. Here we describe the setup, workflow, and image analysis process to maximize the pipeline's output.

2 Materials

2.1 Equipment

1. A camera or a smartphone capable of taking pictures.
2. 64-bit computer with at least 8 GB RAM running either Linux, Windows 10, or Windows 11.
3. Light source with camera/phone mount. Suggested item: ULANZI TH03 Phone stand with ring light, for example, purchased through Amazon.com.
4. Labels for which the exact size is known, or a ruler for scale calibration.

2.2 Plant Specimen

Maize seedlings grown via rolled towel protocol [2, 3].

2.3 Software

1. Google (Colab) account (https://accounts.google.com).
2. ImageJ (https://imagej.net/ij/).
3. Microsoft Photo App with Erase AI tool.
4. Root Painter (https://github.com/Abe404/root_painter).
5. RhizoVision Explorer (https://www.rhizovision.com/).
6. Batch image resizer, recommendation: IrfanView 64 (https://www.irfanview.com/).
7. Software packages for data analysis (i.e., Microsoft Excel, GraphPad Prism, R, or Python).

2.4 Data

A total of 500 (or more) 2D-images consisting of maize seedlings grown in rolled towel assays, evenly spaced to avoid overlap or touching.

3 Methods

3.1 Account and Software Setup

Each of the software programs in the Sect. 2 contains instructions for downloading, installation, and account setup. Follow these instructions. For RootPainter, it is recommended to use a Google Colab account, unless you have access to 8 GB GPU computers.

3.2 Plant Specimen

Instructions on how to prepare maize seedlings are included in previously published rolled towel assay protocols [2, 3]. When setting up, place 8–10 kernels approximately 1 in. (2.54 cm) apart to prevent overcrowding or later seedling overlap. Optional: affix a label (dot or square) to the inside of each rolled towel to use for scale calibration.

3.3 Image Capturing

1. Gather all rolled towels to the image. Set up a ring light with a smartphone focused on a table or tray. For experiments, images must be captured at equidistant positions, using a tripod to ensure that lighting and scale calibrations are consistent and uniform for all comparative images.
2. Open each rolled towel containing maize seedlings on this tray or flat surface, thereby exposing the seedlings and the ToughTag. If not using ToughTag labels, place a ruler next to the seedlings to set the scale (Sect. 3.7).
3. Photograph the maize seedlings using the smartphone setup. Ensure that there is uniform light while taking these pictures (Fig. 1).

3.4 Image Data Processing—Optional

If using Google Colab or a local computer, resizing images can help reduce the computational power needed. Otherwise, full-size original images can be used directly in RootPainter if storage is available.

1. Perform batch conversion in InfranView 64. Open the files in IrfanView by going to > File, then > Batch conversion (Fig. 2A). Set output format to JPG or PNG format.

   Check the box for "advanced options" and select advanced, which opens an option selection menu (Fig. 2B). Select the settings for all images: check the RESIZE box, select "Set new size" and select "set one or both sides to" and enter width: 600 pixels. Check the "Preserve aspect ratio (proportional)" box. This resizes the original image of our examples, which are 4032 × 3024 pixels, to a rectangle of 600 × 800 pixels. Check the "Use Resample function (better quality)" box. In the MISCELLANEOUS section, select "Apply changes to all pages (TIF/PDF saving)." When all changes are made, proceed with the "OK" button (Fig. 2B).

Fig. 1 Maize seedlings grown via the rolled towel assay. Image of 5-day-old W22 maize seedlings grown in the rolled towel assay in a controlled environment chamber. The rolled towel was opened onto a tray and spread out for imaging

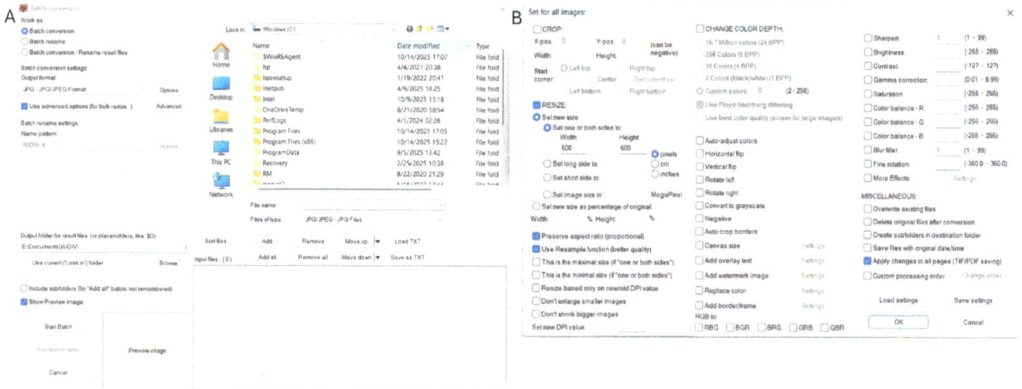

Fig. 2 Batch conversion in InfranView. **A** Screenshot of the Batch conversion menu in InfranView, an image viewer, organizer, and editor tool available for Microsoft Windows. **B** Advanced menu settings for image resizing. For initial image processing, select all boxes shown

Set batch rename settings if applicable in the Output folder for the result files list path to the output folder (Fig. 2A). Click "Start batch".

2. Use data processing language of choice (R, or Python) to randomly split a pool of 500 images into two datasets with the following ratios: ~80% of the images to a Training dataset, which will be automatically split in the RootPainter software into a training and validation set of images and ~20% of the images to a Testing dataset in which the final RootPainter model will be evaluated.

3.5 Model Training

1. Load the Training dataset into the RootPainter datasets.
2. Follow the instructions in RootPainter to start training a new model in a New Project (Create new project). Select how many tiles must be randomly generated by the software (this is necessary unless the images in your dataset are square 600×600 pixel images). Note: When creating the model, there are options for random weights or specifying the model. Random weights are recommended for a new model, while an existing model is recommended for learning from previously trained models, which can reduce the number of images needed for training.
3. On the image tiles that are generated in the RootPainter training dataset, trace the roots thoroughly and highlight a portion of the background. Change brush sizes to better highlight roots and backgrounds. Class imbalance would affect the quality of the model, so do not include too much background annotation.
4. Repeat this for multiple tiles generated in the images of the training dataset. The model will attempt to improve its predicted annotation to match your traced annotation through an iterative process. Over time, after a set number of iterations, the model will not improve its scores any further and can be considered complete for the given dataset.
5. Save the resulting RootPainter model.

3.6 Using a Trained Model for Data Processing

1. Acquire images using the imaging setup outlined in Sect. 3.3. If processing multiple images, capturing images at the same distance is imperative to allow smooth calibration and comparative quantification in later steps.
2. In RootPainter (Fig. 3), select Network>Segment Folder with Testing dataset. Specify the input and output directory (Fig. 3).
3. Select the.pkl model file resulting from Sect. 3.5.

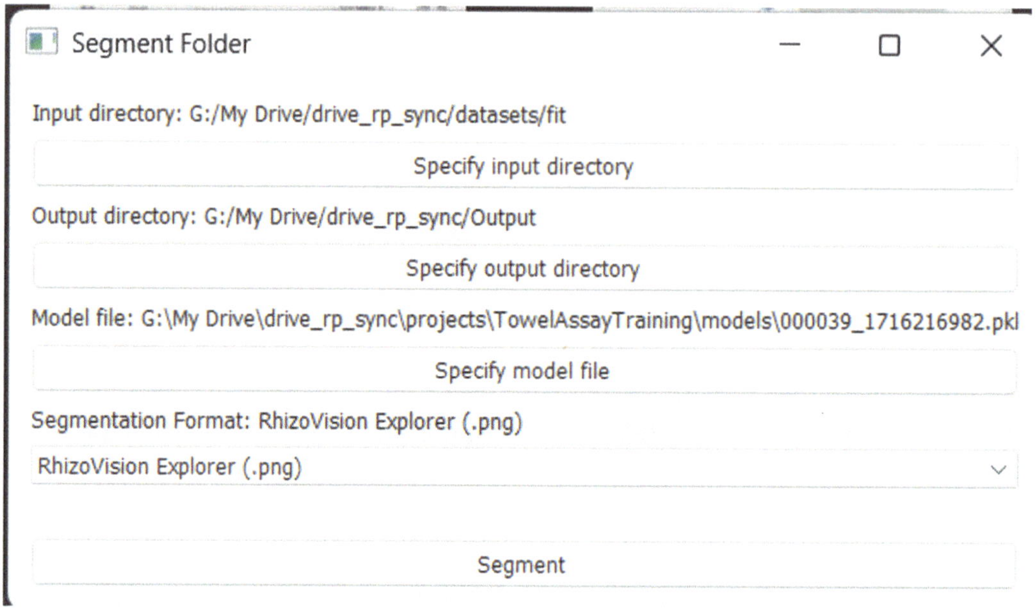

Fig. 3 RootPainter network segmentation settings. Example screenshot of settings for loading an existing model and segmenting images in RootPainter

4. In RootPainter, select Extras > Convert segmentations for RhizoVision Explorer to produce segmentations compatible with RhizoVision Explorer (Fig. 3), and change the segment format to RhizoVision Explorer (.png).
5. Press the segment to proceed with the generation of a segmentation mask.
6. Open and verify the generated images, ensuring that all the images are properly segmented. If images are not properly segmented, there are several causes and options

   (1) Additional cleaning with photo-editing software such as ImageJ or the Microsoft Photo App to remove added noise, overcrowding, or shadows from segmented images.

   (2) The input images may be significantly different in formatting from the images used for training. If these differences are significant enough, a new model will need to be trained to compensate for these variations.

3.7 Measurements, Diameter Thresholding, and Estimation of Gaps in the Segmentation

3.7.1 Calibration Measurements to Set the Scale for the Images

(1) Open one image in ImageJ, select the line tool, and draw a line across the reference label (Fig. 4A).

(2) Select Analyze > Measure or CTRL +M to activate the measurement tool. The resulting length measurement is the number of pixels corresponding to the length (in metric units) of the label. Record this number, because this factor will be used for all calibrations in length units instead of pixels. A second option is to select Analyze > Set Scale, do not change the Distance in pixels, set the Known distance to the length of the label or ruler selection (in metric units). Keep the Pixel aspect ratio at 1.0 and change the Unit of length to a metric unit (e.g., cm). Click Global if you want to measure multiple images in ImageJ and if you are sure that these images are taken at similar settings.

3.7.2 Diameter Thresholding

This step is necessary because the diameter parameter will be used in the RhizoVision software (Sect. 3.9).

1. Use ImageJ to measure the root width of a subset of images. Measure the widths of a variety of roots in 2–3 images (Fig. 4B).
2. Save the width measurements as a .csv or .txt file.

3.7.3 Segmentation Gap Estimation

1. Randomly select and open 10% of the segmented images in ImageJ and look for gaps in the segmentation.

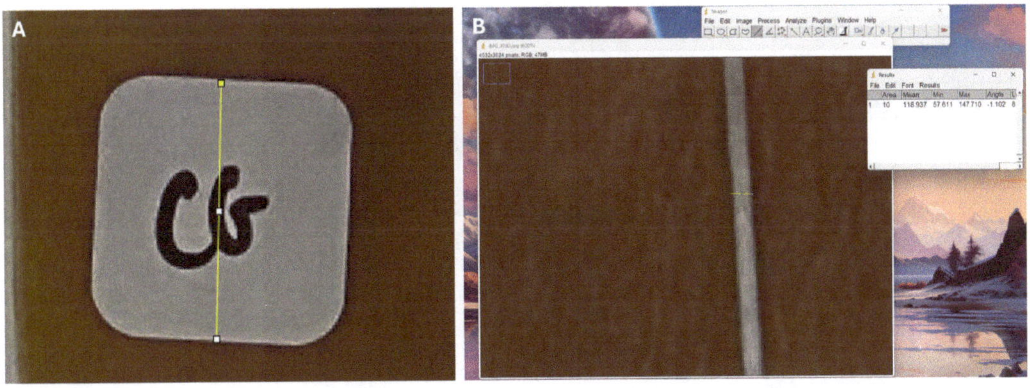

Fig. 4 ImageJ operations for calibration and diameter thresholding. (**A**) Measured label of a known length for pixel conversion in ImageJ. (**B**) Using ImageJ to measure diameters of roots, which will be used subsequently for thresholding in the RhizoVision settings

2. These can be filled using the Process > Binary > Dilate/Close options, but if an estimate is known on the size of these gaps, this can be used in the RhizoVision settings.
3. If segmentation holes are observed, measure the gap estimation length of the segmented roots and note the ranges in measurements.
4. Note these ranges for later use.

3.8 Image Cleanup (Optional)

This cleanup step might not be necessary if the RootPainter model performs well and foreground items (roots) and background are properly segmented. The manual cleanup we describe below is time-consuming, so it might be worth investing more time in the annotations and training of the RootPainter model.

1. RootPainter segmentation might not be perfect and may have additional background objects that require removal. Open the segmented images with a photo editing program such as Microsoft Photo app eraser AI tool (Fig. 5).
2. Use the photo editor to remove additional background objects (Fig. 5C) to result in a cleaner and more accurately masked image (Fig. 5D).
3. Save the cleaned images into a separate folder for the RhizoVision analysis.

3.9 RhizoVision Explorer for Root Parameters

1. Open RhizoVision Explorer and set the following settings for the images (Fig. 6).
2. Image preprocessing
   Set analysis mode to whole roots.
   Check convert pixels to physical units, change drop-down menu to Pixels per.. mm..r. Based on the measurement of Sect. 3.7.1, you can now fill this value (or recalculate it to number of pixels per millimeter).

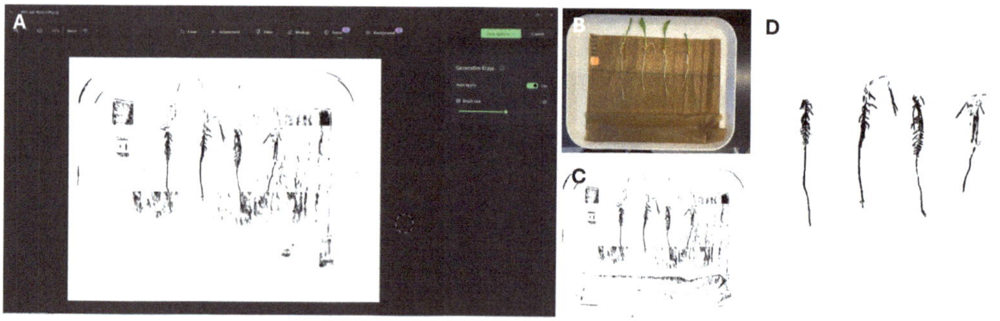

Fig. 5 Example cleanup using the Erase AI tool to remove the wrongly segmented objects. (**A**) Erase the AI tool screenshot of the settings. (**B**) Original rolled towel image. (**C**) Segmentation image from RootPainter that still contains unwanted background. (**D**) Final image post-processing after RootPainter and Erase AI

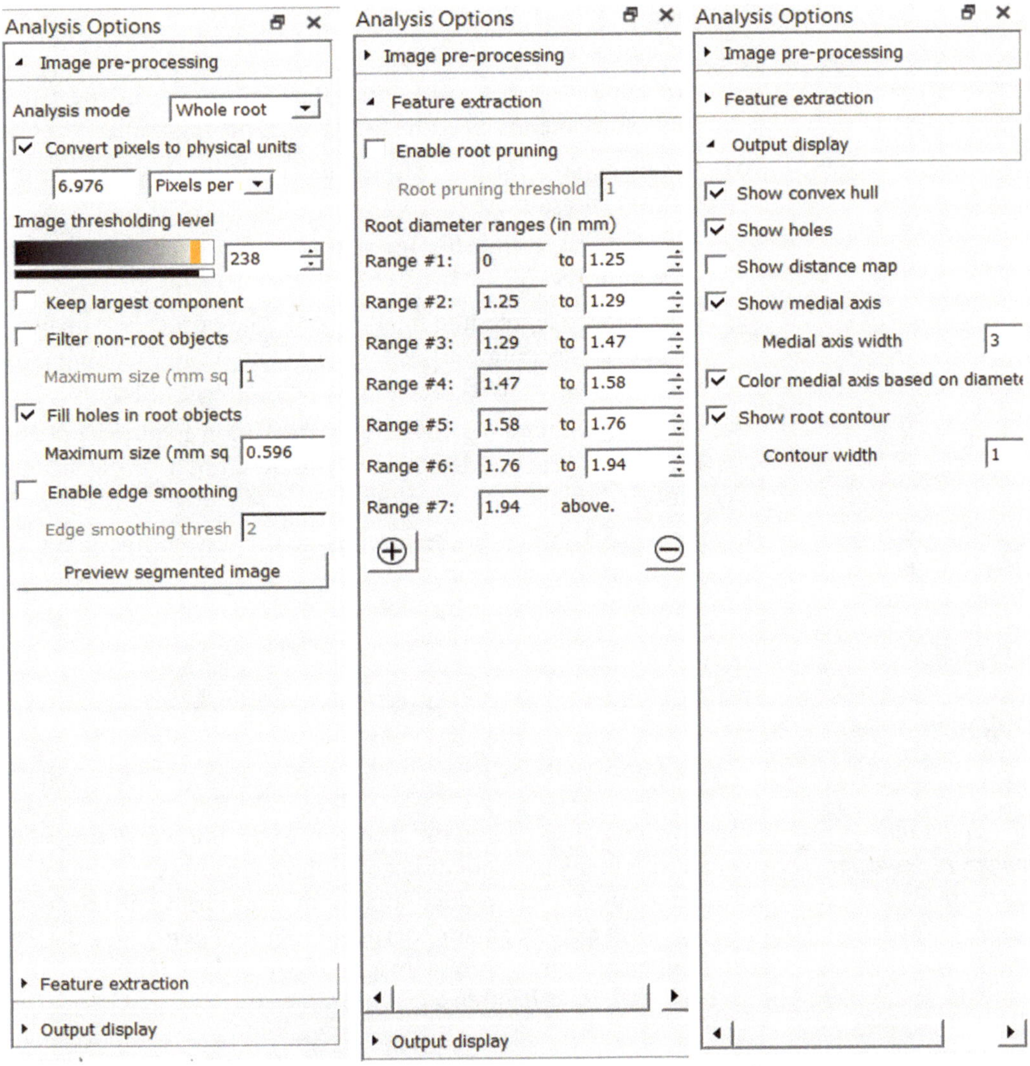

Fig. 6 RhizoVision settings for image analysis. Suggested settings for the Analysis Options for processing root traits of segmented images generated by RootPainter

Deselect "Keep largest component."
(Optional) Select Fill holes in root objects. Use the measurements from Sect. 3.7.3 for an estimate of the hole size to fill in this option. If no gaps in the segmentation were observed, this step can be skipped. Note: the measurement is in mm^2 squared.

3. Feature Extraction
Deselect "Enable root pruning."
Load the diameter ranges in mm for the width of the seedling roots. Thanks to the measurements in Sect. 3.7.2, an estimate of this diameter range can be included.

4. Output Display
   Default settings are acceptable.
5. Process all images using File > Batch Analysis, in the batch processing menu select the folder with the segmented images.
6. Create a folder for the output data and select this folder at the Output location.
7. The features output file and metadata file can be renamed or remain in the default.
8. Select "Save segmented images" and "Save processed images" for record-keeping purposes. Note: The default naming conventions can be used, or this can be modified if desired.
9. Press "Start" and do not close RhizoVision Explorer until all images are processed. The data will be saved as a .csv file and exported to the output folder.
10. The .csv data file can be opened in any .csv compatible program, such as Microsoft Excel, LibreOffice Calc, or Google Sheets, to perform data analysis.

4 Notes

4.1 Optimizing Imaging of Seedlings and Roots

Overcrowding or spreading the seedlings to the borders of the paper towel results in mistaken identification of root features and improper segmentation. If overcrowded images or paper rolls with overlapping seedlings are a regular occurrence, you can generate a dataset specifically with some of these images to overcrowding images for a more accurate RootPainter segmentation model.

4.2 Image Capturing

Images that are too dark may not segment properly. Taking images in a bright room or using a ring light, ensuring uniform light on the seedlings and sufficiently high brightness, may help later processing steps.

4.3 Image Cleanup (Optional)

Microsoft Photos AI eraser has difficulty separating objects close to root tips. Additional manipulation in programs, such as Microsoft Paint, can help to create barriers between the root tip and other objects in the segmented image, to make clean separation between the objects easier.

Acknowledgments

This work was supported by a Department of Defense SMART scholarship to JBC, an USDA NIFA AFRI grant to DRK (Award No. 20206701330914), and State of Iowa Hatch funds to DRK (IOW05745). We wish to thank Luke David for his assistance with training RootPainter during his summer internship.

References

1. Furbank RT, Tester M (2011) Phenomics—technologies to relieve the phenotyping bottleneck. Trends Plant Sci 16:635–644. https://doi.org/10.1016/j.tplants.2011.09.005
2. Draves MA, Muench RL, Lang MG, Kelley DR (2022) Maize seedling growth and hormone response assays using the rolled towel method. Curr Protoc 2:e562. https://doi.org/10.1002/cpz1.562
3. Gonzales CF, Cowling CL, Kelley DR (2024) The rolled towel method for hormone response assays in maize. Cold Spring Harb Protoc https://doi.org/10.1101/pdb.prot108623
4. Smith AG, Han E, Petersen J et al (2022) RootPainter: deep learning segmentation of biological images with corrective annotation. New Phytol 236:774–791. https://doi.org/10.1111/nph.18387
5. Seethepalli A, Dhakal K, Griffiths M et al (2021) RhizoVision Explorer: open-source software for root image analysis and measurement standardization. AoB PLANTS 13:plab056. https://doi.org/10.1093/aobpla/plab056
6. Seethepalli A, Guo H, Liu X et al (2020) RhizoVision Crown: an integrated hardware and software platform for root crown phenotyping. Plant Phenomics 2020:1–15. https://doi.org/10.34133/2020/3074916
7. Schneider CA, Rasband WS, Eliceiri KW (2012) NIH Image to ImageJ: 25 years of image analysis. Nat Methods 9:671–675. https://doi.org/10.1038/nmeth.2089
8. Schindelin J, Arganda-Carreras I, Frise E et al (2012) Fiji: an open-source platform for biological-image analysis. Nat Methods 9:676–682. https://doi.org/10.1038/nmeth.2019
9. Rueden CT, Schindelin J, Hiner MC et al (2017) ImageJ2: ImageJ for the next generation of scientific image data. BMC Bioinform 18:529. https://doi.org/10.1186/s12859-017-1934-z
10. Gaudin ACM, McClymont SA, Holmes BM et al (2011) Novel temporal, fine-scale and growth variation phenotypes in roots of adult-stage maize (*Zea mays* L.) in response to low nitrogen stress. Plant Cell Environ 34:2122–2137. https://doi.org/10.1111/j.1365-3040.2011.02409.x
11. Callwood JB, Cowling CL, Townsend EG et al (2025) Identification of phenotypic and transcriptomic signatures underpinning maize crown root systems. Plant Phenomics 7(1):100008. https://doi.org/10.1016/j.plaphe.2025.100008

Chapter 8

RootHairML: A Simple Machine-Learning Tool for Quantifying Root Hairs

Hans Motte, Joris Jourquin, Kenzo Vereecken, Wouter Smet, Tom Beeckman and Jan Verwaeren

Abstract

Root hairs play pivotal roles in nutrient and water acquisition and in plant–microbe interactions. Consequently, understanding the mechanisms underlying root hair development and their regulatory pathways is an important aspect of plant physiology research. Quantifying root hairs and root hair length is often essential in such studies, but is labor-intensive and prone to subjectivity. The availability of straightforward tools for automated root hair measurements is limited, and existing options are often tailored for specific images or do not measure individual root hairs. To address this, we developed RootHairML, a flexible and simple Python-based machine learning tool designed for efficient quantification of root hair lengths. Based on labeled images and pixel features, RootHairML trains a Random Forest model to enable the detection of root hairs in new images. It provides measurements of individual root hair lengths per image and generates annotated images showing all detected root hairs, allowing for manual verification and adjustments. Here, we describe and showcase the use of RootHairML. Overall, RootHairML offers a valuable tool for root hair analysis, enabling researchers to increase data collection and enhance the reproducibility of root hair studies.

Keywords Root hair, Arabidopsis, Machine learning, Image analysis

1 Introduction

Root hairs are specialized tubular structures that develop from root epidermal cells, playing a crucial role in plant nutrient and water uptake, interactions with microorganisms, and plant anchorage [1]. The dynamic adjustments in root hair growth and density significantly impact the root's surface area and overall plant fitness. Phosphate or nitrogen starvation, for example, typically results in longer and denser root hairs to increase nutrient uptake [2, 3]. Likewise, plants with longer root hairs cope better with drought conditions [4]. Hence, there is a clear interest in understanding factors that impact root hair parameters. Such studies require the quantification of root hair

Steffen Vanneste and Inge Verstraeten (eds.), *Root Development*, Methods in Molecular Biology, vol. 3086,
https://doi.org/10.1007/978-1-0716-5603-7_8,

parameters, often measured by manually tracing all root hairs. This is, however, time-consuming and labor-intensive, while subject to the researcher.

Currently, several automatic tools exist for quantifying root hair parameters. Vincent et al. [5] developed a tool that measures the total two-dimensional area covered by root hairs, while the ImageJ Plugin *Root Hair Sizer* automatically assesses the root hair profile and the root hair length distribution over the root [6]. None of these tools, however, distinguishes individual root hairs. Lu et al. [7] developed a deep learning tool to segment root hairs and measure root hair area from roots grown in microrhizotrons. The use of a specific training dataset might, however, complicate the use of this tool to automatically segment root hairs from roots in a different environment. Moreover, all those tools focus on the assessment of densely grown root hair areas, often in soil conditions, while there is also a need to measure individual root hairs from microscopic images. Recently, *DIRT/μ* is reported to detect individual root hairs, even from roots with dense and crossing root hairs, but because of the complexity of resolving intersecting root hairs, training of the model cannot be performed on personal computers and relies on high-performance computing [8].

As we were in need to measure root hair length or total root hair length from in vitro grown plants in different conditions, we developed a simple and straightforward machine learning tool to detect and measure root hairs, which we call RootHairML. RootHairML uses a Python script to train a Random Forest model on user-provided labeled images, making it a highly flexible and adaptable tool. The trained model is used to detect horizontally emerging root hairs and quantify their length using either fitted lines or skeletonized representations. Although RootHairML is not able to distinguish overlapping root hairs, it is able to measure the total length of these root hairs. Here, we demonstrate a strong correlation between manual and automated measurements, making RootHairML a robust and time-efficient tool for detecting differences in root hair elongation or formation across experimental conditions. We finally illustrate the use of RootHairML by assessing phosphate starvation-induced root hair development in Arabidopsis.

2 Materials

1. Computer with Python and ImageJ/Fiji installed, preferably with the NeuronJ plugin available.

2. User-provided root images: root images (JPEG format) of *Arabidopsis* (or other species) grown on vertically oriented in vitro plates, captured with sufficient resolution to detect individual root hairs (e.g., using a stereomicroscope or digital microscope).

3 Method

RootHairML is a computational tool designed for automated root hair detection. It requires a limited set of user-provided labeled images. From this, it first extracts intensity, edge, and texture features of pixels associated with root hairs to characterize them and their local structural context. Using this feature set, RootHairML trains a random forest model to accurately detect root hairs in new images. The tool is highly flexible, as users can provide custom training data, allowing adaptation to various imaging conditions.

3.1 Installing RootHairML

1. Use the following command in a terminal or command prompt: *pip install git*+https://github.com/jverwaer/RootHairML.git to directly install RootHairML from the GitHub repository.
2. Alternatively, download the *bundle* folder or the zipped version *RootHairML_bundle.zip* from https://github.com/jverwaer/RootHairML. This option provides all-in-one scripts (*train.py* or *analyze.py*) to execute preconfigured pipelines directly from the command line. Sects. 3.2–3.4 explain how to use the all-in-one scripts.

3.2 Preparing Training Data

To train the model, users must supply a set of images along with corresponding vertex files that define the coordinates of all root hairs in each image. A training set of 4–15 images is recommended to achieve reliable model performance.

1. Select 4–15 representative images from the provided set of images, ensuring that the chosen images capture a diversity of phenotypes and root hair morphologies.
2. Manually annotate horizontally emerging root hairs on the images by drawing traces over the center of the root hair and create vertex files that contain for each root hair the X, Y coordinates that jointly form the trace covering the root hairs in a CSV-file formatted as presented in Table 1 (*see* **Note 1**).
3. Create one vertex file per image. Name the vertex file "Image name' vertices.csv."
4. Place all images and corresponding vertex files in the "bundle/trainData" folder.

Table 1
Example of a vertices file to be provided for training of RootHairML

Image;Tracing;X [pix];Y [pix]
Image001;N1;46;625
Image001;N1;45;643
Image001;N1;36;653
Image001;N1;37;670
Image001;N1;32;690
Image001;N1;30;711
Image001;N2;53;651
Image001;N2;42;675
Image001;N2;40;707
...

3.3 Training a Model

The *train.py* script computes pixel-level features and corresponding labels from the training images and fits a random forest classifier on a subset of the data.

1. Specify a model name and the number of estimators in the random forest classifier (*see* **Note 2**) in Step 3 of the step-by-step analysis (see https://github.com/jverwaer/RootHairML/blob/main/README.md) or open the *train.py* file (located in the "bundle/scripts" folder) in an editor to make these changes.
2. Run Step 3 (see https://github.com/jverwaer/RootHairML/blob/main/README.md) or use a terminal to navigate to the "bundle/scripts" folder and run *train.py* via the command *python train.py*. The trained model is saved in the "bundle/models" folder.

3.4 Running Root Hair Analysis

1. Place images to be analyzed in the "bundle/images."
2. Load a model in Step 4 or open the *analyze.py* file (in the "bundle/scripts" folder) in an editor to specify the model.
3. Use a terminal to navigate to the "RootHairML/scripts" folder and run *analyze.py* via the command *python train.py*. A *result.xlsx* output table (*see* **Note 3**) is generated in the "bundle" folder, and annotated images (*see* **Note 4**) are generated in the "bundle/images" folder.

3.5 Step-by-Step Analysis

In the step-by-step analysis, the overall pipeline remains highly similar to the all-in-one scripts, but each step must be executed individually as described in the tutorial (see https://github.com/jverwaer/RootHairML/blob/main/README.md). The folders for loading training data, models, or images are specified in each step and can be easily modified by the user. In addition to the functionalities offered by the all-in-one script, the step-by-step analysis provides additional checks: verifying whether traces in user-provided vertex files correctly match root hairs (Step 1) and assessing whether a given model reliably annotates root hairs on a specific image type (Steps 5 and 6).

3.6 Validation

To evaluate the performance of RootHairML, we conducted a case study comparing the root hair length of Arabidopsis roots grown for 8 days in vitro under high phosphate (HP) and low phosphate (LP) conditions. Images of a 2 mm segment of the primary root—located 6–8 mm from the root tip—were obtained using a Keyence VHX-6000 digital microscope. A total of 42 roots were imaged, with 23 from HP conditions and 19 from LP conditions. The RootHairML model was trained on a diverse set of 12 roots from both treatments.

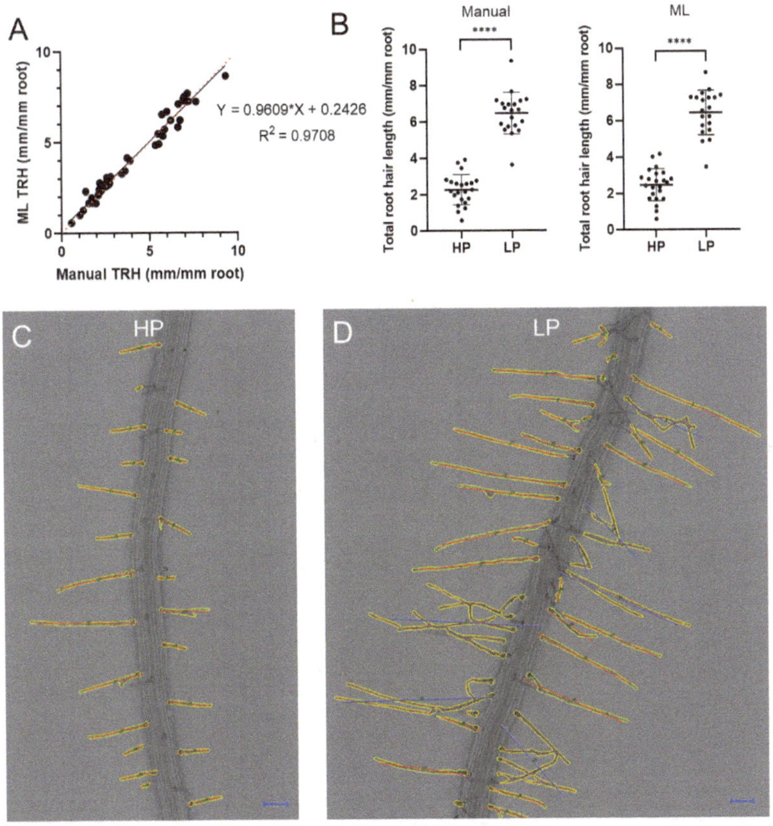

Fig. 1 Total root hair length measurements using manual annotation and RootHairML from Arabidopsis roots grown under high phosphate (HP) or low phosphate (LP) conditions. (**A**) XY scatter plot showing a strong correlation between total root hair length (TRH) values obtained manually or via RootHairML. The black line represents the linear regression fit; the red dashed line indicates the identity line (y = x). (**B**) Scatter dot plots comparing TRH between HP- and LP-grown plants based on manual (left) or RootHairML-based (right) measurements. Horizontal lines indicate group means; error bars represent standard deviation (SD). Asterisks denote statistically significant differences based on unpaired *t*-tests (****$p < 0.0001$). (**C**, **D**) Representative images of Arabidopsis roots grown under high phosphate (HP) (**C**) or low phosphate (LP) (**D**) conditions, analyzed using RootHairML. Note the numerous blue lines in the LP condition, indicating cases where root hairs deviated from linearity due to bending or overlapping. In such cases, RootHairML applied skeletonization to accurately estimate total root hair length. Scale bars: 100 μm

Our analysis revealed a clear difference in root hair length between HP and LP conditions (Fig. 1). Importantly, both manual and RootHairML-based measurements yielded comparable results, demonstrating the reliability of RootHairML in quantifying root hair length across different phosphate treatments. Further examination of the correlation between manual and RootHairML-based measurements showed a strong linear relationship (Fig. 1). The strong linear regression between the two measurement methods underscores the utility of RootHairML in comparative studies between treatments. Importantly, we noted that in a few cases, non-root hair elements such as dirt on the image or the scale bar were mistakenly detected as root hairs. However, the output images in which all root hairs are numbered allowed us to easily identify and manually remove these objects from the results table.

4 Notes

1. Vertex files can be easily generated via NeuronJ [9], an ImageJ plugin designed for 2D tracing and analysis of elongated structures. After indicating all horizontally emerging root hairs in an image, vertex measurements can be displayed from which *Image, Tracing, X [pix],* and *Y [pix]* columns can be copied and saved in a CSV file, as shown in Table 1. Tracings will be transformed into a root-segmentation mask by the script.
2. The number of estimators defines the number of decision trees in the ensemble. A higher number of trees generally improves performance but slows down efficiency. Five hundred gives in our experience a good balance between model accuracy and computational efficiency, but in particular for clearly distinguishable root hairs, the number of estimators can be lowered without large effects on the performance.
3. RootHairML will assess the outline of root hairs in new root images. To measure the length, it will fit a line or calculate the length of the skeleton. The latter will occur in case of bent or aberrant root hairs, when a line is not a good approximation of the root hair length. The script generates an output table (Table 2) that contains for each detected root hair on the provided images (1) the coordinates of the centroid, (2) its orientation, (3) length (assuming linearity), (4) check of the linearity assumption, (5) adjusted length based on the skeleton, (6) adjusted length based on the skeleton, but taking into account diagonal connections between pixels, and (7) the filename of the image. The method in (6) provides a more accurate total length of the skeleton and is preferably used.

Table 2
Example of a RootHairML output table

	(1)		(2)	(3)	(4)	(5)	(6)	(7)
	X	Y	Orientation	Length	Approx_check	Adjusted_ length	Adjusted_length_ with_sqrt2	fname
...								
17	1141.303	601.0037	−1.11824	161.0999	Bad approx	188	222.1371	Image001.jpg
18	956.7066	641.3089	−0.83639	83.68632	OK	83.68632	83.68632	Image001.jpg
19	1145.18	648.9253	−1.41681	87.94558	OK	87.94558	87.94558	Image001.jpg
20	1159.746	685.3198	−1.41437	72.85329	OK	72.85329	72.85329	Image001.jpg
21	1011.38	705.8013	−0.89361	50.35896	OK	50.35896	50.35896	Image001.jpg
...								

(5) and (6) are only calculated if the linearity check gave a bad approximation; otherwhise, the linear length is provided. The output table can be further processed to assess the number of detected root hairs, the average length per root hair, or the total root hair length. We focused on the latter, as it is the most relevant metric for evaluating overall root hair development and its contribution to nutrient uptake. Moreover, undetected very short root hairs or the merging of two or more root hairs into a single detected object will have minimal impact on this metric.

4. The script generates for each input image an annotated image that shows all the detected root hairs, including root hair numbers, outlines, fitted lines, and centroids, and that can be used as a quality control (Fig. 2), and decide to manually add or remove measurements, or increase the number of images used to train the model. This approach is illustrated for a set of images that were captured with a LEICA S9D stereomicroscope and featured a dark background. Initially, we trained the model using only two annotated images, but this resulted in poor detection of horizontally emerging root hairs (Fig. 3). Detection accuracy significantly improved when the training set was expanded to four images, while in particular the model trained on six images demonstrated good performance (Fig. 3).

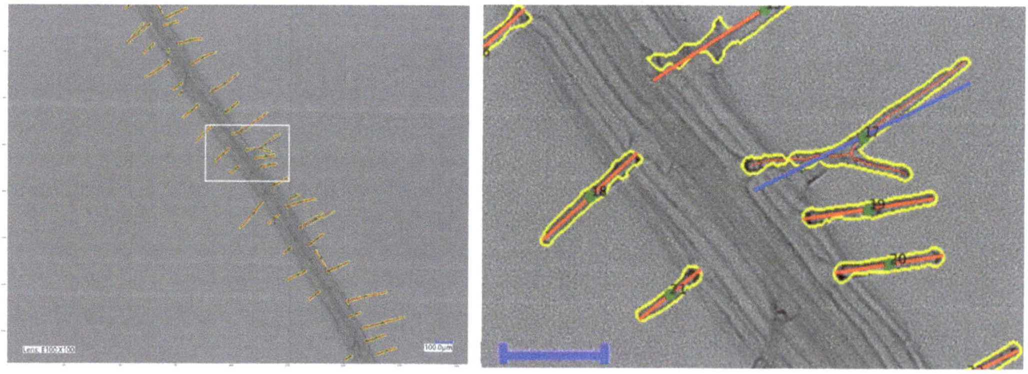

Fig. 2 Output image generated by RootHairML. Detected horizontally emerging root hair outlines are shown in yellow. Fitted lines representing root hairs are displayed in red or blue, with blue indicating cases where a line could not be well fitted, and a skeleton (in red) was used to assess the root hair length. Centroids are marked in green and labeled with the corresponding object number. The right panel provides a magnified view of the region within the white square in the left panel. Blue scale bars represent 100 μm. Object 17 consists of two overlapping root hairs but is detected as a single object. Since the line gives a bad approximation of the root hair length, the length is also calculated based on the skeleton (Table 2)

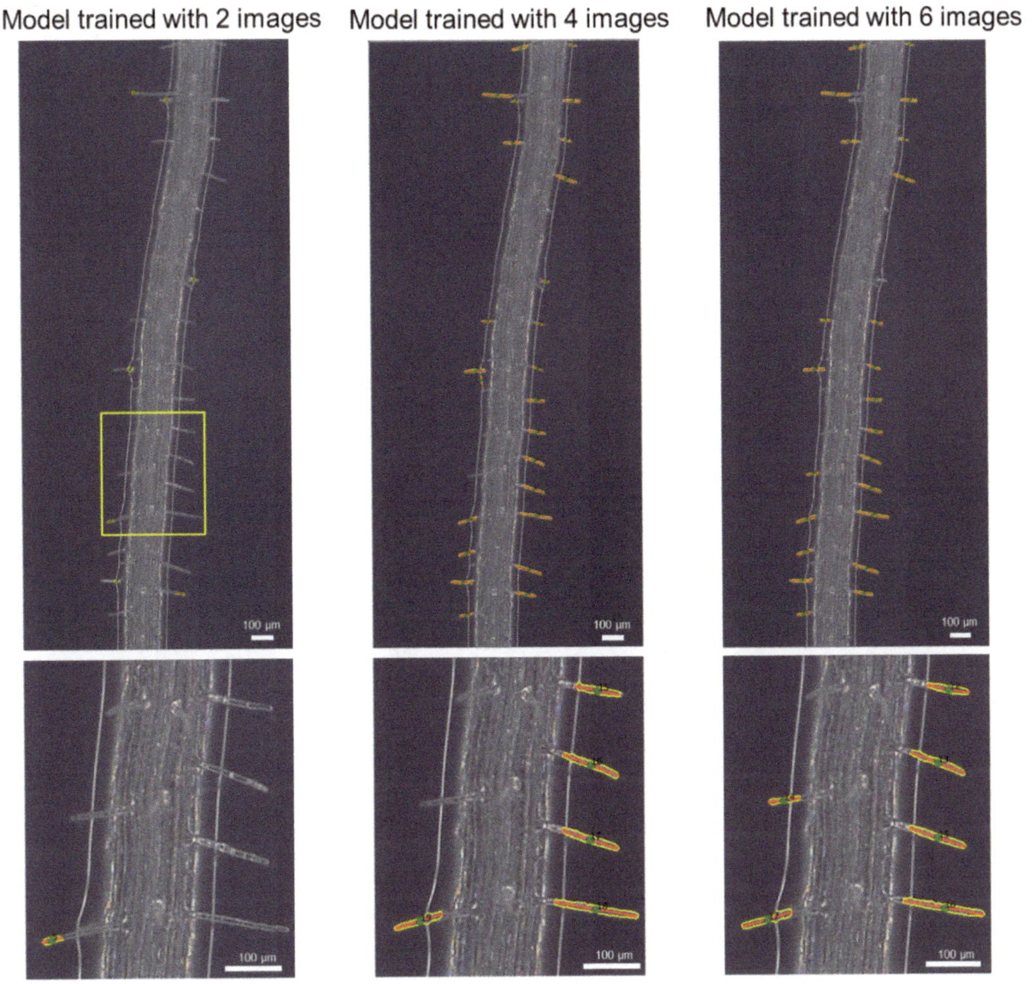

Fig. 3 Detection of horizontally emerged root hairs via RootHairML using a model trained on 2, 4, or 6 images

Competing Interests

The authors declare no competing interests.

References

1. Grierson C, Nielsen E, Ketelaarc T, Schiefelbein J (2014) Root hairs. The Arabidopsis Book 12:e0172. https://doi.org/10.1199/tab.0172
2. Crombez H, Motte H, Beeckman T (2019) Tackling plant phosphate starvation by the roots. Dev Cell 48:599–615. https://doi.org/10.1016/j.devcel.2019.01.002
3. De Pessemier J, Moturu TR, Nacry P et al (2022) Root system size and root hair length are key phenes for nitrate acquisition and biomass production across natural variation in Arabidopsis. J Exp Bot 73(11):3569–3583. https://doi.org/10.1093/jxb/erac118
4. Cai G, Ahmed MA (2022) The role of root hairs in water uptake: recent advances and future perspectives. J Exp Bot 73:3330–3338. https://doi.org/10.1093/jxb/erac114
5. Vincent C, Rowland D, Na C, Schaffer B (2017) A high-throughput method to quantify root hair area in digital images taken in situ. Plant Soil 412:61–80. https://doi.org/10.1007/s11104-016-3016-9

6. Guichard M, Allain J-M, Bianchi MW, Frachisse J-M (2019) Root Hair Sizer: an algorithm for high throughput recovery of different root hair and root developmental parameters. Plant Methods 15: 104. https://doi.org/10.1186/s13007-019-0483-z
7. Lu W, Wang X, Jia W (2022) Root hair image processing based on deep learning and prior knowledge. Comput Electron Agric 202:107397. https://doi.org/10.1016/j.compag.2022.107397
8. Pietrzyk P, Phan-Udom N, Chutoe C et al (2024) DIRT/μ: automated extraction of root hair traits using combinatorial optimization. J Exp Bot 76:285–298. https://doi.org/10.1093/jxb/erae385
9. Meijering E, Jacob M, Sarria JCF et al (2004) Design and validation of a tool for neurite tracing and analysis in fluorescence microscopy images. Cytometry Part A 58A:167–176. https://doi.org/10.1002/cyto.a.20022

Part II

Diagnosis and Manipulation

Check for updates

Chapter 9

Model Systems for Investigating Root Responses to Localized Water Availability

Danielle Colyer, James Donaldson, Jason Banda, Malcolm J. Bennett and Poonam Mehra

Abstract

Water is distributed unevenly in soil, creating fine-scale patterns that shift with soil structure and drying. To navigate this complexity, roots deploy localized adaptive responses that allow them to sense and respond to water availability with high spatial precision. These include alterations in root branching (hydropatterning and xerobranching) and directional growth responses (hydrotropism and xerotropism), all of which collectively optimize water capture. Despite their importance, reproducing these heterogeneous water cues in the laboratory is challenging because soil systems are structurally and hydraulically complex. To address this, a suite of agar-based model systems has been developed to provide reproducible, tractable, and quantifiable platforms for studying root water sensing. These approaches rely on manipulating water potential using solutes or air interfaces to mimic microscale environmental heterogeneity and trigger adaptive responses. In this chapter, we outline the conceptual principles underlying water-driven root adaptive responses and provide detailed protocols for five agar-based experimental systems designed to study hydrotropism, xerotropism, xerobranching, and hydropatterning. These methods support robust mechanistic investigation and can be adapted for both Arabidopsis and crop species.

Keywords Water sensing, Hydropatterning, Hydrotropism, Xerobranching, Xerotropism

1 Introduction

Water availability is a major environmental determinant of root system architecture, shaping where and how roots grow. In natural soils, water is rarely distributed uniformly; instead, it forms highly dynamic and heterogeneous microscale patches. Root systems must therefore continuously sense and respond to local water fluctuations to maintain resource acquisition and survival. This can be achieved through a suite of highly localized and finely tuned root adaptive responses targeting physiological traits such as root branching or growth direction (Fig. 1).

The formation of lateral roots is particularly sensitive to spatial water availability [1]. Under sufficient water availability, lateral roots

Steffen Vanneste and Inge Verstraeten (eds.), *Root Development*, Methods in Molecular Biology, vol. 3086,
https://doi.org/10.1007/978-1-0716-5603-7_9,

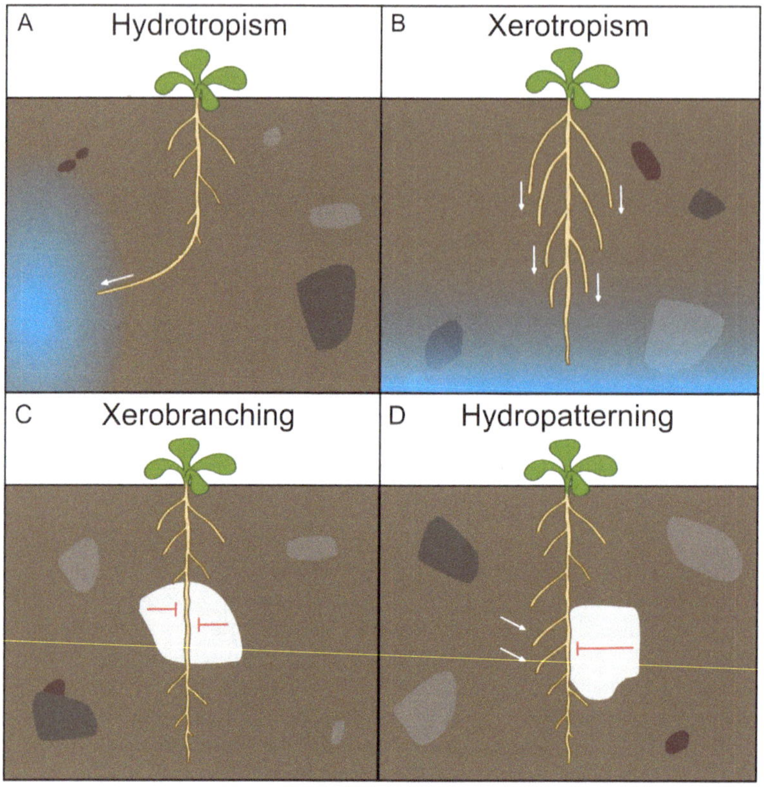

Fig. 1 Root tropic and developmental responses to heterogeneous water availability. (**A**) Hydrotropism and (**B**) xerotropism, direct primary and lateral root growth toward water-rich regions, while (**C**) xerobranching and (**D**) hydropatterning restrict lateral root formation in areas lacking soil contact or moisture

emerge regularly and exhibit diverse growth directions. However, when a primary root experiences partial contact with moist soil along one axis, hydropatterning ensures that lateral roots preferentially emerge from the water-contacted side [2]. Conversely, when the root loses contact with water entirely, such as when crossing an air-filled soil pore, lateral root formation in that region is fully suppressed by an adaptive response termed xerobranching. Lateral root growth resumes when water contact is restored [3,4]. This water-dependent regulation of lateral roots allows the plant to conserve energy by suppressing the formation of roots in regions where there are no resources to benefit from.

Roots also respond directionally to water gradients. The bending and growth of roots towards areas of greater moisture, termed hydrotropism, was first described as early as the 18th century [5] and further investigated by several scientists in the late 19th century [6,7]. During hydrotropism, a lateral moisture gradient induces the reorientation of primary root growth towards the water source, allowing roots to overcome the effects of gravitropism. A slightly different but related response, known as xerotropism, is triggered when the upper soil becomes sufficiently drier than deeper layers. In this case, both primary and lateral roots display enhanced gravitropism, resulting in increased root elongation and downward bending [8]. Together, these directional responses, along with lateral root modulation, allow the root system to efficiently explore and exploit heterogeneous water environments.

These processes provide powerful experimental systems for understanding the mechanisms of root water sensing. They represent local water-sensing responses that precede and differ fundamentally from whole-plant drought responses. Severe drought triggers tissue dehydration, which leads to systemic damage before plants can acclimatize to the reduced water availability. In contrast, mild or spatially localized reductions in water availability elicit adaptive responses at the root tip or along the root axis that are essential for early acclimation and decision-making. Recreating such heterogeneous water distributions in the laboratory is challenging because soil imposes complex and poorly controlled hydraulic conditions. Agar-based systems offer an effective alternative: by manipulating water potential with solutes or by using air interfaces to create sharp moisture contrasts, researchers can simulate localized water cues in a reproducible and quantifiable manner. These systems also facilitate imaging, measurement, and mechanistic dissection of root responses.

Experiments have been designed to reduce water potential in specific regions of the growth medium to mimic heterogeneous water distribution and induce these adaptive responses for further study. Water flows from a high to a low water potential gradient, and these water potential gradients are a key driving force of water movements in both plants and soil. As soil dries, its water potential decreases, reducing the flow of water into the plant roots. Reducing the water potential of agar growth medium can mimic aspects of either drought or heterogeneous water distribution, depending on the experimental design. Several solutes are commonly used to lower the water potential of the growth medium, notably sorbitol, mannitol, polyethylene glycol (PEG), and NaCl. Sorbitol and mannitol will be preferentially used in this chapter, primarily due to the lack of ionic effects and technical challenges compared to other solutes mentioned. For a more detailed discussion of the advantages and disadvantages of these solutes in osmotic stress experiments, see Dubois and Inzé [9].

Inverting agar plates, or creating air gaps in the growth medium, are two approaches to control the distribution of water in an agar-based growth environment through using air, which mimics soil conditions more accurately than osmotic treatments. These approaches can create a hydrotropic stimulus by creating a sharp boundary between moist agar and drier air, or by forcing the root to transiently lose contact with water in a specific region for a controlled amount of time. The inverted plate method uses calcium chloride to further lower the water potential of air by reducing humidity. Using air instead of solutes has the benefit of limiting solute-specific effects.

In this chapter, we describe five agar-based experiments designed to study hydrotropism, xerotropism, xerobranching, and hydropatterning. The following are agar-based experimental procedures for *Arabidopsis thaliana*, but these systems can be adapted to specific crop species and further adapted for analysis in paper rolls and soil. It must be noted that these experiments cannot fully replicate the conditions experienced by roots in soil environments, as several factors, including soil structure and soil microbiome, influence root responses to water stresses [10,11]. Nevertheless, well-designed agar-based experiments are instrumental for uncovering the molecular mechanisms underlying root–water interactions.

2 Materials

2.1 Biological Materials

Seeds of wild type and a mutant defective in the corresponding response: e.g., *miz1* for hydrotropism [12], *abi1-1* for xerobranching [13], *arf7-1* for hydropatterning [14].

2.2 Reagents and Chemicals

1. Half-strength Murashige and Skoog (½ MS) medium: Mix 2.2 g/L Murashige and Skoog Basal Salt Mixture, and 0.5 g/L MES in demineralized water. Depending on the experiment, sucrose is added. 200 mM Mannitol for osmotic stress/xerotropism, 400 mM Sorbitol for hydrotropism and split plate assays. Adjust pH to 5.7 and add 10 g/L plant tissue culture grade agar. Autoclave at 121°C for 30 min, prior to use.
2. 50% Bleach.
3. Ethanol for sterilization.
4. 10% (v/v) Triton X-100 stock solution: Add 1 mL of Triton X-100 to 9 mL of demineralized water. Mix gently to avoid foaming. Store at room temperature.
5. 0.02% (v/v) Triton X-100 working solution: Dilute 200 μL of the 10% (v/v) Triton X-100 stock solution to a final volume of 100 mL with sterile water, followed by gentle mixing before use.
6. For hydrotropism-inverted plate: 4 M calcium chloride.

2.3 Tools and Equipment

1. Microcentrifuge tube (MCT).
2. 12 × 12 cm square sterile plates.
3. Micropore tape.
4. Pipette tip/toothpick.
5. Black permanent marker.

6. Tweezers.
7. For hydrotropism—split plate: No. 22 Sterile surgical blade; Ruler.
8. For hydrotropism—inverted plate: 100 mL measuring cylinder; parafilm; scissors; beaker for liquid collection.
9. For xerobranching—agar air-gap assay: No. 22 sterile surgical blade; 10 cm long flat blade; cable ties; scalpel.
10. For hydropatterning: 5× stereo microscope.

3 Methods

The methods described are optimized for our growth settings and will vary depending on the environment and the material used. Please adjust the timings as necessary; the optimum sizes of seedlings are clarified where important. Unless stated, complete all steps in a flow hood to ensure a sterile working area; otherwise, the experiment risks contamination.

3.1 Osmotic Stress/Xerotropism—Mannitol Plate

1. Sterilize seeds in 1.5 mL of MCT with 500 µl of 50% bleach solution for 6 min, inverting to mix, then add 500 µL of 0.02% Triton X-100 solution. Remove all liquid and rinse with a further 1000 µL of 0.02% Triton X-100 solution. Complete five sterile water washes.
2. Leave the seeds in at least 500 µL of sterile water and place in the dark at 4°C for 2 days for stratification.
3. After stratification, sow seeds onto ½ MS media plates (1% sucrose) using a pipette tip or a toothpick (personal preference). Seal with micropore tape and transfer to a growth space (16/8-h photoperiod cycle at 21°C/18°C, respectively).
4. Transfer the uniformly grown seedlings 5 days after germination to 200 mM mannitol plates and mock plates (Fig. 2). No more than ten seedlings per plate. Mark the tip of the root on the plates for growth measurements. Immediately wrap plates with micropore tape to avoid plants drying out (*see* **Note 1**).
5. Put plates back in the growth space and leave for another 7 days.
6. After 7 days, take images of the plates for analysis.
7. Analysis can include:
   a. Primary root length (total root length) or growth (length from dot to tip).
   b. Lateral root number and density (number of emerged lateral roots/cm of primary root.

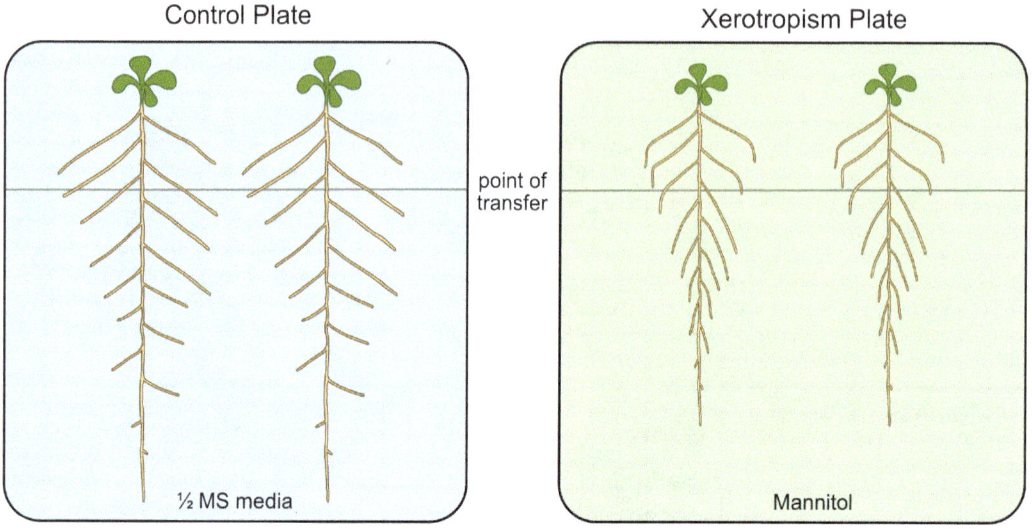

Fig. 2 Schematic of the xerotropism experimental setup. Seedlings are transferred from standard medium to plates containing mannitol to impose osmotic stress (right panel), and the transfer point is marked for later measurement. Under osmotic stress, lateral roots gradually adopt steeper downward angles compared with the more horizontal lateral root growth observed on the control medium (left panel)

c. Lateral root emergence angle: The angle at which a lateral root emerges from the primary root was measured using Fiji (ImageJ) [15] by drawing a line along the primary root axis and a second line along the initial segment of the lateral root, with the angle between the two lines recorded.

d. Lateral root tip angle: The angle of the lateral root tip relative to the gravitational (vertical) axis was measured in Fiji (ImageJ) by drawing a vertical reference line and a line along the distal portion of the lateral root. Lateral roots exposed to osmotic stress typically exhibited steeper bending toward the gravitational vector.

3.2 Hydrotropism—Split Plate

1. Sterilize and germinate seeds as described in Sect. 3.1 (Steps 1–3).
2. Prepare fresh ½ MS media (0% sucrose) and carefully pour 85 mL of the media into 12 × 12 cm square sterile plates using a sterile measuring cylinder (*see* **Note 2**).
3. As soon as the plate has cooled, mark a diagonal line from one corner to the opposite corner on the back of each plate with a ruler and marker pen, separating the plate into two triangular halves (Fig. 3A).
4. Using the scalpel blade, make a straight slice along the line and around the edge of the media and plate wall to separate the lower half of the plate.

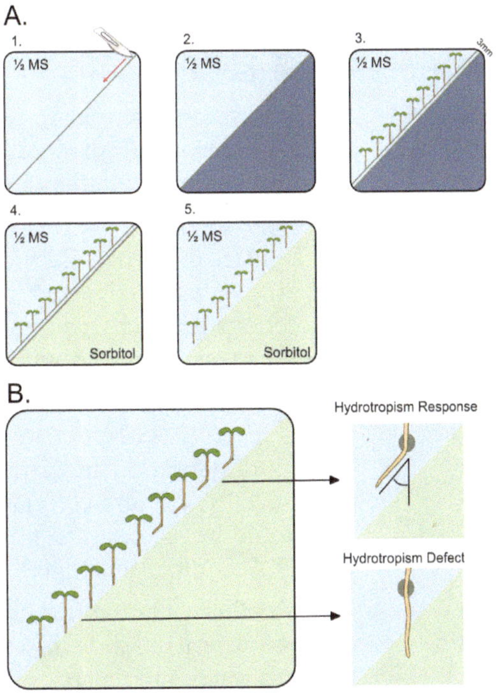

Fig. 3 Schematic of the hydrotropism split-plate experimental setup and analysis workflow. (**A**) The lower portion of ½ MS medium is removed and replaced with sorbitol after positioning seedlings along a marked guideline. (**B**) Root tip position at transfer is recorded, and the deviation of root growth from the vertical axis is measured to quantify hydrotropic curvature. Mutants with impaired hydrotropism grow straight toward the sorbitol interface rather than bending away from it

5. Using a blade or a flat tool (e.g., a scalpel), remove all media from the bottom half of the plate. This will result in half a plate of ½ MS media in a right-angled triangle shape.
6. Place the seedlings vertically on the upper half triangle of media, following an exact 45-degree line 3 mm above the diagonal edge with seedlings 8–10 mm apart. On the back of the plate, mark the location of the root tip with a small dot from a permanent marker (*see* **Note 3**).
7. Once in position, pour freshly autoclaved and lightly cooled 400 mM Sorbitol media onto the now empty side of the plate. Make sure to fill the plate to match the height of the MS media, being careful not to overflow the border.
8. Once the sorbitol media is set, seal the plates with micropore tape and place them vertically in the growth chamber.
9. After 24 h, take an image of the plates using a camera directly above the seedlings. Use a measuring software, such as ImageJ, to determine the hydrotropic response by measuring the angle

of bending of each root tip relative to the vertical direction of gravity (Fig. 3B and *see* **Note 4**).

3.3 Hydrotropism—Inverted Plate

1. Sterilize and germinate seeds as described in Sect. 3.1 (Steps 1–3). Prepare fresh ½ MS medium (0% sucrose) and carefully pour an equal volume (85mL) of the medium into 12 × 12 cm sterile plates using a measuring cylinder.
2. Transfer the 4-day-old seedlings to a fresh ½ MS media plate, using ethanol-sterilized tweezers, separating the plates into sections for each genotype (Fig. 4). Seal with micropore tape and return to the growth space for 24 h of vertical growth.
3. After 24 h, remove the micropore seal and lid from each plate. In spare 12 × 12 cm plate bases, pour 75 mL of 4 M Calcium chloride solution and carefully balance the experimental plates above the calcium chloride solution. The plates with seedlings should be placed in an upside–down position as shown in Fig. 4B. Seal the plates together using parafilm (*see* **Note 5**).
4. After 24 h of horizontal growth, separate the plates and discard the calcium chloride into a beaker. The materials can be rinsed with water and reused.
5. Score the number of seedlings that have grown away (hydrotropism defect) and those that have remained in contact with the media for each plate (*see* **Note 6**).

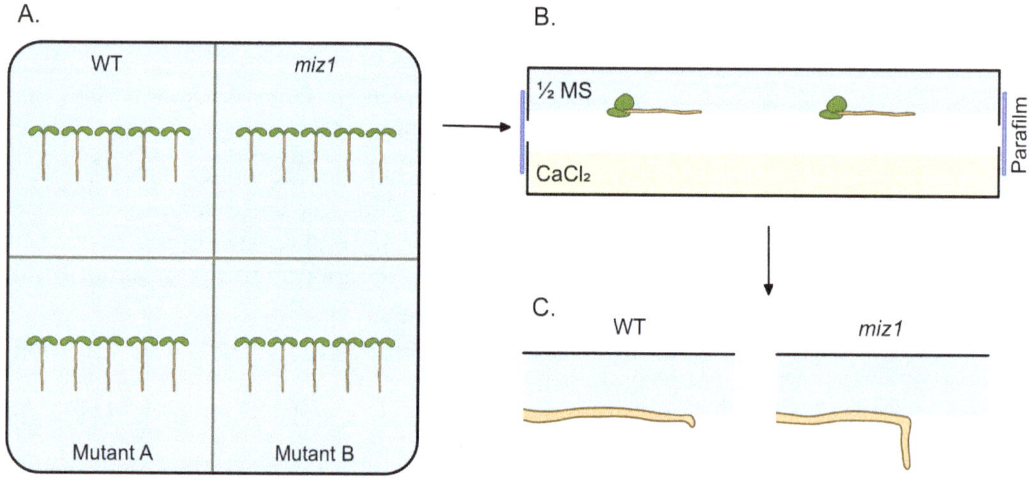

Fig. 4 Schematic of the inverted plate hydrotropism assay and scoring criteria. (**A**) Seedlings are initially grown on ½ MS medium plate. (**B**) The plate is inverted over a 4 M calcium chloride solution to create a strong water potential gradient, with the interface sealed to maintain the air gap. (**C**) Wild-type roots exhibiting hydrotropism continue growing along the medium, whereas hydrotropism-defective mutants (*miz1*) bend into the air gap and grow toward the gravity vector

3.4 Xerobranching—Agar Air-Gap Assay

1. Sterilize and germinate seeds as described in Sect. 3.1 (Steps 1–3).
2. Carefully pour the autoclaved ½ MS media (0.3% sucrose) into 12 × 12cm square sterile plates, which are slanted at 15° by resting one edge on its lid. Fill to 1–2 mm from the edge at the thickest side (*see* **Note 7**).
3. Leave the media to cool, reseal the plates in a sterile bag, and keep at 4°C to ensure the media is firm for cutting.
4. Remove the plates from the fridge and mark on the base two perpendicular lines, 40 mm and 45 mm from the thickest edge of the media (Fig. 5A). These will be the guidelines for cutting the 5 mm air gap.
5. In a spare square plate, sterilize the scalpel blade, a flat blade, and a cable tie (cut to 7–8 cm long) by submerging in 100% ethanol and leaving to air dry in the laminar flow hood.
6. Whilst being careful not to touch the media surface, use the flat blade to cut a 45° angled cut into the media, where the uppermost surface of the cut aligns with the first guideline and will result in a steep overhang for the root tip to grow over.
7. With the scalpel, cut the media at 90° exactly 5 mm below the blade cut using the second guideline for reference and stopping at the 10 cm mark to leave the last 2 cm uncut. Cut two perpendicular lines to connect both previous cuts.
8. Rotate and hold the plate upright, slide the shortened cable tie between the section that has been cut and the plate wall. Using the cable tie, push down lightly on the media strip while tilting the plate so that the excised strip falls out under gravity, leaving a clean space. If any residual media remains in the gap, remove it with the cable tie (*see* **Note 8**).
9. Store plates at 4°C until they are required for seedling transfer (*see* **Note 9**).
10. Transfer the 5-day-old seedlings to the previously prepared xerobranching plates (steps 2–9). Each xerobranching plate will contain six seedlings from a single genotype. The first seedling will be placed on the control region of the media (not containing the air gap), 5 mm above the boundary line. The roots of the other five seedlings will be placed 5 mm above the boundary line in the region containing the air gap, so that their roots will grow across this gap (*see* **Note 10**).
11. Seal each plate with micropore tape and return to the same growth environment as used for germination for 7 days (*see* **Note 11**).

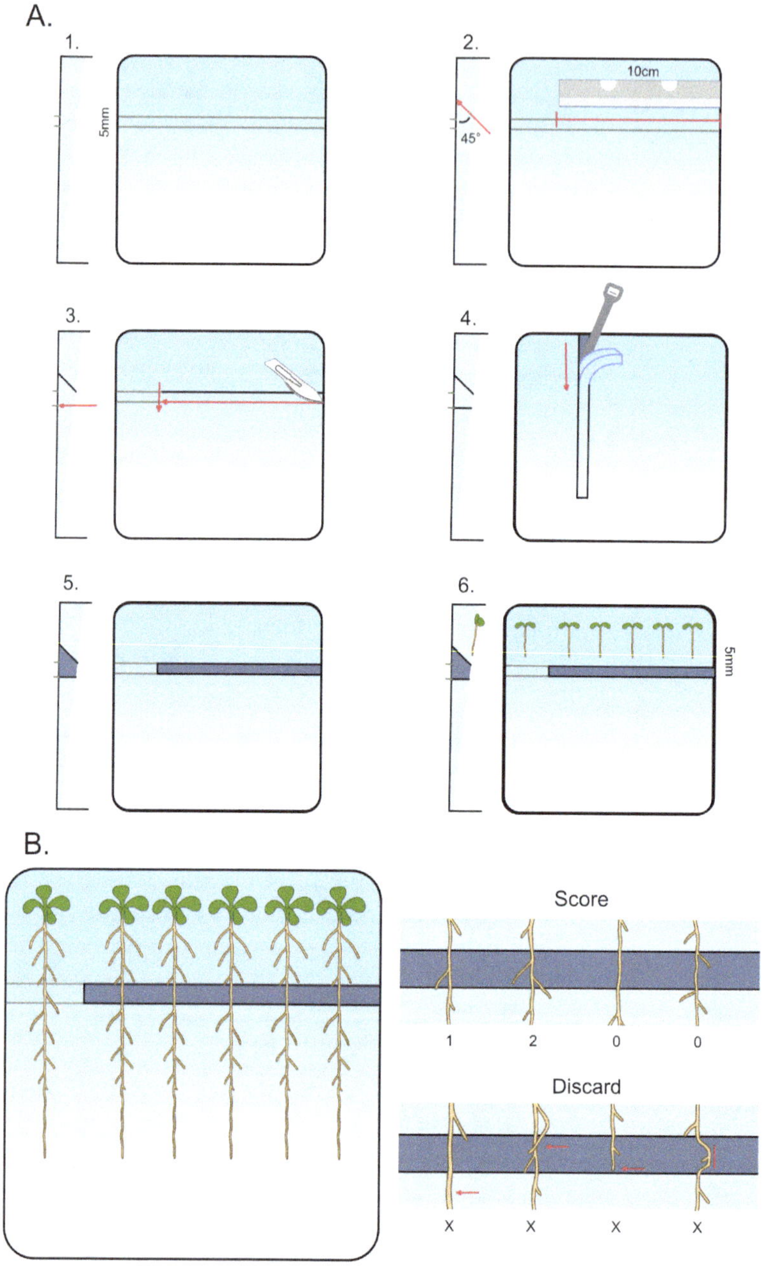

Fig. 5 Schematic of the xerobranching experiment preparation and experiment analysis. (**A**) The step-wise preparation of xerobranching plates with the vertical and side views of the plate. Seedlings are positioned so that most roots grow across the air gap while one seedling remains on intact agar as a branching control. (**B**) Representative outcomes and scoring criteria for xerobranching. Lateral root formation is quantified only in the air gap region. Seedlings are scored based on the number of lateral roots emerging in the air gap, while roots are excluded from analysis if they do not resume branching after crossing the gap, contact another root, fail to cross the gap, or bend into the plate surface

12. Score the number of lateral roots formed on each primary root that successfully passed the 5 mm gap, and the number of lateral roots formed within the corresponding 5 mm region in the control plant (Fig. 5B). The data gathered can be represented as a frequency of the absolute numbers of lateral roots or as a percentage of plants that exhibit a xerobranching defect (*see* **Note 12**).

3.5 Hydropatterning

1. Prepare ½ MS media (0.4% sucrose), autoclave, and pour 75 mL of the media into 12 × 12 cm square sterile plates using a measuring cylinder. The plates can be stored at 4°C or used immediately.
2. Sterilize seeds as described in Sect. 3.1 (Steps 1–2).
3. Sow the seeds in a row ~1.5 cm below the plate edge. Preferred set-up is with an internal plate control, where you have a control on one half of the plate (five seeds), while you sow the mutant you are screening on the other half (another five seeds). Sow the seeds roughly 1 cm apart from each other using a pipette tip or a toothpick (personal preference). Seal with micropore tape and transfer to growth space for 5 days (16/8-h photoperiod cycle at 21°C/18°C, respectively).
4. After twelve days of growth, take images of the plate to assess the general phenotype, such as primary root length. Before taking the plate images, it is wise to number the seedlings on the plate to confirm which seedlings will be measured and analyzed down the line (*see* **Note 13**).
5. For the hydropatterning analysis, remove the micropore tape and put the plate under a binocular microscope with a 5× objective, with the open side of the plate facing upwards. Analyze the direction of initiation of all the laterals on a seedling and mark if they emerge towards you (air) or towards the agar (agar/contact) (as shown in Fig. 6). Use this to calculate the ratio of air emergence vs contact emergence per plant (*see* **Note 14**).

4 Notes

1. The ideal option is to use a split plate setup, in which control (either positive or negative) is together with the mutant(s) you are screening.
2. Ensure the work surface for pouring is completely level. If uneven, it will not be possible to add the sorbitol to a matching level without overflow.
3. Drawing guidelines on the back of the plate or using a template helps maintain the consistency of angle and start

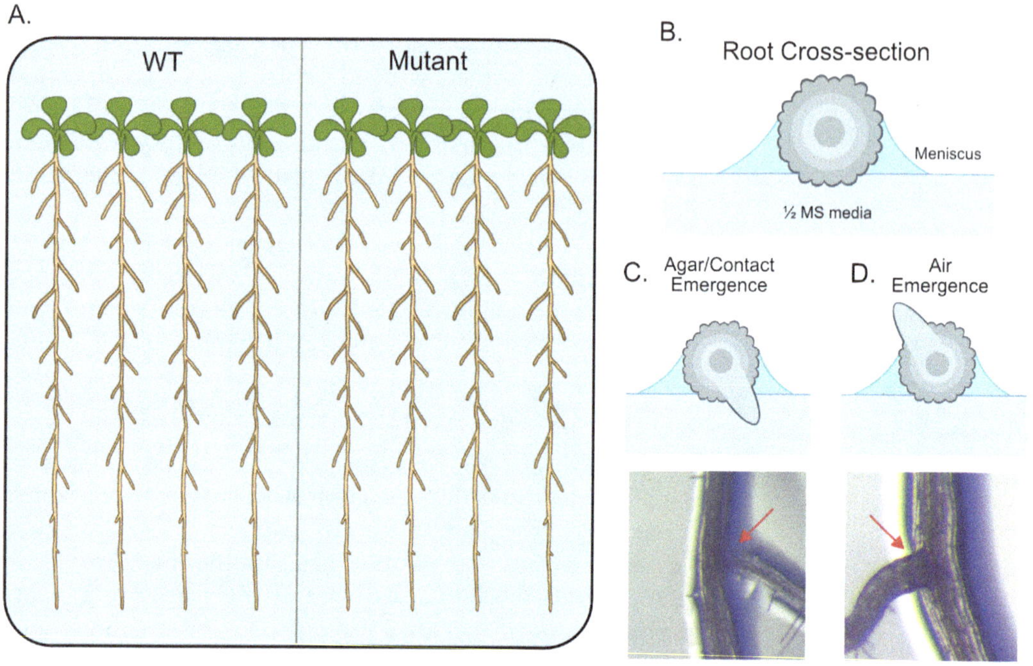

Fig. 6 Schematic of the hydropatterning assay. (**A**) Seedlings are grown vertically on MS agar plates until the lateral roots begin to emerge. (**B**) Cross-section schematic showing the primary root positioned with one side in contact with the agar and the opposite side exposed to air, establishing an asymmetric water environment. (**C**) Scoring of lateral root emergence orientation. In "Agar/Contact Emergence," lateral roots emerge from the agar-contact side, whereas in "Air Emergence," they emerge from the air-exposed side. Representative microscope images illustrate each emergence pattern

location. The marks on the root tip location will be needed for the analysis of growth after the experiment.

4. If the seedlings are too small, they will not grow enough on the split plates. In these cases, allow a further day of growth before transferring. Growth speed will affect the angle perceived in the hydrotropism experiment and is the major challenge in the interpretation of the results. An alternative setup for this experiment is to complete a time-course assay over 12 h, where images are taken every hour.

5. It is best to set up the inversion in the growth room or in a space nearby to reduce the risk of getting calcium chloride on the seedlings during transfer. Flow hoods do not need to be used at this step, as there is insufficient time for bacteria to grow and affect the experiment. We use corner pieces (adaptors) which are used to secure the two plate bases together. The 3D printed corner adapters allow perfectly aligned plates; however, it is possible to carry out the experiment without these, but the plates must be more carefully wrapped in the parafilm, as the risk of slipping is higher.

6. This experiment is a very strong water stress stimulus on the plants. The concentration of calcium chloride can be decreased to 2 M as needed to adjust the stimulus strength. This adjustment is made if most mutants do not display a clear defect at 4 M, allowing a milder stress condition to reveal more subtle phenotypes.
7. Ensure the work surface for pouring is completely level. If uneven, the media will not have a uniform gradient and therefore will be harder to cut. Make sure the media is well mixed before pouring to ensure uniformity of agarose throughout all plates. One liter of media will produce approximately twelve xerobranching plates.
8. A ripple will appear near the top of the vertical cut when pushing down on the media strip. Continue applying gentle pressure until the ripple reaches the end of the cut, indicating that the suction between the media and plate has been released. This helps the strip of media to be removed cleanly and in one piece.
9. Avoid storing the plates for too long, ideally cut all plates within a couple of days before seedling transfer. This is a time-intensive task, and therefore it is easier to split the tasks of cutting the plates and transferring the seedlings to different days.
10. We suggest using more plates than expected (15/20 plates per line) as many seedlings will fail to cross successfully and will need to be discarded from the analysis. Using more than six seedlings will cause the plates to be overcrowded. This can result in lateral roots of neighboring seedlings coming into contact with the primary roots, preventing the xerobranching response in those primary roots.
11. Ensure the plates are supported in a perpendicular position in the growth space. This will improve the number of roots successfully crossing the air gap neatly, as aided by gravitropic growth.
12. Many plants may need to be discarded. Reasons include bending to touch the plate surface, contact with another root during the crossing of the gap, failure to fully cross, or absence of lateral root emergence on the primary root when back in contact with the media after crossing the air gap. A lateral root will often form at the boundary between the media and air gap, which can be challenging to determine if it should be included in the scoring. For a lateral root to be counted as a xerobranching defect, it must be some distance from the media. There is some room for variability in which lateral roots are included or discarded; this can be minimized by using rigorous and consistent criteria to avoid bias.

13. Twelve days of growth is right for our growth conditions; it may take longer or shorter depending on the growth chamber. Preferably, you want the root tip of the longest seedling to be near the bottom edge (1–2 cm away from the edge of the plate). This allows for enough time for the seedling to form a good number of laterals and for the measurement of the primary root length to assess lateral root density. Seedlings that are outliers can be discarded at this point.
14. Scoring is determined by the point of emergence; therefore, be careful to look at the emergence of the laterals from the primary root, not the direction of growth. Some roots will emerge toward the air and quickly bend to the agar, but this will still be scored as an air-emerged lateral. Try adjusting the focus plane on the microscope to assess the depth level of the lateral. If it is not in focus when you focus on the top of the primary root, then it is an agar-oriented root. Furthermore, lateral roots that emerge towards the air will often show a crack in the epidermis that is visible when observing from the top angle. Do not include the small, newly forming laterals, as they cannot be assessed for their emergence direction. Adventitious roots from the hypocotyl of the seedlings should not be included in the analysis.

Acknowledgments

This work was supported by the Biotechnology and Biological Sciences Research Council (BBSRC) Discovery Fellowship (BB/Z514482/1 to PM) and the Horizon Europe ERC Starting Grant (101161820-WATER-BLIND to PM); European Research Council grant HYDROSENSING (101118769 to MJB).

Competing Interests The authors have no conflicts of interest to declare that are relevant to the content of this chapter.

References

1. Mehra P, Fairburn R, Leftley N et al (2023) Turning up the volume: how root branching adaptive responses aid water foraging. Curr Opin Plant Biol 75:102405
2. Bao Y, Aggarwal P, Robbins NE et al (2014) Plant roots use a patterning mechanism to position lateral root branches toward available water. Proc Natl Acad Sci USA 111:9319–9324
3. Babé A, Lavigne T, Séverin JP et al (2012) Repression of early lateral root initiation events by transient water deficit in barley and maize. Philos Trans R Soc B Biol Sci 367:1534–1541
4. Orman-Ligeza B, Morris EC, Parizot B et al (2018) The xerobranching response represses lateral root formation when roots are not in contact with water. Curr Biol 28:3165–3173.e5
5. Bonnet, C, 1754. Recherches sur l'usage des feuilles dans les plantes et sur quelques autres sujets relatifs a l'histoire de la vegetation. Elie Luzac, Goettingen, Leiden
6. Sachs J (1872) Ablenkung der Wurzel von ihrer normalen Wachsthumsrichtung durch feuchte Koerper. In: Sachs J (Ed.)Eds. Arbeiten des

Botanischen Institus in Wuerzburg. Wilhelm Engelmann, Leipzig, 209–222
7. Darwin, C, Darwin, F, 1880. The power of movement in plants. John Murray, London
8. Feng W, Linder H, Robbins NE et al (2016) Growing out of stress: the role of cell- and organ-scale growth control in plant water-stress responses. Plant Cell 28:1769–1782
9. Dubois M, Inzé D (2020) Plant growth under suboptimal water conditions: early responses and methods to study them. J Exp Bot 71:1706–1722
10. Ginnan NA, Custódio V, Gopaulchan D et al (2025) Precipitation legacy effects on soil microbiota facilitate adaptive drought responses in plants. Nat Microbiol 10:2823–2844
11. Pandey BK, George TS, Cooper HV et al (2025) Root RADAR: how "rhizocrine" signals allow roots to detect and respond to their soil environment and stresses. J Exp Bot 76:1500–1509
12. Zhang Y, Bao Z, Smoljan A et al (2025) Foraging for water by MIZ1-mediated antagonism between root gravitropism and hydrotropism. Proc Natl Acad Sci USA 122:e2427315122
13. Mehra P, Pandey BK, Melebari D et al (2022) Hydraulic flux-responsive hormone redistribution determines root branching. Science 378:762–768
14. Orosa-Puente B, Leftley N, von Wangenheim D et al (2018) Root branching toward water involves posttranslational modification of transcription factor ARF7. Science 362:1407–1410
15. Schindelin J, Arganda-Carreras I, Frise E et al (2012) Fiji: An open-source platform for biological-image analysis. Nat Methods 9:676–682

Check for updates

Chapter 10

Quantifying Root Turgor Using Dew Point Hygrometry

Martina Špundová and Jakub Hajný

Abstract

Water potential is the driving force behind the movement of water within the soil–plant–atmosphere continuum. Measuring water potential and its components provides valuable information about plant- and root water relations, including osmoregulation. Here, we describe a straightforward method for determining water and osmotic potential, as well as cell turgor in roots, using dew point hygrometry. The chapter begins with a brief overview of issues related to plant and root water potential, and then proceeds to describe the principles and measurement procedures of dew point hygrometry in more detail, including calibration, equilibration, and sample preparation. Practical recommendations are provided to minimize common error sources. This methodology may be useful for researchers studying plant–water balance, root water uptake, plant stress physiology, and osmoregulation.

Keywords Dew point hygrometry, Thermocouple psychrometry, Osmotic potential, Water potential, Root turgor, Osmoregulation

1 Introduction

1.1 Water Potential in Plants

Water transport in plants is driven by differences in water potential (Ψ). The water potential, expressed in pressure units [MPa], is defined as the free energy of water [J mol^{-1}] divided by the molar volume of water ($V_w = 1.8 \times 10^{-5}$ m^3 mol^{-1}) (for recent reviews of plant water transport and the role of Ψ see, e.g., [1, 2]). For pure water at atmospheric pressure and at sea level, Ψ has been defined as 0 MPa. In different plant compartments, Ψ is negative, mainly depending on the concentration of solutes and the pressure exerted.

Generally, water moves passively from regions of higher (less negative) Ψ to regions of lower (more negative) Ψ. Thus, plant water uptake from the soil to the root and transport to the shoot requires a gradual decrease in Ψ. In detail, water movement along the soil–plant–atmosphere continuum occurs along a gradient of Ψ in the liquid phase (from soil to leaf apoplast) and a difference in water vapor pressure (from intercellular air spaces in leaves to

Steffen Vanneste and Inge Verstraeten (eds.), *Root Development*, Methods in Molecular Biology, vol. 3086,
https://doi.org/10.1007/978-1-0716-5603-7_10,

external air). The mechanism of this movement is explained by the cohesion–tension theory [1, 3], which proposes that transpiration creates a pulling tension that draws a continuous column of cohering water molecules upward through xylem. The size of the Ψ gradient and a difference in vapor pressure determine the rate and extent of water uptake from the soil, water movement through the plant, and water loss by transpiration.

In plants, four components contribute to Ψ: solute (osmotic) potential, pressure potential, gravitational potential, and a matrix-dependent component. The matrix-dependent component originating from the interaction of water molecules with a solid matrix in plant cells (typically cell walls) is usually ignored. Similarly, the gravitational component is usually neglected, as it becomes significant only in very tall plant species. Thus, plant Ψ is typically described as the sum of two main components: osmotic potential (Ψs) and pressure potential or turgor (Ψp). The osmotic potential Ψs is determined by the concentration of solutes. When solutes are present, Ψs becomes more negative because of the increased disorder and thereby lower free energy levels generated by mixing solute and water molecules. The higher the solute concentration, the lower the free energy of the water and the more negative Ψs (and in consequence Ψ).

The pressure component Ψp is determined by the difference between the pressure within the system and atmospheric pressure. A positive pressure (i.e., higher than atmospheric pressure) raises Ψ while a negative pressure (i.e., below atmospheric pressure) reduces it. There are both positive and negative Ψp in plants. In xylem and in cell walls, Ψp is negative due to tension at the evaporating surfaces (or water menisci) within the cellulose microfibril pores of mesophyll cell walls. This tension is transmitted through a continuous water column from the leaves to the root apices and throughout the apoplast [1, 3].

Within well-hydrated living plant cells, Ψp is positive. This is because the solute concentration inside the cells is usually higher than in their surroundings, causing water to flow into cells to "pressurize" them, due to the rigidity of cell walls. Thus, the positive Ψp in cells (typically in the region of tenths of MPa) is the result of mechanical pressure exerted by the cytoplasm on the cell wall and vice versa. Cell turgor is essential for plant mechanical stability [4] and motions [5] as well as for many physiological processes, including growth [6].

The positive pressure is also present in the phloem. According to Münch's theory [7], this Ψp is created by the loading of soluble sugars (especially sucrose) and amino acids from the source tissue (i.e., photosynthetic mesophyll cells). This reduces the local Ψs, resulting in osmotic inflow of water from the surrounding xylem and in an increase of hydrostatic pressure. In sink locations (such as roots, meristems, fruits, or seeds), solutes are removed from the phloem, which causes an increase in Ψs, water outflow, and a decrease in the phloem Ψp.

1.2 Root Water Uptake, Water Potential and Turgor

There are two main physicochemical forces regulating root water uptake: hydrostatic and osmotic forces. Hydrostatic force is generated by the transpiration stream, whereas osmotic force is generated by active import of solutes or biosynthesis of new osmolytes. The current understanding of the mechanism of water uptake by roots is based on the "composite transport model" [8] that describes the main pathways of water movement through root tissues and the contribution of the individual pathways to water uptake and transport depending on the transpiration rate and/or soil water availability.

When water availability in the rhizosphere is limited, i.e., when Ψ of the soil (or other root environment) decreases, the total Ψ of the roots usually decreases too [9]. While in leaves the primary reason for the Ψ decrease is a decrease in the cell Ψp, in roots the situation is more complicated. In principle, two basic scenarios may arise: (1) Root cells actively increase the concentration of osmotically active substances, which reduces Ψs and restores the osmotic inflow of water into the roots. This osmotic adjustment can suppress or postpone the decrease in Ψp of the root cells. Thus, in such a case (typically during soil drying or decreasing osmotic potential of growth medium), the decrease of Ψ is primarily caused by a decrease in Ψs, while Ψp is maintained (Fig. 1) or even increased [10, 11]. Osmotic potential can also be increased in non-stressed plants by stimulation of shoot-to-root sugar transport [12]. (2) If there is no osmotic adjustment of the roots, the decrease in root Ψ is primarily caused by a decrease

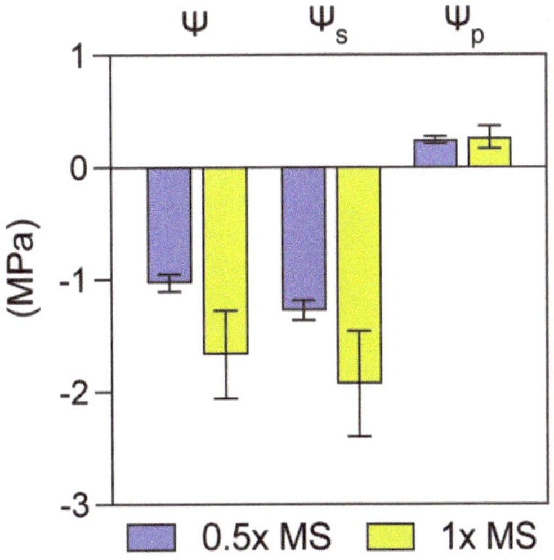

Fig. 1 Water potential (Ψ), osmotic potential (Ψs), and turgor (Ψp) of roots of Arabidopsis seedlings grown in Murashige and Skoog (MS) medium of half strength (0.5× MS; $\Psi s = -0.08$ MPa) and full strength (1× MS; $\Psi s = -0.14$ MPa). With decreasing Ψs of the medium, the root Ψ decreases slightly due to osmotic adjustment, resulting in lower root Ψs. The decrease in root Ψs restores the osmotic inflow of water into the roots, thereby maintaining the Ψp. Means and SD are shown, $n = 5$

of Ψp in root cells [13–15], while a Ψs decrease occurs at later stages of water limitation (similarly to the situation in the leaves).

Due to the complexity of the root structure and functioning [16], Ψs and Ψp, as well as their regulation, may vary by root cell type and location. This is related, among other things, to the existence of Ψ, Ψs, and Ψp gradients in roots, both transverse and longitudinal [17, 18].

Measurement of root Ψ and its components provides essential insights into the plant's water status and its response to stress conditions, such as drought and salinity, as well as osmoregulation [12]. However, targeted measurement of Ψ, Ψs, and Ψp in roots is used much less frequently than in shoots and leaves, and the results are often inaccurate and variable, as the measurement in roots is methodologically more demanding and requires specific approaches. For example, it can be challenging to sample roots without damaging them, and it is hard to clean them thoroughly of soil. Furthermore, there is generally a greater risk of an artificial decrease of Ψ due to evaporative water loss during sampling than for leaves.

There are two main ways in which the root Ψ and/or its components can be measured: (1) measurements using a pressure probe technique and (2) thermocouple psychrometry/dew point hygrometry. The pressure probe method [19, 20] is the only one that directly measures Ψp and related parameters in individual cells. However, the insertion of the probe into a target cell requires extremely precise micromanipulation, which is both technically and operationally demanding and time-consuming. Consequently, this method is not suitable for routine or large-scale measurements. In contrast, the thermocouple psychrometry/dew point hygrometry is a relatively simple and time-efficient approach, enabling measurements of many samples in a relatively short time. A more comprehensive overview of the methodology used to study various aspects of root water uptake and transport in plants can be found in [21].

1.3 Thermocouple Psychrometry/Dew Point Hygrometry

1.3.1 Principle

Generally, psychrometry measurements are based on determining Ψ of a small volume of air, which, after equilibrium is established in a closed chamber with the to-be-measured sample, is the same as Ψ of the sample. The water potential of air is given by the relative humidity (RH) according to the Kelvin equation:

$$\Psi = \frac{RT}{V_w} \ln \frac{p}{P_0} \tag{1}$$

where R is the universal gas constant (8.3143 J mol^{-1} K^{-1}), p is the partial water vapor pressure in equilibrium with the liquid phase in a sample, and p_0 is the saturation water vapor pressure (= the pressure that would result when air is in equilibrium with pure water). T is the absolute temperature (K) and V_w is the molar volume of water ($1.8 \times 10^{-5}\,m^3\,mol^{-1}$). The ratio p/p_0 represents RH.

Thermocouple psychrometry/dew point hygrometry relies on two physical effects: the Seebeck effect and the Peltier effect. The Seebeck effect generates an electric voltage between two different metals when their junctions are at different temperatures. The Peltier effect allows cooling of a thermocouple junction by passing an electrical current; its use for the Ψ measurements was first proposed by [22]. In practice, the Peltier current cools the measuring thermocouple junction below the dew point, causing condensation of water vapor [23]. When held at dew point temperature, the junction neither gains nor loses water. This temperature depends on the RH of the air above the sample, which reflects the sample's Ψ.

Two methods are used to estimate Ψ from RH measurements: the psychrometric (wet bulb) method and the hygrometric (dew point) method [24]. In the psychrometric method, warming of the junction is monitored after the Peltier current stops, which slows transiently due to evaporation of condensed water. The dew point method determines the RH by measuring the dew point temperature [25]. As in the psychrometric method, the thermocouple junction is cooled below the dew point using the Peltier current, which causes condensation of water from the air above the sample onto the junction. The temperature of the junction is then automatically maintained at the dew point using short pulses of Peltier current. These pulses counteract the junction's spontaneous warming, which is caused by the absorption of heat from its surroundings [25].

Using psychrometry/hygrometry, we can not only determine the total Ψ (i.e., the sum of Ψs and Ψp), but also the osmotic potential Ψs, and calculate cell turgor Ψp from the difference between Ψ and Ψs. For this purpose, the sample, for which we first determined the total Ψ, is frozen in liquid nitrogen for several minutes, which destroys cell membranes and eliminates cell turgor. By remeasuring the same sample after thawing at room temperature (in a sample chamber, during equilibration), we determine Ψs and obtain Ψp by subtracting Ψs from Ψ.

1.3.2 The Dewpoint Measurement

1. The sample is placed in a sample holder (a shallow metal cup) and positioned inside a sample chamber, under a miniature thermocouple junction, using a nylon slide.
2. Then the metal mounting block is secured against the sample holder with a nylon screw, and an O-ring seals the chamber, creating a closed microenvironment (Fig. 2).
3. Once temperature and water vapor equilibrium have been reached, a voltage output (in μV) corresponding to the dew-point temperature is determined, and the air's RH and finally Ψ of the measured sample are estimated.

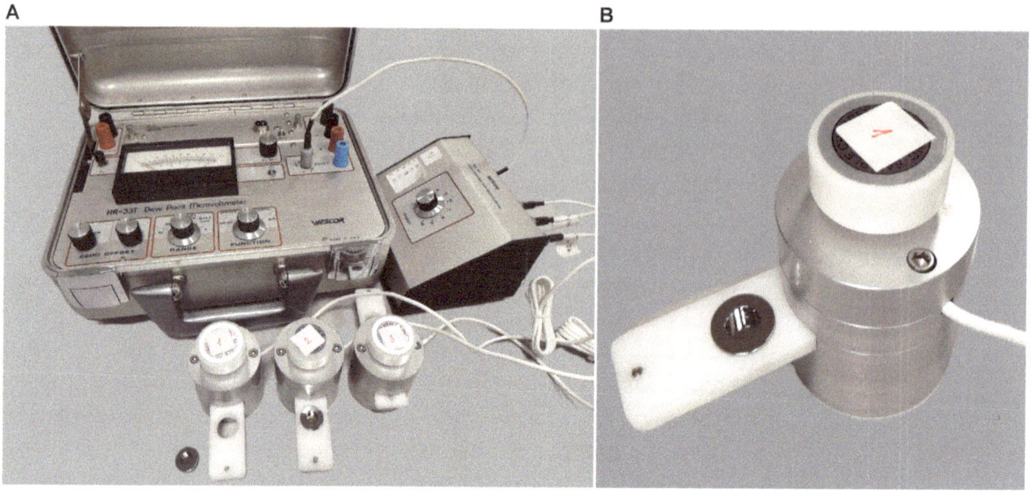

Fig. 2 Instrumentation for measuring water potential: (**A**) HR-33T Dew Point Microvoltmeter, C-52 Psychrometric/Hygrometric Sample Chambers and Switchbox PS-10 (Wescor, Logan, UT, USA), (**B**) C-52 Psychrometric/Hygrometric Sample Chamber in detail

4. While Ψ can be calculated from RH using Eq. 1, in practice, it is determined by calibration.

1.3.3 Calibration

As mentioned, the sample Ψ is typically determined using voltage output calibration. Correct calibration is hence crucial for the accuracy of all subsequent measurements. Due to variability in voltage output characteristics among individual sensors, each sensor (= sample chamber) must be calibrated separately. The calibration is typically performed with a set of salt solutions (NaCl or KCl) of different concentrations and known Ψ to convert the voltage output (in μV) to Ψ (in MPa). It exploits the fact that the Ψ of a solution is equal to its Ψs, which is directly proportional to the solution's concentration. Water potentials of NaCl and KCl solutions for a range of concentrations and temperatures are tabulated, for example, in [26, 27]. Since the sample's Ψ is mainly, but not the only factor affecting the output of the sensor, it is desirable to maintain the conditions during the sample measurement as constant as possible while calibrating.

1.3.4 Equilibration Time

After sample loading into the sample chamber, vapor and thermal equilibrium must be achieved before the reading. The time required for this equilibrium varies for different conditions and samples, so it should be determined empirically, by repeated reading (Fig. 3). Plant tissues require a longer equilibration time than solutions due to slower evaporation. Generally, the equilibrium time of drier samples (with more negative water potential) is shorter than that of samples with higher water potential [28]. The equilibration time required for

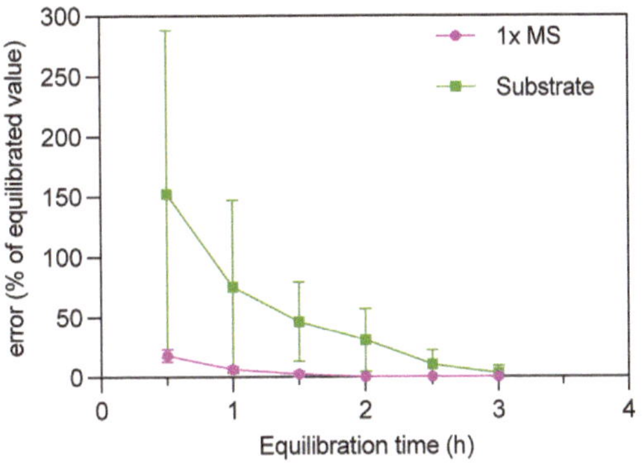

Fig. 3 The dependence of relative error in determining water potential (*Ψ*) on the time of equilibration of roots of Arabidopsis plants grown on standard 1× Murashige and Skoog (MS) medium or in a soil substrate. The error is expressed in % of the *Ψ* value after reaching equilibration (i.e., 2.5 and 4 h for plants from the 1× MS medium and from the substrate, respectively, after placing the root samples in the sample chamber). Means and SD are shown, $n=5$

plant roots grown in soil substrates is longer than for those cultivated on agar medium (Fig. 3), due to differences in their anatomy.

1.3.5 Sample Preparation

During sample preparation, the time between excision and sealing the sample into the chamber should be minimized to prevent water loss, erroneously reducing the *Ψ* readings. The error magnitude is correlated with the size of the sample [29] and the evaporative demand in ambient air.

Given the definition of *Ψ*, its value is independent of the size or volume of the sample. Nevertheless, larger sample volumes provide more accurate *Ψ* measurements, as errors related to water adsorption in thermocouple psychrometer/hygrometer assemblies are minimized when the volume of plant tissue is large relative to the sample chamber size [30].

1.3.6 Contamination

The accurate measurement of *Ψ* with the thermocouple psychrometer/hygrometer requires a uniform evaporation rate of water from the thermocouple junction and a uniform vapor concentration throughout the cavity. It is therefore very important to avoid contamination of both the thermocouple and the interior of the chamber (*see* **Note 1**).

2 Materials

2.1 Instrumentation

1. HR-33T Dew Point Microvoltmeter (Wescor, Logan, UT, USA)

2. C-52 Psychrometric/Hygrometric Sample Chamber(s) with *SUREFAST*^{TM} connector(s) (Wescor, Logan, UT, USA)
3. Switchbox PS-10 with *SUREFAST*^{TM} connector (Wescor, Logan, UT, USA), allowing 10 psychrometric/hygrometric chambers to be connected to a single HR-33T Microvoltmeter
4. Sample holders (Wescor, Logan, UT, USA)

2.2 Calibration

1. A set of calibration solutions (NaCl or KCl) of different concentrations (their water potentials tabulated in [26, 27])
2. Filter paper disks (5 mm diameter)

2.3 Sample Preparation

1. A humidity box
2. Scalpel/surgical scissors, tweezers
3. Safe-lock Eppendorf tubes
4. Liquid nitrogen

3 Methods

3.1 Contamination Check and Cleaning

1. Check the cleanliness of the sample holders, the thermocouple junction, and the inner surfaces of the sample chamber. Clean if necessary (*see* **Note 1**).

3.2 Calibration

1. Prepare standard calibration solutions of NaCl or KCl with a known water potential (*see* **Note 2**). Choose solution concentrations to cover the anticipated range of sample Ψ (*see* **Note 3**).
2. Set a cooling coefficient for the individual sample chambers (*see* **Note 4**).
3. Place a filter paper disk that is thoroughly wetted with the calibration salt solution into the sample holder and seal the chamber. The filter paper matrix should be completely saturated with the solution (*see* **Note 5**).
4. After an equilibration period, switch the device to measuring mode, apply the cooling current for a sufficient period of time (*see* **Note 6**), and measure the voltage output using the dew point method.
5. Using repeated measurements at selected time intervals (without opening the sample chamber), determine the time required for sample equilibration in the chamber (usually minutes). Equilibrium is achieved when two consecutive measurements give the same voltage output.

6. Repeat the voltage output measurement (using the empirically determined equilibration time) for solutions of different concentrations (usually 4–6).
7. Plot the voltage output (in μV) versus Ψ (in MPa) (the resulting dependence is linear) and find the calibration equation for the conversion (it should be approximately 0.75 μV per 0.1 MPa). There will be a specific calibration equation for each individual sample chamber.

3.3 Sample Preparation

1. In a high RH environment (*see* **Note** 7), remove the roots from the soil substrate or medium and gently clean them.
2. Using a scalpel or surgical scissors, excise the root sample, place it as quickly as possible in a sample holder of appropriate size, and seal it in the chamber (*see* **Notes 8–11**).

3.4 Sample Measurement

1. Maintain the conditions during the measurement of the samples, constant while calibrating (*see* **Note 12**).
2. Determine the equilibration time for the measured sample type in the same way as for calibration solutions (this will be minutes to hours, depending on the sample type) (*see* **Note 13**).
3. Measure (using the same procedure as for measuring calibration solutions) the voltage output, which corresponds to the total Ψ of the sample.
4. Remove the sample from the chamber and sample holder (again in a high RH environment), immediately place it using tweezers in a safe-lock Eppendorf tube, and freeze in liquid nitrogen for approximately 10 min.
5. Remove the sample from the Eppendorf tube, place it back into the sample holder as quickly as possible, and seal the chamber.
6. After equilibration, measure the voltage output again (this time it will correspond to the Ψs of the sample).
7. Using the calibration equation, convert the first and second voltage outputs to Ψ and Ψs, respectively.
8. Calculate cell turgor of the sample as $\Psi p = \Psi - \Psi s$.

4 Notes

1. Measurement by the thermocouple psychrometer/hygrometer may be compromised by contamination of the thermocouple and chamber interior, including the surface of the sample holder. Common contaminants include skin oils, plant and plant sap residues, and residues from evaporated liquids. When changing samples, all surfaces within the chamber

must be cleaned thoroughly to remove all contaminants. Additionally, the thermocouple junctions need to be cleaned regularly according to the manufacturer's specifications and subsequently recalibrated. In most cases, sample holders, sample chambers, and thermocouple junctions can be cleaned with deionized water. Most of the water can then be removed by shaking, after which any residual droplets should be carefully blown away with pressurized air or with a stream of gaseous nitrogen to ensure that the water is removed before it can evaporate and leave residues on the cleaned surface. If the thermocouple junction is heavily contaminated, it can be cleaned using a 10% ammonia solution. The thermocouple should then be rinsed several times with deionized water and dried using the above procedure. Extreme care must be taken when cleaning the thermocouple to avoid mechanical damage and scratching of the thermocouple mount and chamber surfaces. For detailed instructions, see the C-52 Sample Chamber Instruction/Service Manual.

2. Standard calibration NaCl solutions are available from the manufacturer of the psychrometric/hydrometric system (Wescor). Generally, calibration solutes may be prepared using distilled/deionized water and reagent-grade NaCl. A one molal solution is made up of 58.44 g of NaCl and 1000 g of water (at 25°C, corresponding to $\Psi = -4.64$ MPa). Once the calibration solutions are prepared, changes in concentration (typically resulting from evaporation or cross-contamination when wetting the paper disks) should be prevented. For routine measurements, a minimum of four calibration solutions should be selected.
3. The C-52 psychrometric/hygrometric sample chamber is designated to measure Ψ roughly between −0.05 MPa and −8 MPa, which covers most biological tissues from very moist to moderately dry.
4. To correctly determine Ψ by the dew point method, a cooling coefficient (Π_V) (reflecting the cooling capability of a thermocouple junction) must be set to the proper value for the actual ambient temperature before the measurement. Then the hygrometric output is independent of the ambient temperature. For detailed setup instructions, see the HR-33T Dew Point Microvoltmeter Instruction/Service Manual.
5. The use of the filter paper disks eliminates accidental splashes of the solutions and contamination of the thermocouple assembly.
6. For the best accuracy, it is desirable to use a consistent cooling period from measurement to measurement, and use the same cooling period for measuring samples as the one used

for calibration. The lower the sample Ψ, the longer the recommended cooling time.

7. It is highly recommended that samples be transferred to the chamber in a humid environment (ideally in a humidity-controlled glove box or at least in a closed space/box with the highest possible RH). This arrangement is particularly important for root measurement because evaporative water loss from roots is usually higher than from leaves [31].
8. When preparing samples, it is desirable to avoid mechanical damage and, if possible, to maintain the smallest possible ratio of cells damaged by cutting to the remaining intact tissue in the sample area. A higher proportion of cut cells can erroneously increase Ψ [28, 32].
9. For accurate measurement, use a sufficiently large sample volume, but before inserting it into the chamber, make sure that the sample does not exceed the edges of the sample holder, as this can lead to contamination or even damage of the thermocouple junction.
10. Try to select samples consistently from the same part of the root system. Try to excise the samples with a sharp cut, do not damage/crush them during sampling, and maintain approximately the same ratio of cut/intact area.
11. In the case of delicate roots such as those from Arabidopsis seedlings, it is best to first collect a larger number of seedlings (using tweezers), gently join the roots of individual seedlings together, and only then cut them.
12. The ambient temperature should not change significantly during the measurement. Slight variations in air temperature will not affect the measurement, but care should be taken to avoid direct solar radiation or a nearby heat or cooling source. It is also desirable to avoid touching the metal parts of the chamber since heat transferred from an operator's hand may have a considerable effect.
13. When the equilibration time is (too) long, a shorter time can be used (consistently) in comparative studies; however, such Ψ values can not be taken as absolute ones [33]. Figure 3 shows a decrease in relative error in determining Ψ with increasing equilibration time in the roots of plants grown in 1× MS medium and in soil substrate.

Acknowledgments

The work was supported by the project TowArds Next GENeration Crops, reg. no. CZ.02.01.01/00/22_008/0004581 of the ERDF Program Johannes Amos Comenius. The work of J.H. was supported by the project JG_2024_003 implemented within the Palacký University Young Researcher Grant and by Junior Star GACR 26-20529M.

Competing Interests The authors have no conflict of interest to declare that is relevant to the content of this chapter.

References

1. Venturas MD, Sperry JS, Hacke UG (2017) Plant xylem hydraulics: what we understand, current research, and future challenges. J Integr Plant Biol 59:356–389. https://doi.org/10.1111/jipb.12534
2. De Swaef T, Pieters O, Appeltans S et al (2022) On the pivotal role of water potential to model plant physiological processes. In Silico Plants 4:diab038. https://doi.org/10.1093/insilicoplants/diab038
3. Dixon HH, Joly J (1895) On the ascent of sap. Philos Trans R Soc Lond B 186:563–576
4. Zhang X, Ramakanth KK, Long Y (2024) The biomechanics of turgor pressure. Curr Biol 34:R986–R991. https://doi.org/10.1016/j.cub.2024.07.013
5. Morris RJ, Blyth M (2019) How water flow, geometry, and material properties drive plant movements. J Exp Bot 70:3549–3560. https://doi.org/10.1093/jxb/erz167
6. Ali O, Cheddadi I, Landrein B et al (2023) Revisiting the relationship between turgor pressure and plant cell growth. New Phytol 238:62–69. https://doi.org/10.1111/nph.18683
7. Münch E (1930) Die Stoffbewegungen in der Pflanze. Gustav Fischer, Jena
8. Steudle E (2000) Water uptake by plant roots: an integration of views. Plant Soil 226:45–56. https://doi.org/10.1023/A:1026439226716
9. Nobel PS, Lee CH (1991) Variations in root water potentials: influence of environmental factors for two succulent species. Ann Bot 67:549–554. https://doi.org/10.1093/oxfordjournals.aob.a088197
10. Schmidhalter U, Evéquoz M, Camp K-H et al (1998) Sequence of drought response of maize seedlings in drying soil. Physiol Plant 104:159–168. https://doi.org/10.1034/j.1399-3054.1998.1040203.x
11. Shabala SN, Lew RR (2002) Turgor regulation in osmotically stressed Arabidopsis epidermal root cells. Direct support for the role of inorganic ion uptake as revealed by concurrent flux and cell turgor measurements. Plant Physiol 129:290–299. https://doi.org/10.1104/pp.020005
12. Hajný J, Trávníčková T, Špundová M et al (2024) Sucrose-responsive osmoregulation of plant cell size by a long non-coding RNA. Mol Plant 17:1719–1732. https://doi.org/10.1016/j.molp.2024.09.011
13. Tyerman S, Oats P, Gibbs J et al (1989) Turgor-volume regulation and cellular water relations of *Nicotiana tabacum* roots grown in high salinities. Aust J Plant Physiol 16:517–531. https://doi.org/10.1071/PP9890517
14. Azaizeh H, Gunse B, Steudle E (1992) Effects of NaCl and $CaCl_2$ on water transport across root cells of maize (*Zea mays* L.) seedlings. Plant Physiol 99:886–894. https://doi.org/10.1104/pp.99.3.886
15. Crabos A, Huang Y, Boursat T et al (2023) Distinct early transcriptional regulations by turgor and osmotic potential in the roots of Arabidopsis. J Exp Bot 74:5917–5930. https://doi.org/10.1093/jxb/erad307
16. Maurel C, Nacry P (2020) Root architecture and hydraulics converge for acclimation to changing water availability. Nat Plants 6:744–749. https://doi.org/10.1038/s41477-020-0684-5
17. Rygol J, Pritchard J, Zhu JJ et al (1993) Transpiration induces radial turgor pressure gradients in wheat and maize roots. Plant Physiol 103:493–500. https://doi.org/10.1104/pp.103.2.493
18. Ball RA, Oosterhuis DM (2005) Measurement of root and leaf osmotic potential using the vapor-pressure osmometer. Environ Exp Bot 53:77–84. https://doi.org/10.1016/j.envexpbot.2004.03.003
19. Hüsken D, Steudle E, Zimmermann U (1978) Pressure probe technique for measuring water relations of cells in higher plants. Plant Physiol 61:158–163.https://doi.org/10.1104/pp.61.2.158

20. Zimmermann U (1989) Water relations of plant cells: Pressure probe technique. In: Methods in enzymology, Vol 174. Academic Press. https://doi.org/10.1016/0076-6879(89)74027-1
21. Boursiac Y, Protto V, Rishmawi L et al (2022) Experimental and conceptual approaches to root water transport. Plant Soil 478:349–370. https://doi.org/10.1007/s11104-022-05427-z
22. Spanner DC (1951) The Peltier effect and its use in the measurement of suction pressure. J Exp Bot 2:145–168. https://doi.org/10.1093/jxb/2.2.145
23. Wiebe HH (1984) Water condensation on Peltier-cooled thermocouple psychrometers: a photographic study. Agron J 76:166–168. https://doi.org/10.2134/agronj1984.00021962007600010041x
24. Briscoe R (1986) Thermocouple psychrometers for water potential measurements. In: Gensler WG (ed) Advanced agricultural instrumentation: design and use. Springer Netherlands, Dordrecht
25. Campbell EC, Campbell GS, Barlow WK (1973) A dewpoint hygrometer for water potential measurement. Agric Meteorol 12:113–121. https://doi.org/10.1016/0002-1571(73)90012-5
26. Campbell GS, Gardner WH (1971) Psychrometric measurement of soil water potential: temperature and bulk density effects. Soil Sci Am J 35:8–12. https://doi.org/10.2136/sssaj1971.03615995003500010011x
27. Lang A (1967) Osmotic coefficients and water potentials of sodium chloride solutions from 0 to 40°C. Aust J Chem 20:2017–2023. https://doi.org/10.1071/CH9672017
28. Nelsen CE, Safir GR, Hanson AD (1978) Water potential in excised leaf tissue: comparison of a commercial dew point hygrometer and a thermocouple psychrometer on soybean, wheat, and barley. Plant Physiol 61:131–133. https://doi.org/10.1104/pp.61.1.131
29. Walker S, Oosterhuis DM, Wiebe HH (1984) Ratio of cut surface area to leaf sample volume for water potential measurements by thermocouple psychrometers. Plant Physiol 75:228–230. https://doi.org/10.1104/pp.75.1.228
30. Bennett JM, Cortes PM (1985) Errors in measuring water potentials of small samples resulting from water adsorption by thermocouple psychrometer chambers. Plant Physiol 79:184–188. https://doi.org/10.1104/pp.79.1.184
31. Oosterhuis DM (1987) A technique to measure the components of root water potential using screen-caged thermocouple psychrometers. Plant Soil 103:285–288. https://doi.org/10.1007/BF02370403
32. Barrs HD, Kramer PJ (1969) Water potential increase in sliced leaf tissue as a cause of error in vapor phase determinations of water potential. Plant Physiol 44:959–964. https://doi.org/10.1104/pp.44.7.959
33. Hanson AD, Nelsen CE, Everson EH (1977) Evaluation of free proline accumulation as an index of drought resistance using two contrasting barley cultivars. Crop Sci 17:720–726. https://doi.org/10.2135/cropsci1977.0011183X001700050012x

Check for updates

Chapter 11

Targeted Flow-Free Proton Delivery to Arabidopsis Seedlings Using a Capillary-Based Organic Electronic Ion Pump (cOEIP)

Julie Ducla, Iwona Bernacka-Wojcik, Eleni Stavrinidou and Peter Marhavý

Abstract

The capillary-based Organic Electronic Ion Pump (cOEIP) is an emerging bioelectronic delivery tool that enables precise and electronically controlled delivery of charged biomolecules to biological targets without inducing bulk fluid flow. Traditional treatment methods often induce mechanical disturbances or secondary signaling that obscure the direct physiological effects of the delivered compounds. In contrast, the iontronic devices enable precise, flow-free ionic delivery, allowing the isolation of responses specifically attributable to the applied biomolecule. Here, we present a detailed protocol for integrating the cOEIP with live confocal imaging of *Arabidopsis thaliana* roots, enabling spatiotemporally resolved studies of root physiology under controlled ionic delivery. As a model application, we demonstrate the localized delivery of protons (H^+) to the root microenvironment, providing a framework for investigating pH-dependent signaling processes. The protocol includes step-by-step instructions for device assembly, preparation, calibration, and experimental setup, accompanied by representative images illustrating the mounting and operational procedures. Importantly, this approach can be adapted for the delivery of a wide range of charged biomolecules, expanding the toolkit for dynamic in vivo studies of plant–environment interactions.

Keywords *Arabidopsis thaliana* , Iontronic delivery, Proton delivery, Root, Root physiology, pH manipulation

1 Introduction

Precise and localized delivery of biomolecules to plant tissues remains a major challenge in experimental plant biology. Conventional treatment methods—such as medium supplementation, droplet application, or microinjection—often lack spatial and dynamic control and can introduce confounding factors, including mechanical stress, convective disturbance of ionic gradients, diffusion artifacts, or unintended signaling cascades. These artifacts make it difficult to distinguish between

---

The authors "Julie Ducla and Iwona Bernacka-Wojcik" contributed equally to this work.

Steffen Vanneste and Inge Verstraeten (eds.), *Root Development*, Methods in Molecular Biology, vol. 3086,
https://doi.org/10.1007/978-1-0716-5603-7_11,

physiological responses that are caused directly by the delivered molecule and those arising from mechanical disturbance or secondary signaling events. To overcome these limitations, the capillary-based Organic Electronic Ion Pump (cOEIP) provides an electronically controlled, flow-free means of delivering charged molecules to living tissues with high spatial and temporal precision [1, 2].

The cOEIP translates electrical signals into controlled ionic fluxes through ion-conducting polymers; consequently, the dynamics of ion transport and the temporal pattern of stimulation can be modulated by adjusting the amplitude and duration of the applied current (e.g., continuous or pulsed delivery at defined frequencies). The cOEIP offers a way to investigate plant responses to specific molecular or ionic cues under strictly controlled conditions, bridging the gap between electrophysiology, chemical delivery, and live imaging.

This unique mechanism enables the modulation of the chemical environment around biological samples without disturbing their physical state or introducing bulk liquid flow. In plant systems, where local ionic changes can trigger complex physiological responses, such precision is particularly valuable.

In OEIP-based delivery, the delivered dose is proportional to the applied delivery current and delivery time, according to Faraday's law, scaled by the transport efficiency of the device. The total amount of substance delivered (n) can be approximated as

$$n = \eta \frac{I \cdot t}{zF} \tag{1}$$

where I is the applied current, t is the delivery time, z is the charge of the transported ion, F is the Faraday constant (96,485 C mol^{-1}), and η represents the delivery efficiency (fraction of current contributing to transport of the target ion). The delivery efficiency can be determined empirically by chemical analysis of the target medium, for example, using pH measurements for proton delivery or analytical techniques such as mass spectrometry or ion-specific sensors for other transported species.

The resulting delivery distribution is diffusion-based and resembles plant-to-plant signaling, with the spatial distribution of the delivered dose determined by the diffusion properties of the surrounding medium.

In recent years, the cOEIP has been successfully applied in plant research to deliver key phytohormones and signaling molecules. For instance, auxin [1] and cytokinin have been delivered in close proximity to *Arabidopsis thaliana* roots to modulate developmental patterning and cellular differentiation [3], while abscisic acid (ABA) has been delivered directly to leaf blade to control stomata aperture and into the plant vasculature to investigate systemic stress responses [2,4]. These studies demonstrated the device's versatility and compatibility with delicate plant tissues, paving the way for its use in more complex and dynamic experimental settings.

Building upon these advancements, we adapted the cOEIP platform for precise manipulation of extracellular pH by controlled delivery of protons (H^+) near *Arabidopsis thaliana* roots [5]. This adaptation enables direct investigation of root physiological responses to localized pH changes, a fundamental parameter influencing nutrient uptake, ion homeostasis, and growth regulation. Furthermore, we combined cOEIP-based proton delivery with real-time, ultra-fast, in vivo confocal imaging, allowing the simultaneous monitoring of rapid cellular and tissue-level responses. This integration provides a powerful approach for studying plant signaling dynamics at unprecedented spatiotemporal resolution.

Beyond proton manipulation, the described protocol can be extended to a wide range of charged biomolecules, enabling dynamic and reversible perturbations of plant physiological states. The method outlined here thus establishes a versatile experimental framework for coupling bioelectronic control with advanced live imaging to explore the molecular basis of plant–environment interactions.

2 Materials

2.1 Seed Sterilization and Plant Growth

1. *Arabidopsis thaliana* seeds.
2. ½ MS medium: Mix 2.2 g of Murashige and Skoog Basal salt mixture including vitamins and 0.1% MES in 1 L of double distilled water, adjust with KOH to pH 5.7, add 8 g of Plant agar. Autoclave, and pour in 50 mL in 8 cm × 8 cm square Petri dishes in the laminar flow hood.
3. 70% ethanol.
4. Sterile water.
5. Eppendorf tubes—1 mL.
6. Pipette and autoclaved pipette tips.
7. Micropore tape.

2.2 cOEIP Mounting

1. Organic Electronic Ion Pump capillary (can be fabricated according to the protocol described in [6] or purchased via iontronics.com). Briefly, the inner surface of polyimide-coated glass capillaries (25/150 µm ID/OD) was surface-treated by sequential flushing with KOH and A174 silane to activate the glass surface and promote covalent anchoring of the polymer network. The capillaries were then filled with an AMPS-based (2-acrylamido-2-methyl-1-propanesulfonic acid) polyelectrolyte precursor containing crosslinker and photoinitiators and photo-polymerized under blue light to form a stable ion-conducting channel.
2. 37% hydrochloric acid (HCl), (w/w), commercial grade.

3. Electrolyte source solution (0.1 M HCl): Add 0.83 mL of 37% HCl in 80 mL of distilled water, and add water to a total volume of 100 mL.
4. Storage solution (0.1 M NaCl): Dissolve 0.586 g of NaCl in 100 mL of distilled water.
5. Adhesive Lined Heat Shrink Tubing (example: Dual Wall, PO-X, 3:1, 0.118″, 3 mm, Transparent, Farnell).
6. Ag/AgCl electrodes (*see* **Note 1**).
7. Electrode preparation solution: Add 8.3 mL of 37% HCl to 80 mL of distilled water, and add water to a total volume of 100 mL.
8. Tweezers.
9. Heat gun or Bunsen burner.

2.3 Sample Mounting and Microscope Stage Setup

1. Source-measure unit (Keithley generator).
2. Micromanipulator.
3. Open glass chamber adapted to the microscope stage (here 45 mm × 65 mm; this chamber was specifically 3D printed and modified using the service of Labstuce (https://labstuce.com/) and various editable models are also available online (e.g., https://www.tinkercad.com/things/0WptAiyt1sz) (*see* **Note 2**).
4. A square Petri dish with solid ½ MS medium.
5. Soft tweezers.
6. Spray bottle of sterile water.
7. Parafilm.
8. Inverted confocal microscope.

3 Methods

3.1 Seed Surface Sterilization and Seedling Growth

1. Place Arabidopsis seeds in a microcentrifugation tube.
2. Add 70% ethanol and vortex the seeds for 30 s followed by manual shaking for 10 min.
3. Remove the ethanol and wash the seeds with sterile water three times.
4. Add 200—300 μL of sterile water with the freshly sterilized seeds using a P1000 pipette and sterile pipetting tips.
5. Place individual seeds on a solid ½ MS agar media (50 mL) in a square plate, e.g., by pipetting.
6. Seal the plate with micropore tape.
7. Place the sealed plate at 4°C for 48 h for stratification.

8. Incubate the plates in a growth room until seedlings have roots of 1–2 cm long.

3.2 Electrode Preparation

1. Cut two pieces of 10 cm silver wire.
2. Place the end of the silver wire in the electrode preparation solution for 30 min to oxidize the wire until the tip of the silver wire turns dark.
3. Rince the oxidized silver wire with 0.1 M NaCl.
4. Store in the storage solution until use (Fig. 1a; see Sect. 3.7).

3.3 Pump and Reservoir Preparation

1. Cut a 5–8 cm long piece of shrink tubing as well as a 5–10 cm long piece of OEIP capillary.
2. Place 3–5 cm of OEIP capillary into the tubing, and using heated metal tweezers, clamp the tubing end onto the capillary to create a hermetic junction.
3. Fill the tubing with the electrolyte source solution (H^+ source), ensuring that the capillary tip is in direct contact with the solution (*see* **Note 4**).

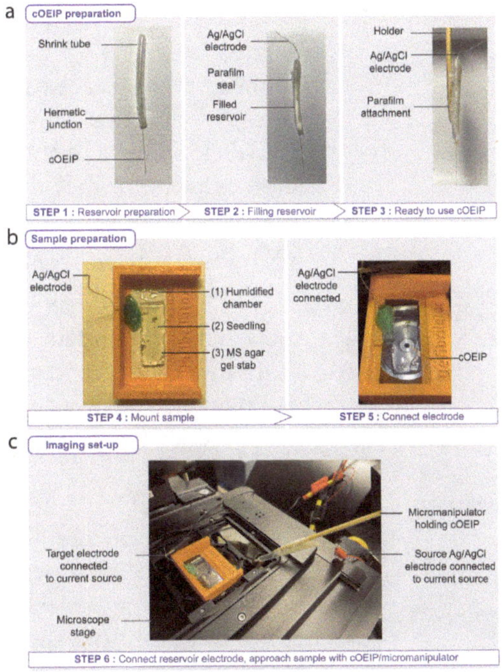

Fig. 1 Representation of the main steps involved in the preparation of cOEIP for the live confocal imaging of *A. thaliana*. seedling. (**a**) Preparation of the cOEIP following the three-step protocol described in Sect. 3.3. (**b**) Preparation and mounting of the seedling sample as described in Sect. 3.4. (**c**) Connection of the cOEIP electrode to the anode (high potential terminal) and the target electrode to the cathode (low potential terminal) of the source meter, as described in Sect. 3.5, after which proton delivery and imaging are initiated (see Sect. 3.6)

4. Insert the oxidized part of the Ag wire into the tube, ensuring contact with the solution.
5. Once the Ag/AgCl electrode and HCl solution are placed in the tube, the interface can be hermetically sealed with a piece of parafilm (Fig. 1a).

3.4 Mounting Seedlings

1. Glue the Ag/AgCl electrode directly on the coverslip at the bottom of the slide to facilitate the manipulation.
2. Prehumidify the coverslip using a spray bottle and place the seedling in the middle of the coverslip.
3. Carefully place a slab of solid ½ MS agar medium on top of the seedling to cover the seedling as well as the electrode (5 mm thick slab obtained from the ½ MS medium prepared in a square Petri dish).
4. Add an additional slab of ½ MS medium on top of the glued electrode to close the electrolytical circuit, to maintain the humidity, and to avoid movement of the sample when approaching the cOEIP capillary to the root (Fig. 1b).

3.5 cOEIP Device Setup and Ion Loading

1. Attach the prefilled and hermetically sealed cOEIP to the micromanipulator arm and secure it firmly with parafilm (Fig. 1c).
2. Connect the source-measure unit to the cOEIP and sample electrodes. For cation delivery, connect the anode to the cOEIP electrode and the cathode to the sample electrode. For anion delivery, reverse the polarity.
3. Position the OEIP delivery tip in the solid ½ MSagar in close proximity to the root (here: 100 μm from the root) and maintain this position throughout loading, control, and delivery.
4. Apply your intended delivery current (here 400 nA) for 10–15 min to load and stabilize the device. Record the operating voltage at the set current (*I*–*V* behavior) as a quality control reference for device performance and reuse (*see* **Note 3**).

3.6 Proton Delivery

1. Maintain the cOEIP delivery tip at the same position (here: 100 μm). Due to the absence of convective transport, cOEIP-mediated delivery generates highly localized concentration profiles, with ion concentration decreasing sharply as a function of distance from the delivery site. Spatiotemporal concentration profiles of OEIP delivery can be estimated based on the diffusion properties of ions in the surrounding medium (see examples in [5] and [1]).
2. Apply 0 nA for 10 min to record a control response of the plants before the cOEIP delivery. The OEIP loading step and passive H^+ leakage may slightly decrease the local pH during this

control period; however, under these conditions, this decrease was insufficient to trigger a measurable plant response [5].

3. Apply a current of 400 nA for 10 min, recording the voltage response.
4. Monitor proton delivery via fluorescent root reporters (e.g., pH-dye such as Fluorescein-5- (and -6)- Sulfonic Acid, Trisodium Salt (FS) [7]; Invitrogen F1130).
5. If dose dependency is tested, extend the delivery to 15–20 min or increase the operating current. The operating current can be increased up to the limiting current. The limiting current can be determined empirically by increasing the applied current until the current–voltage response deviates from the initial linear behavior [8].

3.7 Post-experiment Disassembly and Storage

After completion of the experiment, discard the agar slab and seedling according to institutional biological waste guidelines. Ag/AgCl wire electrodes are stored in storage solution and may be reused provided that the characteristic dark Ag/AgCl layer remains visible at the electrode tips; electrodes should be reoxidized when this coating is diminished or lost.

For cOEIP storage, remove the parafilm seal and completely empty the source reservoir. The OEIP capillary and tubing should be stored in 0.1 M NaCl solution to prevent capillary degradation due to dehydration. Devices may be reused as long as their current–voltage characteristics remain stable and reproducible. This can be verified by operating the device in a standardized dummy setup (agar medium without plant material) at the intended delivery current and confirming that the steady-state operating voltage is comparable to those obtained during initial experiments. Additionally, delivery efficiency can be assessed periodically by chemical analysis of the target medium, for example, using pH measurements for proton delivery.

4 Notes

1. Ag/AgCl wire electrodes are recommended due to their high conductivity and electrochemical stability. To avoid unintended local ionic or pH effects, the Ag/AgCl electrode should be positioned several millimeters away from the seedling and not in direct proximity to the tissue.
2. The chamber used is open on one side, which allows to keep the humidity of the sample while leaving space to approach the capillary from the open side for an easy access.
3. If the system does not reach the desired current, ensure that the ionic circuit is closed: the source electrode and capillary

inlet must be immersed in the source electrolyte, while the capillary outlet and target electrode must be immersed in the same ½ MS agar compartment. If necessary, carefully trim the capillary tip to restore ionic contact.

4. If air bubbles prevent contact, gently tap the tubing or refill the reservoir to remove trapped air.

Conflict of Interest

IBW is a shareholder in the small, researcher-controlled intellectual property company OBOE IPR AB (oboeipr.com). OBOE IPR AB owns patents related to OEIP research and is a holding company for subsidiary Iontronics AB.

References

1. Poxson DJ, Karady M, Gabrielsson R et al (2017) Regulating plant physiology with organic electronics. Proc Natl Acad Sci USA 114:4597–602. https://doi.org/10.1073/pnas.1617758114
2. Bernacka-Wojcik I, Talide L, Abdel Aziz I et al (2023) Flexible organic electronic ion pump for flow-free phytohormone delivery into vasculature of intact plants. Adv Sci 10:2206409. https://doi.org/10.1002/advs.202206409
3. Pařízková B, Antoniadi I, Poxson DJ et al (2022) iP & OEIP—cytokinin micro application modulates root development with high spatial resolution. Adv Mater Technol 7:2101664. https://doi.org/10.1002/admt.202101664
4. Bernacka-Wojcik I, Huerta M, Tybrandt K et al (2019) Implantable bioelectronics: implantable organic electronic ion pump enables ABA hormone delivery for control of stomata in an intact tobacco plant. Small 15:1970233. https://doi.org/10.1002/smll.201970233
5. Ducla J, Di Fino L, Mentzelopoulou A et al (2024) H^+ drives ultra-fast root-to-root responses to wounding. BioRXiv https://doi.org/10.1101/2024.10.04.616436
6. Armiento S, Bernacka-Wojcik I, Dar AM et al (2025) Powering a molecular delivery system by harvesting energy from the leaf motion in wind. Bioinspir Biomim 20:016023. https://doi.org/10.1088/1748-3190/ad98d3
7. Serre NB, Wernerová D, Vittal P et al (2023) The AUX1-AFB1-CNGC14 module establishes a longitudinal root surface pH profile. eLife 12:e85193. https://doi.org/10.7554/eLife.85193
8. Seitanidou M, Tybrandt K, Berggren M, Simon DT (2019) Overcoming transport limitations in miniaturized electrophoretic delivery devices. Lab Chip 19:1427–35. https://doi.org/10.1039/C9LC00038K

Chapter 12

Quantitative Measurement of Basipetal Auxin Transport in Arabidopsis Roots Via ^{3}H-IAA Labeling

Yulin Gan, Yakun Peng and Shutang Tan

Abstract

The polar transport of the phytohormone auxin regulates plant organogenesis, morphogenesis, and gravitropism, which are coordinately regulated by auxin influx and efflux carriers. To measure the basipetal (shootward) transport of auxin in roots, we applied radiolabeled IAA to the root tips of 7-day-old seedlings in the model plant *Arabidopsis thaliana*. A scintillation counter was then used to measure the radioactivity of the samples. Here, we describe a method for measuring auxin transport in *A. thaliana* roots via ^{3}H-indole-3-acetic acid (^{3}H-IAA).

Keywords Auxin transport, *Arabidopsis thaliana* , Isotope labeling, ^{3}H-IAA

1 Introduction

The phytohormone auxin is crucial for plant growth and development, such as phototropism and gravitropism, as well as the formation of roots, stems, and leaves [1–3]. The polar transport of auxin relies on the auxin efflux carriers PIN-FORMEDs (PINs) and the auxin influx carrier AUXIN RESISTANT1/LIKE AUX1s (AUX1/LAXs) [4–7]. The quantitative measurement of auxin transport in plants relies on examining the distribution and dynamics of auxin reporter gene expression signals, including the DR5::GFP/GUS reporter and DII-VENUS auxin degradation sensors [8–10]. In addition, radiolabeled auxin transport assays are widely used for quantitatively measuring auxin transport in plant roots and hypocotyls, which offers a more sensitive and easier method than transgenic markers do [11]. We applied agar droplets containing ^{3}H-labeled indole-3-acetic acid (^{3}H-IAA) to the root tips of 7-day-old Arabidopsis seedlings [12]. The root tissues were subsequently collected and incubated in scintillation solution. The radioactivity in the root tissues was measured by a scintillation counter. To validate the application of this method in the analysis of auxin transport mechanisms, we applied the auxin transport inhibitor

Steffen Vanneste and Inge Verstraeten (eds.), *Root Development*, Methods in Molecular Biology, vol. 3086, https://doi.org/10.1007/978-1-0716-5603-7_12,

N-1-naphthylphthalamic acid (NPA) to ^{3}H-IAA agar droplets and detected the inhibition of auxin transport in plants [4, 13–15]. We confirmed the sensitivity and reliability of our method and demonstrated the ability of the system to detect suppressed auxin transport.

Here, we established an optimized quantitative detection protocol for auxin transport on the basis of the isotope labeling method. Using ^{3}H-labeled IAA as a tracer, we quantified auxin transport in Arabidopsis roots by detecting radioactive signals with a scintillation counter. Compared with traditional methods, this protocol provides a more precise and quantifiable approach, which is suitable for the analysis of auxin transport in plants.

2 Materials

2.1 MS Medium for Seedling Growth and Agar Droplet Preparation

The following 1.25% agar ½ MS medium was prepared for all steps.

1. MS salts (2.3 g/L) added to 1 L deionized water, 0.5 g MES monohydrate. The mixture was stirred until all ingredients dissolved completely.
2. Subsequently, 10 g/L sucrose was added to the above mixture, and again the mixture was stirred until this sucrose completely dissolved.
3. Adjust the pH to 5.8–5.9 using approximately 75 μL 10 M KOH solution.
4. 12.5 g/L plant agar was next added to the mixture.
5. The medium was sterilized in an autoclave at 121°C for 20 min and then maintained at 65°C to prevent solidification.

2.2 Chemicals

^{3}H-IAA (250 μCi/9.25 MBq, 1 mCi/mL, in ethanol, ART-0340-250, PerkinElmer) was dissolved in ethanol to a stock concentration of 3.45 M as a radioactive tracer solution. NPA was dissolved in DMSO to a concentration of 10 mM as a stock solution. Scintillation solution was purchased from Revvity: OptiPhase HiSafe 3.

2.3 Materials and Equipment

1. Petriplates: 9 cm diameter
2. 1.5 mL Eppendorf tubes
3. 2 mL Eppendorf tubes
4. Metal beads for tissue disruption
5. Scintillation counter (Revvity, Tri-Carb)

2.4 Plant Material

Seven-day-old *Arabidopsis thaliana* Col-0 seedlings were grown as follows:

1. Arabidopsis Col-0 seeds were sterilized by incubating them in 75% ethanol solution for 15 min.

2. These surface-sterilized Arabidopsis Col-0 seeds were then spread evenly on ½ MS media in petriplates.
3. The seeds in petriplates were stratified at 4°C for 3 days, and were then placed vertically in a growth chamber at 22°C under a 16 h light/8 h dark photoperiod for 7 days.

3 Methods

3.1 Medium Preparation

1. Transfer seven-day-old Col-0 seedlings to a new MS plate, with 20 seedlings used as a single sample for which four replicates were performed (Fig. 1A).
2. Add ^{3}H-IAA of the stock solution at a concentration of 500 μM to the 1.25% agar ½ MS medium cooled to 65°C (*see* **Note 1**). This means, add 1.45 μL of ^{3}H-IAA stock solution to 10 mL of ½ MS medium (*see* **Note 2**).
3. 1 mL aliquots of this ½ MS medium containing 500 μM ^{3}H-IAA were divided into 1.5 mL centrifuge Eppendorf tubes and stored at 65°C to prevent solidification (*see* **Note 3**).
4. 1 μL DMSO (as a control) was added to one of these aliquots of 1.25% agar ½ MS medium supplemented with 500 μM ^{3}H-IAA.

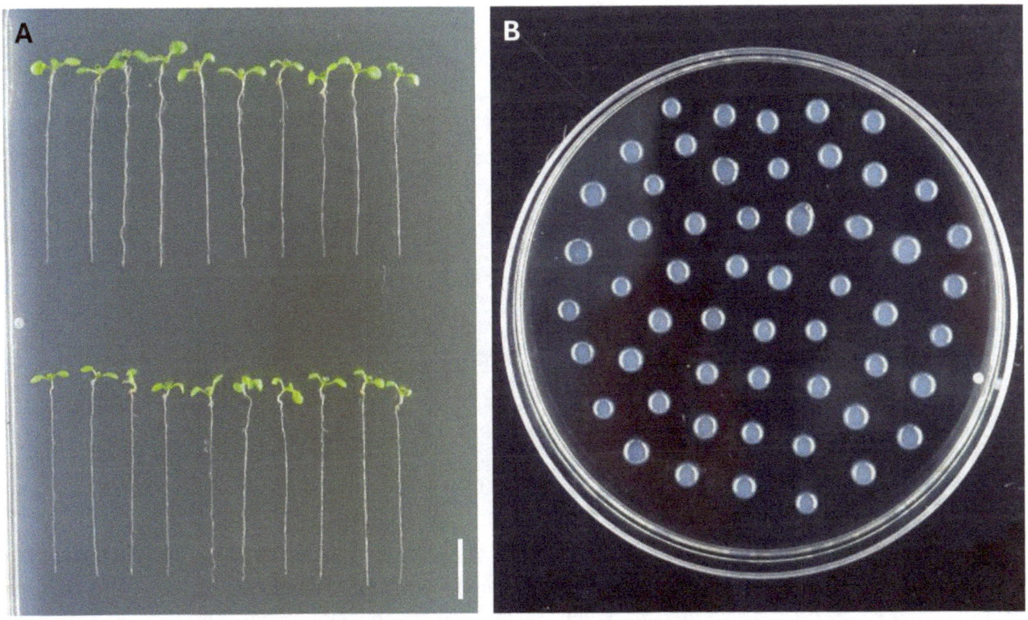

Fig. 1 Preparation of 7-day-old Arabidopsis Col-0 seedlings and ^{3}H-IAA droplets. (**A**) Twenty 7-day-old Col-0 seedlings were placed in ½ MS medium as one sample, with three replicates for each experimental group. Scale bars, 1 cm. (**B**) Dispensing medium containing ^{3}H-IAA into small droplets on a 9 cm diameter round plate

5. 1 μL of 10 mM NPA (in DMSO) was added to another aliquot of 1.25% ½ MS medium in a 1.5 mL centrifuge tube, resulting in a final concentration of 10 μM NPA. This would serve as the validation treatment.

3.2 Droplet Preparation

1. Dispense the ½ MS + 500 μM ^{3}H-IAA medium (including the control DMSO and validation NPA containing tubes) tube in 5 μL droplets (*see* **Note 4**). Quickly form these droplets on a 9 cm diameter round petriplate (*see* **Note 5**) (Fig. 1B).
2. Allow these droplets to solidify in darkness to prevent IAA degradation (*see* **Note 6**).

3.3 Seedling Treatment

1. During ^{3}H-IAA droplet solidification, position the root tips of the 20 seedlings carefully aligned to each other (Fig. 1A).
2. Place a droplet beneath the root tips of the seedlings and position this droplet carefully to achieve approximately 0.5 mm overlap with the root tip (*see* **Note 7**) (Fig. 2).
3. Incubate the seedlings in contact with the different treatments in the dark for 6 h to prevent IAA degradation (*see* **Note 8**).

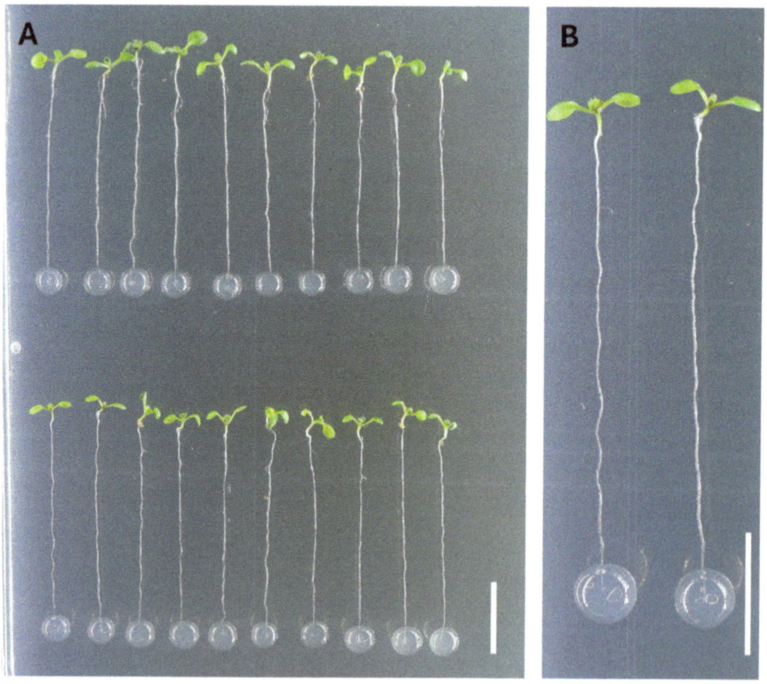

Fig. 2 Application of ^{3}H-IAA droplets to Arabidopsis roots. (**A**) A representative photo of ^{3}H-IAA droplets applied to Arabidopsis root tips. The droplets were carefully transferred onto the root tips of the seedlings via tweezers. (**B**) The ^{3}H-IAA droplets only contacted the apical root tips, overlapping the apex by approximately 0.5 mm. Scale bars, 1 cm

3.4 Sampling for Measurements

1. Cut the apical 2 mm of the root tips, then excise an apically located 5 mm root segment from the remaining primary root as indicated in Fig. 3 (*see* **Note 9**).
2. Place the collected samples in a 2 mL centrifuge tube (with 2 iron beads inside) and immediately freeze the samples in liquid nitrogen.
3. Subsequently, the samples were ground completely with a precooled grinder, spun at 14,000 rpm for 1 min, and then 1 mL of scintillation solution was added (*see* **Note 10**).
4. The mixture was shaken and incubated overnight to allow the radioactivity to evenly diffuse into the whole volume of scintillation solution. One empty vial (no plant material) of scintillation solution was added as a background control.
5. The radioactivity of the samples was measured in a scintillation counter, according to the user manual, with three replicate measurements of each sample (*see* **Notes 11 and 12**) (Fig. 4).

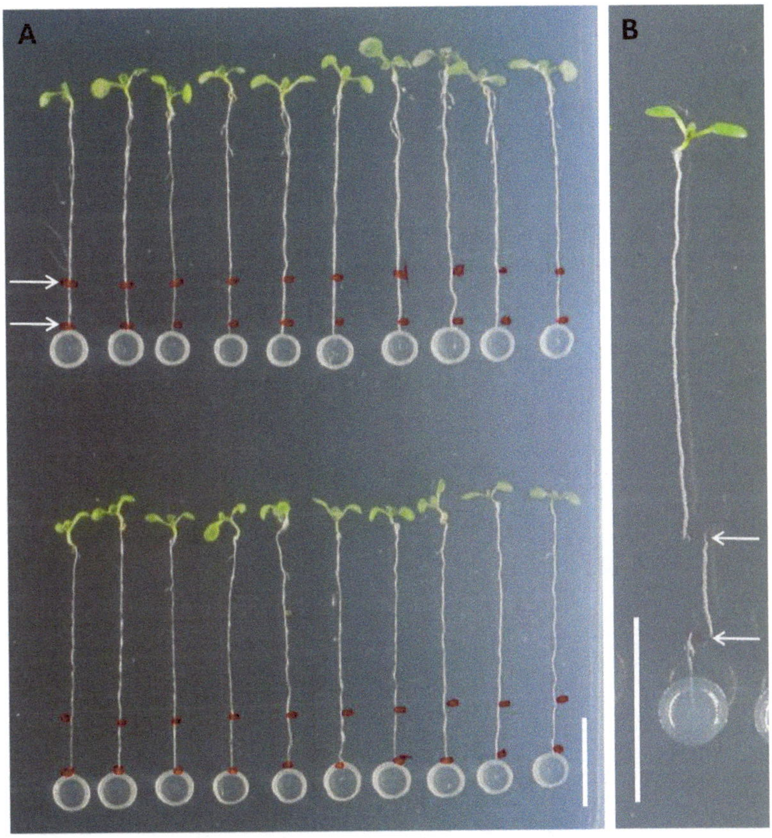

Fig. 3 Sectioning of sample tissues. **(A)** The white arrow indicates the end of the segment where auxin was quantified. **(B)** The apical 5 mm root segment above 2 mm from the root tip was cut. Scale bars, 1 cm

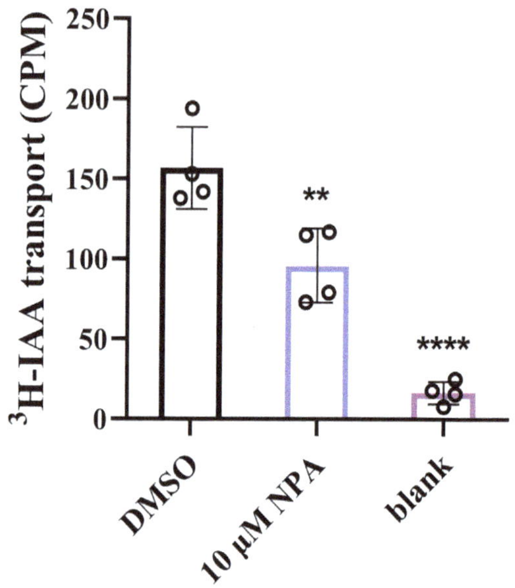

Fig. 4 NPA inhibits auxin transport in Arabidopsis roots. The radioactivity of ^{3}H-IAA in samples treated with NPA-^{3}H-IAA droplets is significantly lower than DMSO-^{3}H-IAA control. The blank control only measured the radioactivity of the scintillation solution. CPM: counts per minute. $^{**}p < 0.01$; $^{****}p < 0.0001$

Table 1

The average radioactivity values (of three repeated measurements) measured by a scintillation counter in the independent replicate samples of blank scintillation solution, in the control treatment (500 μM ^{3}H-IAA and DMSO), and in the validation treatment (500 μM ^{3}H-IAA and NPA). Measurements in counts per minute (CPM)

Replicates	Col-0-DMSO—control treatment	Col-0-NPA—validation treatment	Blank
Sample 1	153	115	8
Sample 2	142	73	25
Sample 3	192	117	16
Sample 4	138	79	18

6. The resulting counts per minute (Fig. 4 and Table 1) demonstrate basipetal transport of ^{3}H-IAA in the primary root segments, while in the 10 μM NPA solution, scintillation counts were significantly reduced, confirming the role of NPA in inhibiting auxin transport in the primary root.

4 Notes

1. ^{3}H-IAA was added after the 1.25% agar MS medium cooled to 65°C at room temperature to prevent the high temperature from affecting its auxin activity.
2. It is necessary to confirm the physiological activity of ^{3}H-IAA compared to a known IAA-containing agar medium to confirm that the auxin concentration is correct. This highlights the importance of quality control and excludes the influence of ^{3}H-IAA degradation.
3. The mixture was incubated in a 65°C water bath to prevent solidification.
4. The ½ MS + ^{3}H-IAA aliquots were divided into 5 μL droplets using yellow pipette (2–200 μL) tips. Work quickly during this process to prevent droplet solidification inside the tips.
5. It is better to have some water in the round plate to increase the humidity and minimize droplet evaporation.
6. Wrap the round plate in aluminum foil to allow droplet solidification in a dark environment, preventing IAA from light-induced degradation.
7. The ^{3}H-IAA droplet only contacts the extreme root tip, overlapping with the root tip by approximately 0.5 mm.
8. Wrap the media plate in aluminum foil and place the plate vertically in the dark for incubation. Do not move the plate to prevent the droplets from falling off.
9. Discard the droplet-contacted apical root segment during sampling to avoid compromising the results by contaminating with high ^{3}H-IAA concentrations.
10. The samples were ground repeatedly to ensure that the tissue was completely homogenized.
11. All samples could be repeatedly measured three times to minimize experimental error.
12. Set up four samples with only the scintillation solution as blank controls. The measured values for these vials will be used to subtract the background radioactivity.

Conflict of Interest

The authors have no conflicts of interest to declare that are relevant to the content of this chapter.

References

1. Vanneste S, Friml J (2009) Auxin: a trigger for change in plant development. Cell 136(6):1005–1016. https://doi.org/10.1016/j.cell.2009.03.001
2. Ötvös K, Marconi M, Vega A, O'Brien J, Johnson A, Abualia R, Antonielli L, Montesinos JC, Zhang Y, Tan S, Cuesta C, Artner C, Bouguyon E, Gojon A, Friml J, Gutiérrez RA, Wabnik K, Benková E (2021) Modulation of plant root growth by nitrogen source-defined regulation of polar auxin transport. EMBO J 40(3):e106862. https://doi.org/10.15252/embj.2020106862
3. Li Y, Wang Y, Tan S, Li Z, Yuan Z, Glanc M, Domjan D, Wang K, Xuan W, Guo Y, Gong Z, Friml J, Zhang J (2020) Root growth adaptation is mediated by PYLs ABA receptor-PP2A protein phosphatase complex. Adv Sci 7(3):1901455. https://doi.org/10.1002/advs.201901455
4. Yang Z, Wei H, Gan Y, Liu H, Cao Y, An H, Que X, Gao Y, Zhu L, Tan S, Liu X, Sun L (2025) Structural insights into auxin influx mediated by the *Arabidopsis* AUX1. Cell 188(15):3960–3973.e3915. https://doi.org/10.1016/j.cell.2025.04.028
5. Adamowski M, Friml J (2015) PIN-dependent auxin transport: action, regulation, and evolution. Plant Cell 27(1):20–32. https://doi.org/10.1105/tpc.114.134874
6. Yang Z, Xia J, Hong J, Zhang C, Wei H, Ying W, Sun C, Sun L, Mao Y, Gao Y, Tan S, Friml J, Li D, Liu X, Sun L (2022) Structural insights into auxin recognition and efflux by *Arabidopsis* PIN1. Nature 609(7927):611–615. https://doi.org/10.1038/s41586-022-05143-9
7. Zhang X, Adamowski M, Marhava P, Tan S, Zhang Y, Rodriguez L, Zwiewka M, Pukyšová V, Sánchez AS, Raxwal VK, Hardtke CS, Nodzyński T, Friml J (2020) Arabidopsis flippases cooperate with ARF GTPase exchange factors to regulate the trafficking and polarity of PIN auxin transporters. Plant Cell 32(5):1644–1664. https://doi.org/10.1105/tpc.19.00869
8. Friml J, Vieten A, Sauer M, Weijers D, Schwarz H, Hamann T, Offringa R, Jürgens G (2003) Efflux-dependent auxin gradients establish the apical–basal axis of *Arabidopsis*. Nature 426(6963):147–153. https://doi.org/10.1038/nature02085
9. Brunoud G, Wells DM, Oliva M, Larrieu A, Mirabet V, Burrow AH, Beeckman T, Kepinski S, Traas J, Bennett MJ, Vernoux T (2012) A novel sensor to map auxin response and distribution at high spatio-temporal resolution. Nature 482(7383):103–106. https://doi.org/10.1038/nature10791
10. Sabatini S, Beis D, Wolkenfelt H, Murfett J, Guilfoyle T, Malamy J, Benfey P, Leyser O, Bechtold N, Weisbeek P, Scheres B (1999) An auxin-dependent distal organizer of pattern and polarity in the *Arabidopsis* root. Cell 99(5):463–472. https://doi.org/10.1016/S0092-8674(00)81535-4
11. Goldsmith MHM (1977) The polar transport of auxin. Ann Rev Plant Biol 28(1977):439–478. https://doi.org/10.1146/annurev.pp.28.060177.002255
12. Lewis DR, Muday GK (2009) Measurement of auxin transport in *Arabidopsis thaliana*. Nat Protoc 4(4):437–451. https://doi.org/10.1038/nprot.2009.1
13. Abas L, Kolb M, Stadlmann J, Janacek DP, Lukic K, Schwechheimer C, Sazanov LA, Mach L, Friml J, Hammes UZ (2021) Naphthylphthalamic acid associates with and inhibits PIN auxin transporters. Proc Natl Acad Sci 118(1):e2020857118. 10.1073/pnas.2020857118
14. Sussman MR, Goldsmith MHM (1981) The action of specific inhibitors of auxin transport on uptake of auxin and binding of N-1-naphthylphthalamic acid to a membrane site in maize coleoptiles. Planta 152(1):13–18. https://doi.org/10.1007/BF00384978
15. Ung KL, Winkler M, Schulz L, Kolb M, Janacek DP, Dedic E, Stokes DL, Hammes UZ, Pedersen BP (2022) Structures and mechanism of the plant PIN-FORMED auxin transporter. Nature 609(7927):605–610. https://doi.org/10.1038/s41586-022-04883-y

Check for updates

Chapter 13

EdU-Based Quantitative Analysis of Cell Proliferation in Arabidopsis Root Meristems

Tilman Jacob, Haiyue Ai and Marcel Quint

Abstract

The division of a mother cell into two daughter cells is a fundamental process in biology, involving the transfer of genomic information to the next generation of cells. In plant roots, cell divisions occur in the root apical meristem, a specialized tissue located at the plant's root tip. Cell proliferation rates in the root apical meristem shape root growth, thereby contributing to general development as well as the acclimatization to stressful environmental conditions through phenotypic plasticity. Here, we present a protocol for the microscopical analysis of cell proliferation in *Arabidopsis thaliana* root meristems by 5-ethynyl-2′-deoxyuridine (EdU) staining. EdU, a thymine-analog, is incorporated into newly synthesized DNA of proliferative cells during the S-phase of the cell cycle. Subsequently, EdU-containing nuclei can be labeled with a fluorophore and detected by confocal laser scanning microscopy. The combination of the EdU assay with cell wall staining enables the quantification of root meristem architecture parameters and cell division activity from microscopic images. Together, this can elucidate how root meristems respond to changes in environmental conditions or address fundamental questions of developmental plant biology.

Keywords Root growth, Root development, Root meristem, Cell cycle, Cell division, S-phase, EdU, *Arabidopsis thaliana* , Microscopy

1 Introduction

Cell division is a fundamental biological process enabling organismal growth, maintenance of tissue integrity, and the inheritance of genetic identity. In plants, cell division is largely confined to specialized tissues termed meristems, including the apical meristems located at the tips of primary shoot and root. The root tip of *Arabidopsis thaliana* has a simple, radially organized structure composed of concentric layers that form single-cell files in longitudinal sections. As individual cells progress through these files, they transition through successive developmental stages. Therefore, a single two-dimensional image of the Arabidopsis root tip essentially captures both spatial and temporal information on organ development, making it an ideal model for the study of growth dynamics [1].

Steffen Vanneste and Inge Verstraeten (eds.), *Root Development*, Methods in Molecular Biology, vol. 3086,
https://doi.org/10.1007/978-1-0716-5603-7_13,

Within the root apical meristem, young cells undergo several rounds of cell division before exiting the meristematic zone, where they begin to elongate and acquire their characteristic anisotropic shape. Normal growth as well as acclimatization to changing environments depend on the precise coordination of cell division and elongation [2, 3]. In the meristem, cell division follows a tightly regulated, unidirectional process known as the cell cycle, which consists of four key phases. During the G1-phase, the cell prepares for DNA replication by synthesizing organelles and protein complexes. This is followed by the S-phase, in which chromosomes are actively replicated. In G2-phase, the cell undergoes genome integrity checks before entering mitosis (the M-phase), during which the cell physically divides into two daughter cells [4]. As cells elongate and differentiate, several rounds of endoreduplication occur, a variant of the cell cycle that bypasses mitosis, resulting in increased ploidy as the cells progress through differentiation [5].

To investigate fundamental questions of plant development as well as root growth plasticity, robust methods are required to quantify both cell division dynamics and the structural organization of the root apical meristem. In the 5-ethynyl-2′-deoxyuridine (EdU) assay, which was first developed for mammalian cell culture [6], proliferating cells in the S-phase are detected based on metabolic labeling of DNA. EdU is a thymine-analog, which is fed to seedlings and incorporated into newly synthesized DNA during a defined pulse period. After fixation, EdU-containing nuclei can be labeled with a fluorophore in a click-chemistry reaction, which can then be detected and quantified using a confocal microscope [6]. Previously, similar assays relied on radioactive ^{3}H-labeled thymidine or 5-bromo-2′-deoxyuridine (BrdU) precursors. ^{3}H-thymidine assays were tedious due to the radioactivity, and BrdU assays were time-consuming and potentially damaging to samples, as they required enzymatic digestion of DNA for antibody-based detection [7, 8]. In contrast, all reaction components required for EdU labeling easily penetrate plant tissues, allowing for fast and convenient detection. Therefore, the EdU assay has quickly been adopted by plant scientists studying root development [8–10].

Here, we describe a workflow for the EdU-based quantitative analysis of cell proliferation in Arabidopsis root meristems together with the analysis of meristem architecture parameters. For this, seedlings are cultivated under experimental conditions of choice and then exposed to EdU for incorporation into newly synthesized DNA. After fixation and permeabilization, EdU-containing nuclei of proliferative cells, as well as cell walls, are stained with suitable fluorophores and imaged by confocal microscopy. The resulting image sets can be used to extract quantitative data on cell division activity and meristem architecture. The entire assay can be performed within a single day following seedling cultivation. Although this workflow was optimized for Arabidopsis seedling roots, it can serve as a starting point to extend the analysis to other species as well.

2 Materials

2.1 Sterile Culture of Arabidopsis Seedlings

Here, we give an example for the sterile cultivation of Arabidopsis seedlings on *A. thaliana* salts (ATS) medium [11] after liquid surface sterilization on vertically oriented plates. However, any standard cultivation protocol can be applied to grow seedlings prior to the EdU assay. In case the researcher wishes to investigate the interaction of cell proliferation with any external treatments (such as light or temperature changes, salt, drought, or nutrient stress, or the application of chemical effectors), these treatment conditions have to be included in the cultivation protocol individually.

1. 70% ethanol in H_2O.
2. 4% (v/v) NaOCl solution (sodium hypochlorite solution, also referred to as "bleach") in H_2O, supplemented with 0.01% (v/v) Triton X-100.
3. Sterile, deionized water.
4. ATS solution: 5 mM KNO_3, 2.5 mM KPO_4, 2 mM $MgSO_4$, 2 mM $Ca(NO_3)_2$, 0.05 mM Fe-Na-EDTA, 70 μM H_3BO_3, 14 μ M $MnCl_2$, 0.5 μM $CuSO_4$, 1 μM $ZnSO_4$, 0.2 μM Na_2MoO_4, 10 μM NaCl, and 0.01 μM $CoCl_2$ (*see* **Note 1** for the preparation of stock solutions).
5. Sucrose (optional, *see* **Note 2**).
6. Gelling agent suitable for plant cell culture (such as Phytagel™, Sigma Aldrich).
7. Sterile petri dishes for cultivation: Can be either square (120 × 120 × 17 mm, Greiner Bio-one) or round (94 × 16 mm, Greiner Bio-one), with vents.
8. Plaster tape that allows gas exchange (such as Leukopor® tape, Leukoplast).
9. Sterile 1.5 mL reaction tubes and micropipettes with corresponding sterile pipette tips.
10. Sterile hood for seed sterilization, plate preparation, and seed sowing.
11. Growth cabinet, phytotron, or otherwise controlled environment equipped for plant cultivation on vertically or near vertically oriented plates.

2.2 EdU Incorporation and Fixation

1. Sterilized liquid plant cultivation medium of identical composition as the solid plate medium used for seedling cultivation (including all optional chemical treatments).
2. 10 mM EdU stock solution (part of commercially available EdU-based cell proliferation kits, such as the ClickTech EdU Cell Proliferation Kit, Carl Roth/baseclick).

3. 1× Phosphate-buffered saline (PBS, pH 7.4): 137 mM NaCl, 2.7 mM KCl, 10 mM Na_2HPO_4, 1.8 mM KH_2PO_4 (*see* **Note 3** for preparation of stock solution).
4. 4% (w/v) Paraformaldehyde (PFA) solution in 1× PBS with 0.5% (v/v) Triton X-100. Dissolve 4 g of PFA in 100 mL 1× PBS. Under a fume hood, heat to 60°C and stir for 30 min, until the solution becomes transparent (*see* **Note 4**). Freshly add Triton X-100 to a concentration of 0.5% (v/v), e.g., add 50 μL Triton X-100 to 10 mL of 4% PFA in 1× PBS.
5. 2 mL reaction tubes or 12- or 24-well plates.

2.3 EdU Labeling and Cell Wall Staining

1. Kit components to prepare the EdU detection solution. For the ClickTech EdU Cell Proliferation Kit, Carl Roth/baseclick, this includes: 10× Reaction buffer, Reactor System, 10 mM Dye Azide (for example, 6-carboxyfluorescin (6-FAM)), 10× buffer additive, and deionized water.
2. 1× PBS buffer (see Sect. 2.2).
3. Cell wall staining dye, such as 1 g/L Calcofluor White solution supplemented with 0.5 g/L Evan's Blue, Sigma (*see* **Note 5** on the compatibility of EdU dyes and cell wall stains).
4. Optional: ClearSee solution [12]: 10% (w/v) xylitol, 15% (w/v) sodium deoxycholate, and 25% (w/v) urea in water (*see* **Note 6**). For 100 mL ClearSee, dissolve 10 g xylitol, 15 g sodium deoxycholate, and 25 g urea in 100 mL of deionized H_2O.

2.4 Root Meristem Imaging by Confocal Laser Scanning Microscopy

1. Confocal laser scanning microscope system equipped with laser lines and detection filters suitable for the EdU labeling and cell wall staining dyes of choice.
2. Microscope slides and cover slips suitable for confocal laser scanning microscopy.
3. Tweezers for sample mounting.
4. 1× PBS buffer (see Sect. 2.2).

3 Methods

3.1 Sterile Culture of Arabidopsis Seedlings

1. Aliquot Arabidopsis seeds into sterile 1.5 mL reaction tubes. Sterilize the seed surfaces by incubation in 70% ethanol for 5 min followed by incubation in 4% NaOCl solution with 0.01% Triton X-100 for 8 min while shaking (*see* **Note 7**). Rinse the seeds 3–5 times with sterilized, deionized water.
2. Stratify the seeds in water at 4°C and in darkness for 2–7 days (*see* **Note 8**).

3. For media plates, prepare the ATS solution from stocks. Optionally, add sucrose at 0.5%–1% (w/v) (*see* **Note 2**). Fill the ATS solution into autoclavable bottles (pre-filled with gelling agent at 0.5% (w/v)) and autoclave this ATS medium according to the instructions of the autoclave's manufacturer. Once the autoclaved media has cooled down to approximately 60°C, pour the medium into petri plates under a sterile hood. Use roughly 50 mL for the large square plates and 30 mL for the smaller round plates. Let the ATS medium solidify and leave the plate lids open for a while to avoid excessive condensation. Plates can be stored at 4°C for several weeks.
4. Sow the stratified seeds directly on the ATS plates, evenly dispersed on a horizontal line within the top part of the plate. Close the plates with a single layer of tape. Prepare at least one plate with 15–20 seeds per investigated genotype and treatment combination.
5. Grow the seedlings on vertically oriented plates under controlled environmental conditions for 5–10 days (or shorter/longer, depending on the specific growth conditions).

3.2 EdU Incorporation and Fixation

1. Prepare the EdU incorporation solution: In 2 mL reaction tubes or a 12- or 24-well plate, supplement 1–2 mL of the liquid plant cultivation medium with EdU stock solution to a final EdU concentration of 10 µM (*see* **Note 9**).
2. Carefully harvest the seedlings from the media plates with tweezers and submerge them in the EdU incorporation solution. Take care that especially the root tips are properly submerged in the solution.
3. Incubate seedlings in EdU incorporation solution for 1 h under the same environmental conditions in which they were cultivated (*see* **Note 9**).
4. After EdU incorporation, transfer the seedlings with tweezers into new reaction tubes or wells for fixation. Add 1 mL 4% PFA solution freshly supplemented with 0.5% (v/v) Triton X-100. Incubate for 20 min at room temperature (*see* **Note 10**).
5. Wash fixed seedlings twice with 1× PBS.

Preferably, researchers should immediately proceed with the staining and imaging steps. If this is not possible, fixed seedlings can also be stored in 1× PBS at 4°C for up to one week.

3.3 EdU Labeling and Cell Wall Staining

1. Prepare the EdU detection solution from kit components as indicated in the kit manufacturer's manual. Prepare the solution immediately before application and keep it protected from light to avoid photodegradation of the labeling dye.
2. Incubate the fixed and washed seedlings in the EdU detection solution for 30 min at room temperature, protected from light.

DNA-incorporated EdU is now labeled with the fluorescent dye.

3. Wash the seedlings three times in 1× PBS.
4. If a clearing step is required before imaging, clear the seedlings for a desired time in ClearSee solution (*see* **Note 6**). If no clearing is required, proceed immediately with cell wall staining.
5. To stain the cell walls with Calcofluor White, incubate the seedlings in Calcofluor White staining solution for 30 min at room temperature, while keeping everything protected from light.
6. Wash the seedlings with 1× PBS for 30 min at room temperature once, followed by three quick washing steps with 1× PBS. The seedlings are now ready for imaging.

3.4 Root Meristem Imaging by Confocal Laser Scanning Microscopy

1. Mount the fixed and stained seedlings with 1× PBS on a microscope slide and cover the sample with a coverslip. If several seedlings are mounted on the same slide, make sure that the individual root tips are separated.
2. Adjust the microscope settings according to the EdU labeling dye and cell wall stain of choice. Ensure that the fluorescence signals from both dyes are well-separated in different channels, so that each dye can be clearly distinguished. In our example (Fig. 1), Calcofluor White-stained cell walls are imaged with an excitation of 405 nm and emission filters set to 410–470 nm, and 6-FAM-labeled EdU-incorporating nuclei are imaged at 488 nm excitation and emission filters set to 410–617 nm.
3. Image the primary root's apical meristem. Set the z-focus on a central longitudinal section of the root, so that the quiescent center and at least one cortical cell file are clearly visible for later extraction of meristem architecture parameters, using the channel for the cell wall dye. In the EdU channel, image either single central sections or whole z-stacks per meristem (*see* **Note 11**). While the settings in the cell wall channel can be optimized individually for each root tip to obtain an optimal image, the settings in the EdU channel should remain the same across all samples to ensure comparability during the analysis.

3.5 Image Analysis

Extracting quantitative data from microscopic images is a common challenge in cell biology. While there are various tools available for specific quantification tasks, in our experience, manual quantification—albeit tedious for large sample sizes—produces the most robust data. However, with recent advancements in AI-supported image analysis, we anticipate that this task will become easier to automate in the near future [14]. Here, we describe a typical Fiji/ImageJ [15] workflow to manually extract cell proliferation and

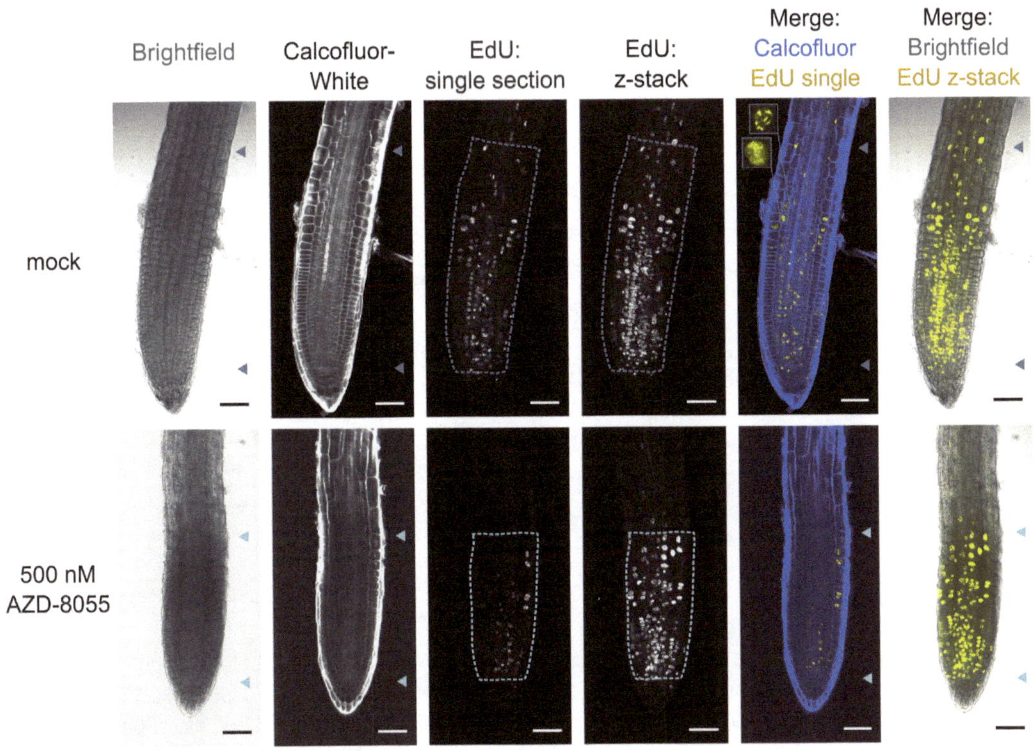

Fig. 1 Example images of Arabidopsis root tips after EdU assay and cell wall staining. Seedlings were grown on ATS medium +1% sucrose at 20°C and long-day white light conditions (16 h lights on/8 h off) for 3 days and then shifted to ATS plates +1% sucrose supplemented with 500 nM of AZD-8055 (TARGET OF RAPAMYCIN (TOR) kinase inhibitor that represses cell division [13]) or DMSO mock for 2 more days under the same light conditions. Images were acquired with a Zeiss LSM900 confocal laser scanning microscope. Brightfield images were recorded via T-PMT detector. Cell wall-staining Calcofluor-White was excited at 405 nm and detected at 410–470 nm. EdU-containing nuclei labeled with 6-FAM were excited at 488 nm and detected at 410–617 nm. The EdU z-stack is represented as a maximum intensity projection. Insets in the upper merged panel show magnified nuclei exhibiting whole and speckled EdU staining patterns, representing early and late S-phase, respectively. Triangles mark the upper and lower borders of the root meristem (first elongated cell and quiescent center, respectively, Fig. 2). Dashed lines depict the meristem area. Scale bars: 50 μm. Quantification: Fig. 2

meristem architecture parameters from an image set acquired as described above.

1. Load the microscopic image to Fiji/ImageJ and split the channels so that you have two separate images for cell wall stain and EdU-labeled nuclei (*see* **Note 12**).
2. Define the root apical meristem area as a region of interest (ROI) in the cell wall image: we define the root apical meristem as the region between the quiescent center and the last cortex cell that is not yet elongated, with the first elongated cortex cell being the first cell which is twice as long as it is wide

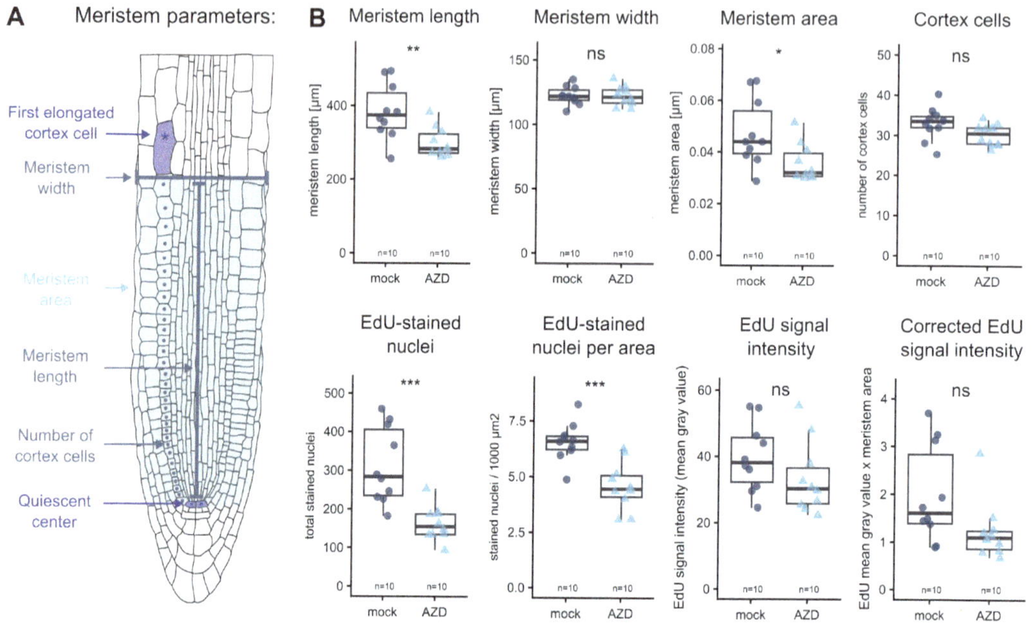

Fig. 2 EdU-based quantitative analysis of TOR inhibition effects on meristem parameters and cell division. (**A**) Definition of meristem parameters. The root apical meristem extends from the quiescent center to the first elongated cortex cell, which is twice as long as it is wide. (**B**) TOR inhibition effects on meristem parameters and cell division. Seedlings were grown on ATS medium +1% sucrose at 20°C and long-day white light conditions (16 h lights on/8 h off) for 3 days and then shifted to ATS plates +1% sucrose supplemented with 500 nM of AZD-8055 (AZD; TOR kinase inhibitor that represses cell division [13]) or DMSO mock for 2 more days under the same light conditions. Meristem parameters were quantified according to (**A**). For quantification of cell division activity, maximum intensity projections of z-stacks were used. For comparison, EdU-stained nuclei were either counted manually or EdU signal intensity was quantified as a rapid proxy. Note that only the manual quantification results in statistically significant differences (*see* **Note 16**). $n = 10$ seedlings per treatment. Significance was tested with Welch's *t*-test (***$p < 0.001$, **$p < 0.01$, *$p < 0.05$, ns $p > 0.05$). Boxes show median (line), 25th–75th percentiles (box), and 1.5× interquartile range (whiskers). Example images: Fig. 1

(Fig. 2; *see* **Note 13** for alternative approaches). Transfer the ROI to the EdU image and crop everything outside the ROI in both images.

3. Quantify the meristem area, length, and width (Fig. 2). Measure the meristem width at the shoot-ward end of the ROI. In the cell wall image, additionally count the number of cortex cells in the meristem.

4. In the EdU image, quantify the number of nuclei, which were labeled by the fluorescent dye. All labeled nuclei contain DNA that has been synthesized while the seedlings were incubated in the EdU incorporation solution and serve as a proxy for the number of cells undergoing the S-phase during this period (*see* **Notes 14** and **15**). Depending on the number and overlap of stained nuclei, they can be quantified with the "Analyze

particles" function of ImageJ or have to be counted manually (*see* **Note 16**).

4 Notes

1. Individual components of ATS solution can be prepared as stock solutions to facilitate easier preparation of the medium. For stock solutions, prepare individually: 1 M KNO_3 (101 g/L), 1 M KPO_4 (for pH 5.5: 15.79 g/L K_2HPO_4 and 123.8 g/L KH_2PO_4), 1 M $MgSO_4 \times 7H_2O$ (246.5 g/L), 1 M $Ca(NO_3)_2$ (236 g/L), 20 mM Fe-EDTA $\times$ $3H_2O$ (8.42 g/L) and 1000× micronutrient solution (4.32 g/L H_3B_3, 2.77 g/L $MnCl_2 \times 4H_2O$, 80 mg/L $CuSO_4$, 288 mg/L $ZnSO_4 \times 7H_2O$, 48 mg/L $Na_2MoO_4 \times 2H_2O$, 580 mg/L NaCl, 2.38 mg/L $CoCl_2 \times 6H_2O$).

   For 1 L of 1× ATS solution, mix: 5 mL of 1 M KNO_3, 2.5 mL of 1 M KPO_4, 2 mL of 1 M $MgSO_4$, 2 mL of 1 M $Ca(NO_3)_2$, 2.5 mL of 20 mM Fe-EDTA and 1 mL of 1000× micronutrient solution.

2. Some experimental setups include 0.5%–1% (w/v) sucrose in the growth media. Adding sucrose accelerates seedling development and leads to more uniform germination and growth during the first days. However, keep in mind that the presence of external sucrose affects the cell cycle regulation [16].
3. PBS buffer can be prepared as 10× stock solution. For this, dissolve 80 g NaCl (1.37 M), 2 g KCl (27 mM), 14.4 g Na_2HPO_4 (100 mM), and 2.4 g KH_2PO_4 (18 mM) in 800 mL deionized H_2O. Adjust the pH to 7.4 with HCl, then add H_2O to 1 L and autoclave.
4. PFA is the polymerized, solid form of formaldehyde. When dissolved in 1× PBS buffer and heated to 60°C, PFA de-polymerizes and releases monomeric formaldehyde, which is the actual fixing agent that crosslinks proteins and nucleic acids to preserve the structure of the sample. Over time, the formaldehyde degrades, oxidizes, or re-polymerizes, indicated by cloudiness or yellowing of the solution. Therefore, PFA solution should always be prepared fresh if possible. Alternatively, it can be aliquoted and stored at −20°C for up to one year.
5. Commercially available EdU assay kits employ click chemistry-based conjugation to attach a fluorescent dye to EdU, which is incorporated into newly synthesized DNA. This approach provides flexibility in selecting the fluorophore so that its excitation and emission properties can be matched to the specific requirements of the imaging setup and any additional fluorescent markers that may be present in the sample. Different kits are available that include different dyes with

different spectra. In our experiments, EdU was labeled with 6-FAM, whose emission in the green-yellow spectral range is readily distinguishable from the blue fluorescence of the cell wall stain Calcofluor White (Fig. 1).

6. Whether sample clearing is required prior to imaging depends on the specific experimental design. Clearing is generally not required for seedling root tissue up to approximately one to two weeks of age, but it can improve image quality in older plants or in tissues with reduced light penetration.
7. These incubation times should not be exceeded, as prolonged exposure to ethanol or bleach might harm the embryo and lead to reduced germination rates.
8. 3–4 days of stratification are ideal, but everything between 2 and 7 days works as well. It is also possible to sow the seeds onto plates immediately after sterilization and place the plates at 4°C and in darkness for several days to facilitate stratification.
9. Both the working concentration of EdU in the incorporation solution as well as the incorporation duration can be adjusted in both directions depending on the requirements of the experiment. However, it has been shown for mammalian cell cultures that EdU can have cytotoxic effects if applied in very high concentrations or over periods of several hours [17]. In our hands, incubation with 10 μM EdU for 1 h consistently produces a sufficient number of labeled nuclei to enable robust quantitative analysis.
10. Many fixation protocols incorporate a vacuum infiltration step to facilitate improved penetration of the fixative into the tissue. However, when working with young seedling roots, which contain minimal intercellular air spaces, this step is generally unnecessary.
11. Imaging single optical sections is faster and allows for a larger number of samples to be processed. In contrast, maximum intensity projections of z-stacks capture a more complete representation of the sample and are less susceptible to variability between replicates. In our experience, EdU labeling in the meristem can vary considerably depending on the z-plane of the section. Consequently, maximum intensity projections provide more consistent and reliable quantification, even though fewer meristems can be imaged per session.
12. As manual quantification of microscopic images can involve subjective decisions, appropriate measures should be implemented to reduce confirmation bias, such as blinding file names before analysis.

13. Although the Arabidopsis primary root has served as a model tissue to study plant development for several decades, there is no consensus regarding the description of its longitudinal zonation pattern [18]. Alternative approaches for determining the shoot-ward boundary of the meristem include different ways of cell shape analysis—for example, identifying the first cell that is 1.5 or 2 times longer than its predecessor in the cell file—the use of molecular markers, or reference to the columella. Depending on the EdU concentration and incubation duration (*see* **Note 9**), EdU incorporation itself can also serve as an indicator of the meristematic region.
14. Upon careful examination of the stained nuclei, two distinct patterns are observable: a "whole" and a "speckled" pattern (s. inset in Fig. 1). These patterns have been associated with early and late stages of the S-phase, respectively [19].
15. Although the EdU assay is conceptually a pulse-chase experiment, it is important to consider that the proportion of actively proliferating cells may already differ between experimental conditions at the onset of the EdU pulse. Consequently, variation in EdU labeling cannot be attributed solely to differences in cell cycle activity during the pulse period itself but instead reflects the cumulative effects of the experimental conditions on meristematic activity throughout the entire growth period.
16. In case the observed difference in the number of stained nuclei between the different samples is sufficiently large, it might be feasible to simply measure the mean fluorescence intensity of the EdU labeling dye in the ROI as a rapid approximation. In this case, signal saturation during imaging must be avoided. However, direct counting of individual nuclei represents a more objective approach, as variations in per-cell signal intensity, arising from multiple technical or biological factors, can bias intensity-based measurements. Accordingly, in the example presented here, no significant differences were detected in mean EdU fluorescence intensity, whereas manual counting of individual nuclei revealed clear and statistically significant differences between the treatments (Fig. 2).

Conflict of Interest

This work was funded by Deutsche Forschungsgemeinschaft [grant 141/12-1 to MQ] and a stipend of the China Scholarship Council to HA. The authors have no conflicts of interest to declare that are relevant to the content of this chapter.

References

1. Scheres B, Wolkenfelt H (1998) The *Arabidopsis* root as a model to study plant development. Plant Physiol Biochem 36:21–32. https://doi.org/10.1016/S0981-9428(98)80088-0
2. Shimotohno A, Aki SS, Takahashi N, Umeda M (2021) Regulation of the plant cell cycle in response to hormones and the environment. Annu Rev Plant Biol 72:273–296. https://doi.org/10.1146/annurev-arplant-080720-103739
3. Beemster GTS, Baskin TI (1998) Analysis of cell division and elongation underlying the developmental acceleration of root growth in *Arabidopsis thaliana*. Plant Physiol 116:1515–1526. https://doi.org/10.1104/pp.116.4.1515
4. Dewitte W, Murray JAH (2003) The plant cell cycle. Annu Rev Plant Biol 54:235–264. https://doi.org/10.1146/annurev.arplant.54.031902.134836
5. Breuer C, Braidwood L, Sugimoto K (2014) Endocycling in the path of plant development. Curr Opin Plant Biol 17:78–85. https://doi.org/10.1016/j.pbi.2013.11.007
6. Salic A, Mitchison TJ (2008) A chemical method for fast and sensitive detection of DNA synthesis *in vivo*. Proc Natl Acad Sci USA 105:2415–2420. https://doi.org/10.1073/pnas.0712168105
7. Echevarría C, Gutierrez C, Desvoyes B (2021) Tools for assessing cell-cycle progression in plants. Plant Cell Physiol 62:1231–1238. https://doi.org/10.1093/pcp/pcab066
8. Kotogány E, Dudits D, Horváth GV, Ayaydin F (2010) A rapid and robust assay for detection of S-phase cell cycle progression in plant cells and tissues by using ethynyl deoxyuridine. Plant Methods 6: 5. https://doi.org/10.1186/1746-4811-6-5
9. Vanstraelen M, Baloban M, Da Ines O, et al. (2009) APC/C^{CCS52A} complexes control meristem maintenance in the *Arabidopsis* root. Proc Natl Acad Sci USA 106:11806–11811. https://doi.org/10.1073/pnas.0901193106
10. Perilli S, Perez-Perez JM, Di Mambro R et al. (2013) Retinoblastoma-related protein stimulates cell differentiation in the *Arabidopsis* root meristem by interacting with cytokinin signaling. Plant Cell 25:4469–4478. https://doi.org/10.1105/tpc.113.116632
11. Lincoln C, Britton JH, Estelle M (1990) Growth and development of the axr1 mutants of *Arabidopsis*. Plant Cell 2:1071–1080. https://doi.org/10.1105/tpc.2.11.1071
12. Kurihara D, Mizuta Y, Sato Y, Higashiyama T (2015) ClearSee: a rapid optical clearing reagent for whole-plant fluorescence imaging. Development 142:4168–4179. https://doi.org/10.1242/dev.127613
13. Montané M-H, Menand B (2013) ATP-competitive mTOR kinase inhibitors delay plant growth by triggering early differentiation of meristematic cells but no developmental patterning change. J Exp Bot 64:4361–4374. https://doi.org/10.1093/jxb/ert242
14. Goh T, Song Y, Yonekura T et al. (2023) In-depth quantification of cell division and elongation dynamics at the tips of growing Arabidopsis roots using 4D microscopy, AI-assisted image processing and data sonificiation. Plant Cell Physiol 64:1262–1278. https://doi.org/10.1093/pcp/pcad105
15. Schindelin J, Arganda-Carreras I, Frise E et al. (2012) Fiji: an open-source platform for biological-image analysis. Nat Methods 9:676–682. https://doi.org/10.1038/nmeth.2019
16. Xiong Y, McCormack M, Li L et al. (2013) Glucose-TOR signalling reprograms the transcriptome and activates meristems. Nature 496:181–186. https://doi.org/10.1038/nature12030
17. Ligasová A, Strunin D, Friedecký D et al. (2015) A fatal combination: a thymidylate synthase inhibitor with DNA damaging activity. PLoS One 10:e0117459. https://doi.org/10.1371/journal.pone.0117459
18. Ivanov VB, Dubrovsky JB (2013) Longitudinal zonation pattern in plant roots: conflicts and solutions. Trends Plant Sci 18:237–243. https://doi.org/10.1016/j.tplants.2012.10.002
19. Hayashi K, Hasegawa J, Matsunaga S (2013) The boundary of the meristematic and elongation zones in roots: endoreduplication precedes rapid cell expansion. Sci Rep 3: 2723. https://doi.org/10.1038/srep02723

Chapter 14

Measuring Membrane Protein Dynamics in Plant Roots Using Fluorescence Correlation Spectroscopy

Jasim Basheer and Petra Marhava

Abstract

Quantitative methods using imaging and spectroscopy offer valuable insights into cell biology and molecular biophysics by elucidating complex mechanisms across various cellular environments. Fluorescence correlation spectroscopy (FCS) is particularly advantageous for quantifying the dynamics of fluorescent particles in low-concentration samples. As a high-resolution spectroscopic method with single-molecule sensitivity, FCS is well-suited to investigating membrane proteins, molecular interactions, aggregation, and conformational changes in living systems, including plant cells. However, challenges such as background fluorescence and artifacts arising from biological sample properties require careful experimental design, thereby limiting the routine application of FCS in plant tissues. Recent advancements in technology, such as confocal optics and highly efficient photon detectors, have substantially enhanced FCS's sensitivity and expanded its capacity for single-molecule detection. These advancements now empower plant biologists to explore the dynamics of plant proteins under various conditions. This chapter focuses on methodologies in fluorescence autocorrelation spectroscopy (FACS/FCS) and their application to plant research, with a particular emphasis on plasma membrane proteins in Arabidopsis root cells.

Keywords *Arabidopsis thaliana* , Plasma membrane, FCS, Fluorescence correlation spectroscopy, Root, Protein diffusion, Confocal microscopy

1 Introduction

Fluorescence correlation spectroscopy (FCS) is a high-resolution microscopy technique that measures the spatial and temporal dynamics of biomolecules at extremely low concentrations. It analyzes fluctuations in fluorescence intensity caused by physical, chemical, or biological changes affecting a fluorophore of interest within a specific observation volume [1]. These fluctuations, driven by processes like Brownian motion, provide information about the average number of fluorescent molecules within the detection volume and their diffusion dynamics over time. The resulting data are analyzed further using theoretical models to extract parameters such as fluorophore size, concentration, solution viscosity, and diffusion coefficient [2].

Steffen Vanneste and Inge Verstraeten (eds.), *Root Development*, Methods in Molecular Biology, vol. 3086,
https://doi.org/10.1007/978-1-0716-5603-7_14,

Mathematically, FCS is related to dynamic light scattering (DLS). However, DLS relies on weak, nonspecific signals from scattered photons and therefore typically requires a concentrated, purified sample. In contrast, FCS is an extremely sensitive analytical tool as it can measure particles at nano- to picomolar concentrations within a subfemtoliter observation volume [1].

FCS relies on spontaneous fluctuations in fluorescent intensity arising from deviations from equilibrium, rather than changes in mean fluorescence intensity. These fluctuations reflect changes in local concentrations, shifts in mobility coefficients (including active transport, translational, and rotational diffusion), and alterations in the rates of intramolecular and photophysical reactions [3]. In addition, dual-color fluorescence cross-correlation spectroscopy (dcFCCS) extends FCS to assess co-localization and correlated motion between two distinct fluorescent labels, allowing detection of binding reactions, dynamic co-localization during intracellular transport, and cleavage events. Although FCS is often referred to as a single-molecule technique, it is more accurately described as a single-molecule-sensitive technique, since individual molecules are not directly tracked but are inferred statistically from fluorescence fluctuations [4].

The primary application of FCS is the characterization of temporal fluctuations in fluorescence intensity and concentration arising from the motion of individual fluorescent molecules within the effective detection volume. The diffusion coefficient obtained from these measurements can be used to interpret the size of the diffusing fluorescent particles, given the medium's viscosity and organization. The common phenomena that can be understood using FCS are:

1. *Concentration and mobility* are the two primary outputs of FCS measurements.
2. *Conformational changes* of macromolecules, such as DNA compaction, which have a higher diffusion due to Brownian motion in a free solution [5].
3. *Detection of protein interactions* is key in understanding many cellular processes [6–10].
4. Diffusion changes due to *molecular aggregation*, including the determination of micellization processes and their critical concentrations [11, 12].
5. Changes in diffusion caused by the *lateral reorganization* of biological membranes.
6. Assessment of the role of *concentration in oligomerization* studies [13–15].
7. Determination of the *absolute concentration* of a molecule of interest within different locations in the sample [16, 17].

The foundation of FCS was established in the early 1970s through studies of diffusion and chemical dynamics, including DNA-drug intercalation [18–21]. However, the initial measurements were limited because of poor signal-to-noise ratios due to high background intensities caused by scattering in wide-field optical systems. The integration of confocal optics with highly efficient photon detectors, high numerical aperture (NA) objectives, and emission filters significantly improved the sensitivity of FCS and reduced the observation volume to the subfemtolitre range. These advances eventually increased the detection range to the single-molecule level [22].

Traditionally, FCS measurements rely on the law of free diffusion, which is primarily affected by biological membranes due to their lateral organization and complex, multicomponent assemblies of diverse lipids and proteins. In recent years, FCS has been increasingly used to study plasma membrane (PM) proteins and their roles in multiple physiological processes such as endocytosis and exocytosis, which are essential for PM protein turnover and function. Studies by Cui et al. (2018) [23] and Xing et al. (2019) [24] demonstrated the role of FLS2 endocytosis via clathrin- and sterol-associated pathways during flg22-induced plant defense responses by studying the transport of PLDδ using FCS. Similarly, Yu et al. (2020) [25] and Xing et al. (2022) [26] used FCS to quantify the density of receptor-like kinase FERONIA (FER) at the PM of *Arabidopsis thaliana*, allowing them to investigate FER localization and recycling upon flg22 peptide stimulation.

As outlined above, FCS is a powerful tool widely used to address numerous biophysical and biological phenomena, primarily because of its high sensitivity. Nevertheless, its application in plant cells is challenging due to interference from cell walls, chloroplasts, and other autofluorescent components, can interefere and mask signals from target fluorescent particles. However, some of these limitations can be mitigated by selecting appropriate bright fluorescent markers, which is a critical consideration before conducting the practical experiment. In this chapter, we present a comprehensive, stepwise protocol for performing FCS analysis to quantitatively measure the dynamic properties of biomolecules at the plasma membrane of *Arabidopsis* root cells.

FCS data analysis involves two steps. First, from the measured fluorescence intensity time-trace ($I(t)$), the autocorrelation function ($G(\tau)$) is calculated. Second, $G(\tau)$ is analyzed using fitting models. The first step is straightforward, using a simple autocorrelation equation to generate graphs. Understanding the autocorrelation function is crucial, as it quantifies the correlation between a signal and a delayed copy of itself, representing signal values at successive times. Calculating autocorrelation functions (ACF) is vital for studying molecules, since the diffusion coefficient is inversely proportional to the width of the ACF, and the fluorophore concentration is inversely proportional to the amplitude of the ACF. The primary parameter measured by FCS is the diffusion rate of a fluorescent molecule in and out of the observation volume [19, 21]. For FCS, the fluorescence intensity time trace ($I(t)$), recorded in a small detection volume (V_0), yields the autocorrelation function ($G(\tau)$) via the following equation.

$$G(\tau) = \frac{<I(t)I(t+\tau)>}{<I(t)>^2}$$

The second part requires user input to select a suitable model for interpreting $G(\tau)$, which can be performed using software such as QuickFit, Origin, or MATLAB. Although some users have developed complex models that account for real detection volumes and particle motion, these are rarely used. Instead, standard models of $G(\tau)$ are reviewed, including the triplet state in photophysical transitions on a microsecond timescale, with an average fraction T [3, 27–29].

$$G(\tau) = [1 - \tau + \tau e^{\frac{-\tau}{\tau T}}] \frac{1}{N(1-\tau)} \frac{1}{1 + (\tau/\tau_D)} \sqrt{\frac{1}{1 + (\tau/\tau_D)(\omega_0/\omega_z)^2}}$$

Secondly, diffusion within biological systems differs significantly from Brownian free diffusion, which follows the simple diffusion law. This discrepancy arises from interactions between fluorophores and other molecules or supramolecular structures that impede diffusion, potentially resulting in anomalous diffusion—or anomalous subdiffusion—where a nonlinear diffusion model becomes necessary for precise fitting [3, 27–29].

$$G(\tau) = [1 - \tau + \tau e^{\frac{-\tau}{\tau_\tau}}] \frac{1}{N(1-\tau)} \frac{1}{1 + (\tau/\tau_D)^\alpha} \sqrt{\frac{1}{1 + (\tau/\tau_D)^\alpha (\omega_0/\omega_z)^2}}$$

If the fluorescent particles of interest are not only diffusing but also influenced by a directed flow with velocity v, $G(\tau)$ is modeled using an alternative fitting approach.

$$G(\tau) = \frac{1}{N} \frac{1}{1 + (\tau/\tau_D)} \sqrt{\frac{1}{1 + (\tau/\tau_D)(\omega_0/\omega_z)^2}} \exp\left(\left(-\frac{\tau V}{\omega_0}\right)^2 \frac{1}{1 + (\tau/\tau_D)}\right)$$

In many biological protein samples, such as biomolecules moving within biological membranes, movement is primarily restricted to two dimensions, confined to the membrane plane. In such cases, the model used is:

$$G(\tau) = \frac{1}{N}\frac{1}{1 + (\tau/\tau_D)}$$

2 Materials

2.1 Plant Growth and Preparation

1. *Arabidopsis thaliana* seeds expressing the fluorescent marker(s) of interest.
2. Sterile ½ MS medium: Dissolve 2.2 g/L of Murashige and Skoog Basal salt mixture, including vitamins, 3 g/L of Sucrose, 0.5 g/L of MES in 800 mL of ultrapure water, adjust to pH 5.7 with 1 M KOH. For solid ½ MS medium, add 8 g/L of Plant agar. Autoclave and pour 50 mL into 12 × 12 cm square Petri dishes in the laminar flow hood.
3. Sterile 12 × 12 cm square Petri-dishes.
4. Laminar air flow hood.
5. Sterilization solution: 75% ethanol with 0.1% Tween 20.
6. Ethanol, absolute.
7. Sterile water.
8. Sterile Eppendorf tubes.
9. 1 mL micro-pipette and sterile tips.
10. Sterile toothpicks.
11. Micropore tape.
12. Forceps (fine-tip and soft-tip).
13. Lab-Tek™ II Chambered Cover glass 1.5 H (Catalog number 155360PK).
14. Plant growth chamber set at 21°C, 70% relative humidity, with a 16 h photoperiod (120 μmol/m^2/s).

2.2 Reference Dye for Calibration

1. ATTO488 Carboxy. The list of different reference dyes is summarized in Table 1.

2.3 Microscope Components

For a microscope with FCS functionality (Fig. 1):

1. Laser source and optics: Select either a continuous wave or a pulsed laser source compatible with your fluorophore of interest (e.g., 488 nm for GFP, 561 nm for RFP/mCherry, 633 nm for Cy5). When available, use Pulsed lasers to perform advanced techniques such as FLCS (fluorescence lifetime correlation spectroscopy). Select appropriate filters and mirrors

Table 1
Summary of reference dyes along with their excitation/emission spectra and their respective diffusion coefficient

Fluorophore	Excitation maximum (nm)	Emission maximum (nm)	Diffusion coefficient at 25°C ($\mu m^2/s$)	References
Atto488-carboxylic acid	488	523	400 ± 10	[32]
eGFP	488	525	93 ± 4	[30]
Oregon Green 488	488	550	411 ± 6 410 ± 8	[30, 33]
Fluorescein	498	520	425 ± 1	[32]
Rhodamine 110	500	535	470 ± 40	[34]
Rhodamine 123	508	530	460 ± 40	[34]
Rhodamine 6G	525	550	414 ± 5 430 ± 40 414 ± 1	[30, 33–35]
Rhodamine B	546	560	450 ± 40 427 ± 4	[34, 35]
Alexa 633	633	647	300 ± 10	[36]
Alexa 647	594 or 633	665	330 ± 10	[37]
Cy5	633	670	360 ± 10	[38]
Atto655-maleimid	650	686	407 ± 10 406 ± 9 409 ± 7	[33, 39]
Atto655-NHS ester	663	685	425 ± 6	[40]
Atto 655-carboxylic acid	663	685	426 ± 8	[33, 40]

based on the microscope's laser, fluorophore, and beam path for a pinhole-based confocal microscope (**Note 1**).

2. Microscope objective: A high numerical aperture (NA) objective lens (NA ≥1.2, often water-immersion) is crucial for achieving a femtoliter-sized (fL) detection volume and efficient photon collection.
3. Detector*:* Use highly sensitive single-photon detectors, such as avalanche photodiodes (APDs) or photomultiplier tubes (PMTs), capable of counting individual photons over time.

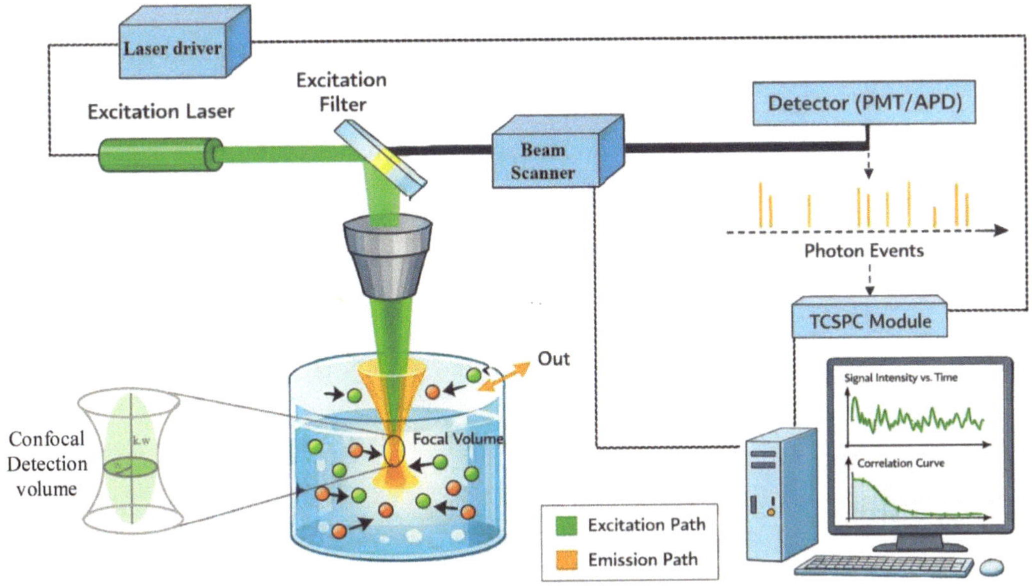

Fig. 1 Scheme of FCS instrumentation

4. Correlator or data acquisition system: Use a time-correlated single photon counting (TCSPC) unit to generate the autocorrelation function (ACF).
5. Computer and software: A dedicated computer workstation equipped with specialized acquisition and analysis software to fit autocorrelation curves for extracting parameters such as diffusion coefficients and molecular concentrations. E.g., microscope-specific software or third-party software such as Origin or MATLAB.

3 Methods

3.1 Seed Surface Sterilization and Plant Growth

1. Place *Arabidopsis thaliana* seeds in a microcentrifuge tube.
2. Add sterilization solution and incubate for 10 min with gentle shaking.
3. Remove the sterilization solution and briefly wash the seeds with 100 % EtOH.
4. Remove the EtOH, spread the seeds along the side of the tube, and air-dry in a laminar airflow hood.
5. Transfer the sterile seeds to 12 × 12 cm square Petri dishes containing solid ½ MS medium using a sterile toothpick or pipette.

6. Seal the plate with micropore tape.
7. Incubate the plates with seeds at 4°C for 48 h before transferring them to the plant growth chamber.

3.2 Mounting Seedlings in the Chamber

1. Add 50 µL of ½ MS liquid medium to a Lab-Tek™ II Chambered Cover glass, then carefully transfer and position a 5-day-old seedling (**Note 4**) (Fig. 4B).
2. Cut a small slab of solid ½ MS medium and place it gently over the root to keep it close to the coverslip.
3. Cover the chamber with the lid and allow the sample to stabilize for 5–10 min in the growth chamber before mounting it on the microscope for measurements.

3.3 Data Acquisition

3.3.1 Setting up the Microscope and Software for Acquisition

1. Launch the FCS software. In this protocol, we use SymPhotime 64 (Picoquant) and ZEN software (Fig. 2).
2. Define the light path and select the appropriate filters/mirrors based on the dyes and sample used in the ZEN software (Fig. 2B).
3. Open and configure the Picoquant FLIM+FCS handshake software to connect the ZEN black software and SymPhoTime64 software.
4. In the Picoquant FLIM+FCS handshake software, set the data acquisition time in the FCS tab according to the duration required for each replicate (Fig. 2C).
5. In SymPhoTime64, define the workspace using "File\open Workspace" from the main menu, and save the data with appropriate names (Fig. 3A, B).
6. Activate "Test Mode" and bring the dye into focus by checking the change in the number of photons received (Fig. 3C).
7. An increase in photon counts indicates that the reference fluorophore solution is in focus (Fig. 3E).

3.3.2 Calibration

1. Prepare a reference dye solution (listed in Table 1) at a final concentration of 10 nM, and place a drop of the solution onto a chambered glass slide with a cover glass (1.5 H) (**Note 2**).
2. To begin calibration, set the acquisition mode to spot-scanning.
3. Activate test mode in the Picoquant remote software and adjust the focus by checking the photon count in the acquisition software (Fig. 3C, D).
4. After achieving focus, the reflection from the glass can be reduced by adjusting the refractive index to maximize the number of photons from the samples.

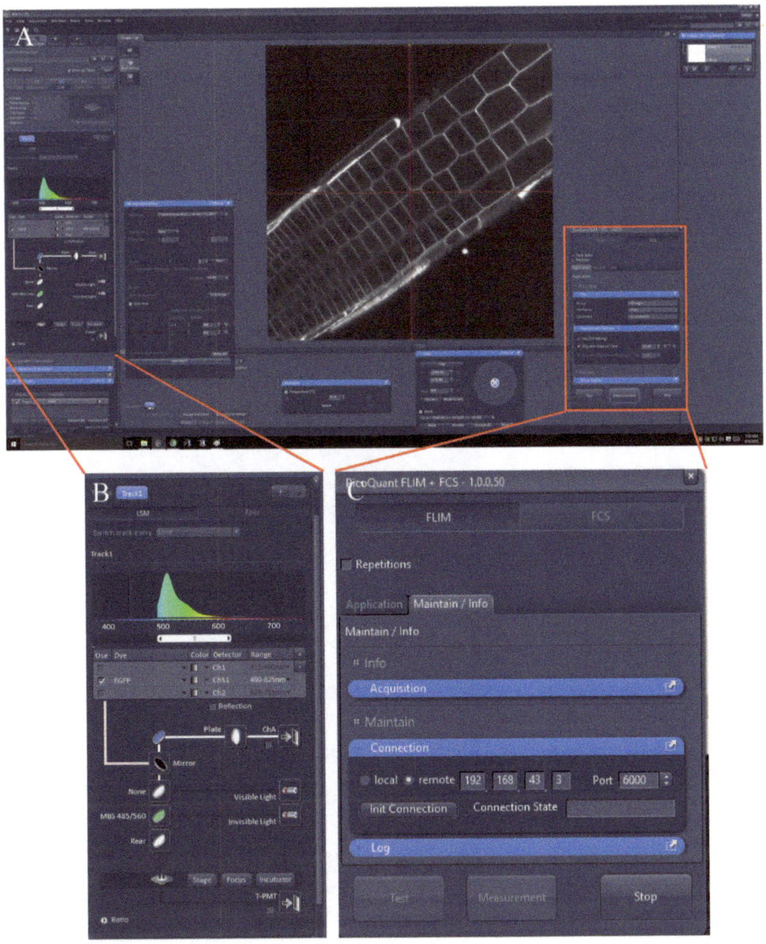

Fig. 2 Initializing the ZEN black microscope software, FCS light path, and initalizing picoquant FLIM+FCS handshake software. (**A**) Zeiss ZEN Black software used for controlling the microscope and modifying the light path. (**B**) Zoomed inset of the light path window of the ZEN black software. (**C**) Zoomed inset of PicoQuant FCS+FLIM handshake software

5. Once the refractive index is fixed, acquire data from 10-s acquisitions (this duration may vary with experimental setup) and repeat the acquisition at least ten times (Fig. 3F).

3.3.3 Measurement of Protein Diffusion in Root Cells

1. Create a new workspace to store the samples in the SymPhotime 64.
2. Adjust the microscope mirror to align the light path with the confocal detector, then enable Live/Continuous mode to view the sample.
3. Bring the sample into focus and identify the appropriate measurement location.

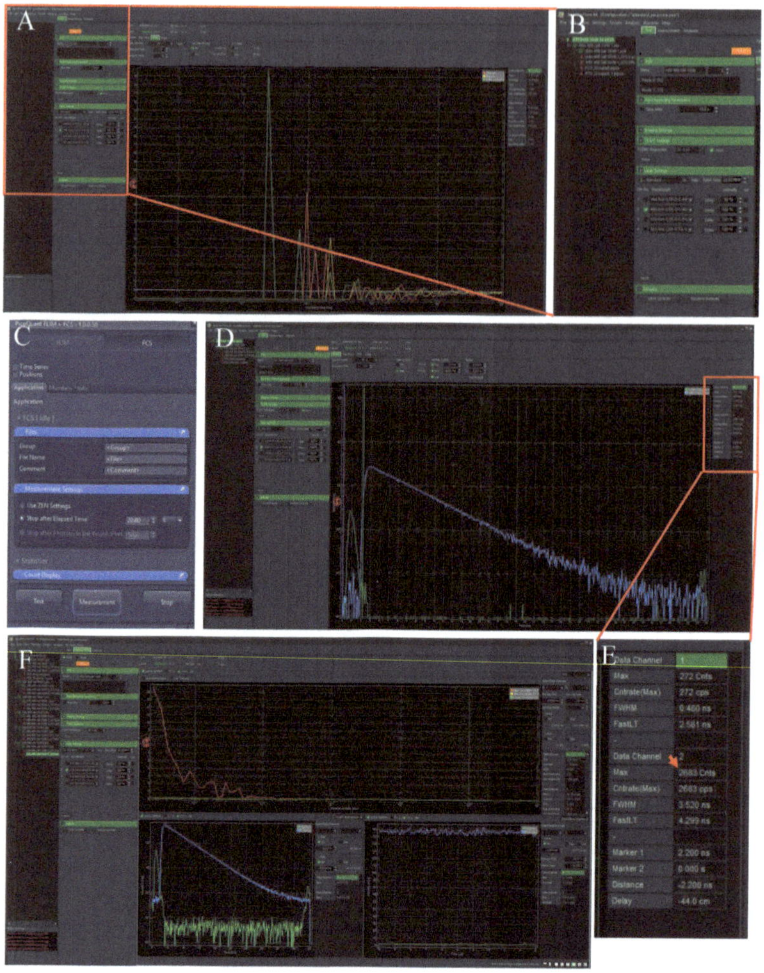

Fig. 3 FCS confocal detection volume, pinhole, refractive index calibration and recording of FCS data for reference fluorophore. (**A**) PicoQuant SymPhotime64 Acquisition software used for FCS measurements. (**B**) Zoomed inset of the PicoQuant SymPhotime64 software workspace and SymPhotime64 acquisition settings. (**C**) Zoomed inset of PicoQuant FCS+FLIM handshake software showing the selection of acquisition time and tabs used for test/measure. (**D**) SymPhotime64 Acquisition software used for FCS measurements, showing the decay peak once the detector collects photons while the sample comes into focus. (**E**) Zoomed inset of photon counting tab in the SymPhotime64 acquisition software. (**F**) PicoQuant Acquisition software showing acquired FCS measurements from the reference dye

4. Select the region of interest (ROI) by positioning the crosshair over the desired spot with your mouse (Fig. 4C).
5. Switch the mirror to guide the light path to the PicoQuant detector.

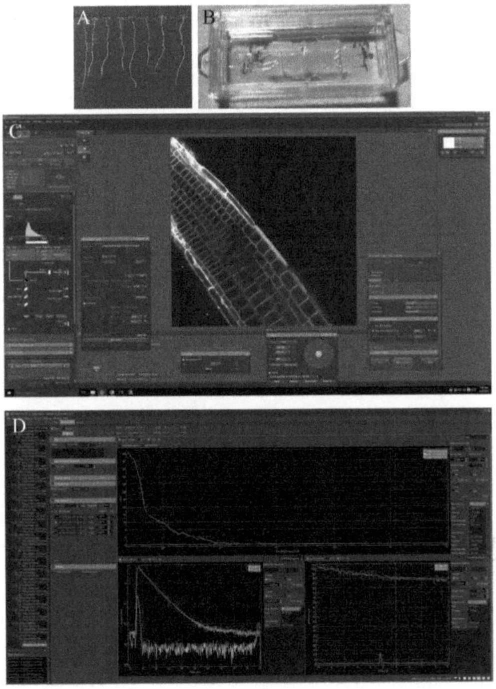

Fig. 4 Preparation of plant samples for FCS and fluorescence fluctuation data collection. (**A**) Plants grown in ½MS plates for performing the FCS on the root cells. (**B**) Plants transferred to Lab-Tek™ II Chambered Cover glass stabilized with a slab of ½ MS agar for FCS measurement. (**C**) Zeiss ZEN software is used to generate the confocal image for selecting the ROI using the crosshair for FCS measurement. (**D**) SymPhotime64 Acquisition software showing acquired FCS measurements from the plant plasma membrane samples

6. Perform a prebleach using the Test tab in the PicoQuant FLIM+FCS Handshake software.
7. After prebleach, measurement was performed using the Measure tab in the PicoQuant FLIM+FCS handshake software remote (**Note 3**).
8. The measurement is obtained multiple times for each sample (Fig. 4D).

3.4 Data Analysis and Fitting

3.4.1 Calibration and Setting up the Confocal Volume

1. Use data obtained from the reference dye, with a known diffusion coefficient.
2. Open the data file, select the appropriate channel (Fig. 5A, B), and generate the autocorrelation graph from the raw data using the Single Calculation tab (Fig. 5C, D).

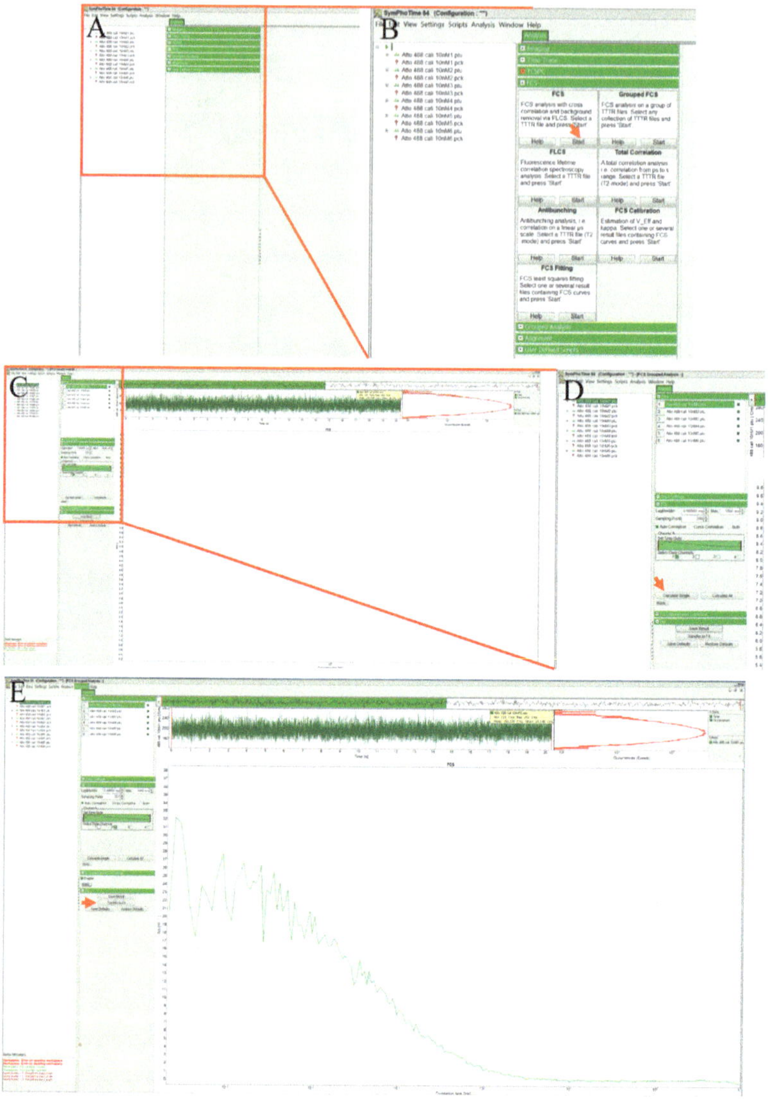

Fig. 5 Generation of the autocorrelation graphs using the reference fluorophore data and transferring the generated autocorrelation to the fitting software. Preparation of plant samples for FCS and fluorescence fluctuation data collection. (**A**) Acquired data opened in PicoQuant SymPhotime 64 Software for further analysis. (**B**) Zoomed inset of SymPhotime64 software with red arrow showing the tab used for FCS analysis. (**C**) After opening the FCS tab for the sample, the generation of the autocorrelation graph was done using SymPhotime 64. (**D**) Zoomed inset of SymPhotime64 software with red arrow showing the tab used for autocorrelation graph generation. (**E**) SymPhotime64 software showing autocorrelation graph with the red arrow showing the tab for transferring the graph to the fitting window

3. The initial portion of the measurement is usually excluded due to disturbances.
4. Transfer the autocorrelation graph to the Fitting window (Fig. 5E).
5. Fitting is performed using an appropriate fitting equation, and the quality of the fit is evaluated (Fig. 6A, B).
6. Adjust the fitting parameters to improve the fit. Once a satisfactory fit is achieved and the diffusion coefficient approaches the predicted value, save the results (Fig. 6C, D).
7. Open the saved data in the Calibration tab and enter the correct diffusion coefficient in the software (Fig. 7A, B).
8. A message box with "__fL set as effective focal volume (V_EFF) and __ set as kappa (based on the objective, pinhole, and laser wavelength)" will be displayed, confirming your calibration (Fig. 7C).
9. Close the workspace before loading the sample of interest.

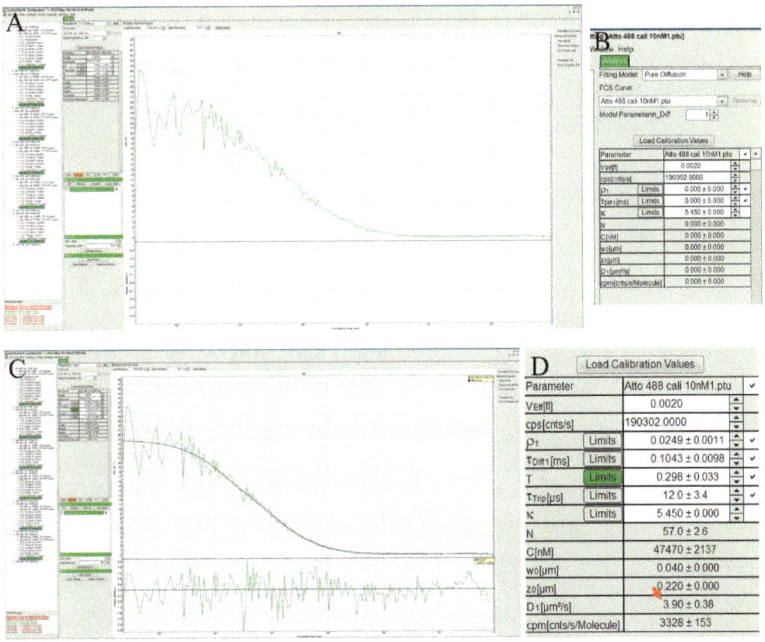

Fig. 6 Fitting of the generated autocorrelation graph using a pure diffusion fitting model since we used a solution of Atto488 carboxy reference dye. (**A**) PicoQuant SymPhotime 64 Software showing the autocorrelation graph in the fitting window. (**B**) Zoomed inset of SymPhotime64 software showing the tab with the fitting parameters to be selected for fitting based on the samples. (**C**) SymPhotime 64 software showing the fitted autocorrelation graph along with the diffusion data. (**D**) Zoomed inset of SymPhotime64 software showing the dynamics of the reference dye with red arrow showing the diffusion coefficient

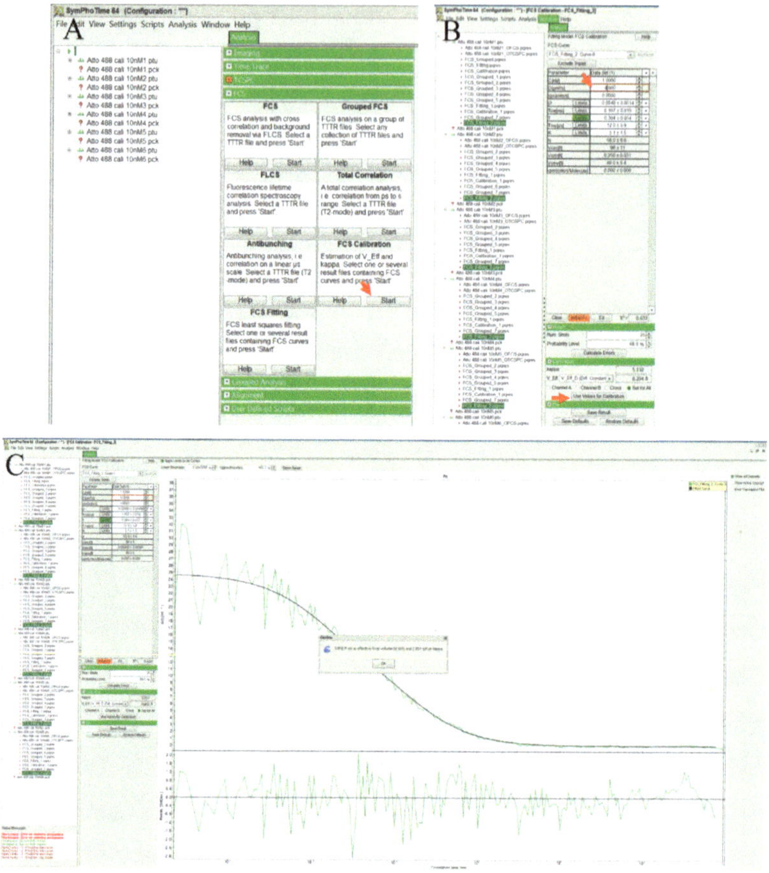

Fig. 7 Recording the effective focal volume (V_EFF) to the analysis software to avoid artifacts in the analysis. (**A**) PicoQuant SymPhotime64 Software showing the tab for setting up the calibration data from the reference dye using a red arrow. (**B**) SymPhotime64 software showing the tabs where the fitted data must be provided and saved. (**C**) SymPhotime 64 software showing the effective focal volume (V_EFF) is successfully added

3.4.2 Data Analysis

1. Open the data file for the sample of interest you measured, along with the reference dye.
2. Select the proper channel and generate the autocorrelation graph (Fig. 8).
3. Transfer the autocorrelation data to the Fitting window for parameter estimation.
4. Exclude areas with large perturbations or variations in the autocorrelation graphs.
5. Perform the fitting using the initial fit function; assess the overall fit and adjust the parameters as needed (**Note 5**).

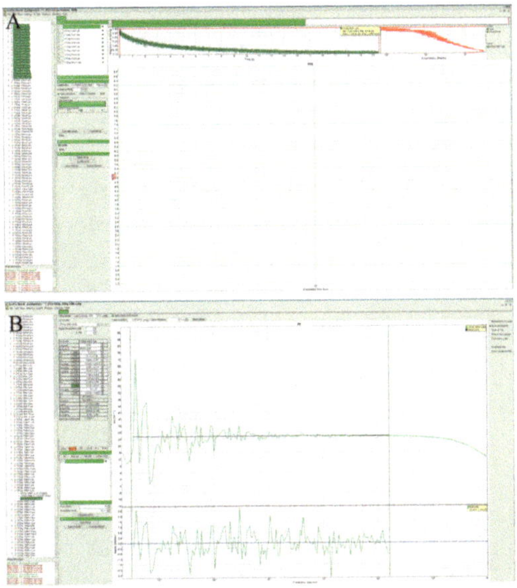

Fig. 8 Analysis of the diffusion coefficient from the FCS data of the plant samples. (**A**) PicoQuant SymPhotime 64 Software showing the data from the plant samples. (**B**) SymPhotime64 software showing fitted autocorrelation graphs along with their diffusion coefficients

6. Once the fit is satisfied, apply the Fit All function to confirm the fit, and save the diffusion coefficient values in Excel format for statistical analysis.

4 Notes

1. Before collecting FCS data, carefully align the microscope's optical path. FCS accuracy relies on the shape and size of the detection volume. Maximize photon collection by optimizing the emission path to the detector. Improve the detection volume shape by aligning the confocal pinhole with a reference fluorophore (Table 1). For example, the pinhole size for the 488 nm excitation wavelength may vary between the 40× and 64× objectives. Likewise, even when the objective is set to the same magnification (e.g., 40×), the pinhole size can differ between the 488 and 561 nm wavelengths. Using a reference dye allows you to account for changes in pinhole size when switching wavelengths or objectives during measurements. Additionally, adjust the objective correction collar to account for the cover-glass thickness and the sample's refractive index.
2. Calibration should be performed after any change to the experimental setup, such as a change in excitation wavelength or

confocal pinhole size, that may affect the size and shape of the detection volume. The calibration measurement is ideally performed at least once per day during the experiment. Defining the confocal volume using a reference dye solution (Table 1) is a necessary step in FCS measurements and a critical prerequisite for determining the diffusion coefficients/concentrations of the sample of interest without artifacts. The selection of the reference dyes is based on the excitation/emission wavelength of the sample of interest. The list of different reference dyes is summarized in Table 1 along with their respective diffusion coefficients and excitation-emission wavelength. The FCS can be used across a wide range of concentrations, but optimal calibration concentrations typically fall between 1 and 20 nM.

3. FCS acquisition times vary significantly, ranging from milliseconds to minutes, depending on the dynamics being studied. Fast molecular motions require shorter acquisition time (microseconds), whereas slower biological processes require longer acquisitions (seconds to minutes). Typically, approximately 10 s is used for reference measurements, as it balances statistical accuracy with considerations such as photobleaching and cell movement. Shorter acquisition times are suitable for capturing rapid events, such as EGFP in solution, whereas longer times are required to observe slower diffusion or binding processes in membranes.
4. FCS is highly sensitive to sample quality and to the cover glass used in sample preparation (Fig. 4A, B). The samples should be positioned close to the glass cover, with minimal refractive-index difference. Because ROI is measured at a single point, the selected location within the root sample should account for growth movement. The point of measurement tends to bleach the ROI; therefore, pre-bleaching it before measurement is preferable to reduce variability in the values.
5. Artifacts in FCS can arise from various factors, introducing errors into the results and potentially leading to misinterpretation of the data. These artifacts can be divided into three major categories:

   a. Artifacts resulting from the limited accuracy of the approximations used in the theoretical derivation models.
   b. Artifacts resulting from detector background and parasitic signal components.
   c. Artifacts associated with calibrating the effective detection volume can lead to incorrect interpretation of τD and N values obtained by fitting [28, 30, 31].

Conflict of Interest

The authors have no competing interests to disclose.

References

1. Wang Z, Zhang H, Jian L et al (2022) Principles of fluorescence correlation spectroscopy applied to studies of biomolecular liquid–liquid phase separation. Biophys Rep 8:100. https://doi.org/10.52601/BPR.2022.210047
2. Bag N, Ng W, Sankaran J et al (2020) Data fitting. An introduction to fluorescence correlation. Spectroscopy 1:8–1-8–30. https://doi.org/10.1088/978-0-7503-2080-1CH8
3. Macháň R, Hof M (2010) Recent developments in fluorescence correlation spectroscopy for diffusion measurements in planar lipid membranes. Int J Mol Sci 11:427–457. https://doi.org/10.3390/IJMS11020427
4. Enderlein J, Gregor I, Patra D, Fitter J (2005) Statistical analysis of diffusion coefficient determination by fluorescence correlation spectroscopy. J Fluoresc 15:415–422. https://doi.org/10.1007/S10895-005-2633-0
5. Adjimatera N, Kral T, Hof M, Blagbrough IS (2006) Lipopolyamine-mediated single nanoparticle formation of calf thymus DNA analyzed by fluorescence correlation spectroscopy. Pharm Res 23:1564–1573. https://doi.org/10.1007/S11095-006-0278-6
6. Schwille P, Heinze KG (2001) Two-photon fluorescence cross-correlation spectroscopy. Chem Phys Chem 2:269–272. https://doi.org/10.1002/1439-7641
7. Bierbaum M, Bastiaens PIH (2013) Cell cycle-dependent binding modes of the ran exchange factor RCC1 to chromatin. Biophys J 104:1642–1651. https://doi.org/10.1016/j.bpj.2013.03.024
8. Caballero-George C, Sorkalla T, Jakobs D et al (2012) Fluorescence correlation spectroscopy in drug discovery: study of Alexa532-endothelin 1 binding to the endothelin ETA receptor to describe the pharmacological profile of natural products. Sci World J 2012:524169. https://doi.org/10.1100/2012/524169
9. Rose RH, Briddon SJ, Hill SJ (2012) A novel fluorescent histamine H 1 receptor antagonist demonstrates the advantage of using fluorescence correlation spectroscopy to study the binding of lipophilic ligands. Br J Pharmacol 165:1789–1800. https://doi.org/10.1111/J.1476-5381.2011.01640.X
10. Karpińska A, Pilz M, Buczkowska J et al (2021) Quantitative analysis of biochemical processes in living cells at a single-molecule level: a case of olaparib-PARP1 (DNA repair protein) interactions. Analyst 146:7131–7143. https://doi.org/10.1039/d1an01769a
11. Yu L, Tan M, Ho B et al (2006) Determination of critical micelle concentrations and aggregation numbers by fluorescence correlation spectroscopy: aggregation of a lipopolysaccharide. Anal Chim Acta 556:216–225. https://doi.org/10.1016/J.ACA.2005.09.008
12. Nag M, Das D, Bandyopadhyay D, Basak S (2015) Unusual denaturation trajectory of bovine gamma globulin studied by fluorescence correlation spectroscopy. Phys Chem Chem Phys 17:19139–19148. https://doi.org/10.1039/c5cp00387c
13. Herrick-Davis K, Grinde E, Cowan A, Mazurkiewicz JE (2013) Fluorescence correlation spectroscopy analysis of serotonin, adrenergic, muscarinic, and dopamine receptor dimerization: the oligomer number puzzle. Mol Pharmacol 84:630–642. https://doi.org/10.1124/MOL.113.087072
14. Ilien B, Glasser N, Clamme JP et al (2009) Pirenzepine promotes the dimerization of muscarinic M1 receptors through a three-step binding process. J Biol Chem 284:19533–19543. https://doi.org/10.1074/JBC.M109.017145
15. Nguyen TTT, Kläsener K, Zürn C et al (2017) The IgM receptor FcμR limits tonic BCR signaling by regulating expression of the IgM BCR. Nat Immunol 18:321–333. https://doi.org/10.1038/ni.3677
16. Erokhova L, Horner A, Kügler P, Pohl P (2011) Monitoring single-channel water permeability in polarized cells. J Biol Chem 286:39926–39932. https://doi.org/10.1074/jbc.M111.291864
17. Yu L, Lei Y, Ma Y et al (2021) A comprehensive review of fluorescence correlation spectroscopy. Front Phys 9: 644450. https://doi.org/10.3389/FPHY.2021.644450
18. Ehrenberg M, Rigler R (1974) Rotational brownian motion and fluorescence intensify fluctuations. Chem Phys 4:390–401. https://doi.org/10.1016/0301-0104(74)85005-6
19. Elson EL, Magde D (1974) Fluorescence correlation spectroscopy. I. Conceptual basis and theory. Biopolymers 13:1–27. https://doi.org/10.1002/BIP.1974.360130102
20. Magde D, Elson E, Webb WW (1972) Thermodynamic fluctuations in a reacting system—measurement by fluorescence correlation

spectroscopy. Phys Rev Lett 29:705. https://doi.org/10.1103/PhysRevLett.29.705
21. Magde D, Elson EL, Webb WW (1974) Fluorescence correlation spectroscopy. II. An experimental realization. Biopolymers 13:29–61. https://doi.org/10.1002/BIP.1974.360130103
22. Hess ST, Webb WW (2002) Focal volume optics and experimental artifacts in confocal fluorescence correlation spectroscopy. Biophys J 83:2300–2317. https://doi.org/10.1016/S0006-3495(02)73990-8
23. Cui Y, Li X, Yu M et al (2018) Sterols regulate endocytic pathways during flg22-induced defense responses in arabidopsis. Development 145:dev165688. https://doi.org/10.1242/dev.165688
24. Xing J, Li X, Wang X et al (2019) Secretion of phospholipase Dδ functions as a regulatory mechanism in plant innate immunity. Plant Cell 31:3015–3032. https://doi.org/10.1105/tpc.19.00534
25. Yu M, Li R, Cui Y et al (2020) The RALF1-FERONIA interaction modulates endocytosis to mediate control of root growth in Arabidopsis. Development 147:dev189902. https://doi.org/10.1242/dev.189902
26. Xing J, Ji D, Duan Z et al (2022) Spatiotemporal dynamics of FERONIA reveal alternative endocytic pathways in response to flg22 elicitor stimuli. New Phytol 235:518–532. https://doi.org/10.1111/nph.18127
27. Widengren J, Mets Ü, Rigler R (1995) Fluorescence correlation spectroscopy of triplet states in solution: a theoretical and experimental study. J Phys Chem 99:13368–13379
28. Ries J, Schwille P (2012) Fluorescence correlation spectroscopy. BioEssays 34:361–368. https://doi.org/10.1002/BIES.201100111
29. Wohland T, Maiti S, Macháň R (2020) An introduction to fluorescence correlation spectroscopy. IOP Publishing. https://doi.org/10.1088/978-0-7503-2080-1
30. Petrášek Z, Schwille P (2008) Precise measurement of diffusion coefficients using scanning fluorescence correlation spectroscopy. Biophys J 94:1437–1448. https://doi.org/10.1529/BIOPHYSJ.107.108811
31. Sýkora J, Kaiser K, Gregor I et al (2007) Exploring fluorescence antibunching in solution to determine the stoichiometry of molecular complexes. Anal Chem 79:4040–4049. https://doi.org/10.1021/AC062024F
32. Abbandonato G, Hoffmann K, Resch-Genger U (2018) Determination of quantum yields of semiconductor nanocrystals at the single emitter level via fluorescence correlation spectroscopy. Nanoscale 10:7147–7154. https://doi.org/10.1039/C7NR09332B
33. Müller CB, Loman A, Pacheco V et al (2008) Precise measurement of diffusion by multi-color dual-focus fluorescence correlation spectroscopy. EPL 83:46001. https://doi.org/10.1209/0295-5075/83/46001
34. Gendron PO, Avaltroni F, Wilkinson KJ (2008) Diffusion coefficients of several rhodamine derivatives as determined by pulsed field gradient–nuclear magnetic resonance and fluorescence correlation spectroscopy. J Fluoresc 18:1093–1101. https://doi.org/10.1007/S10895-008-0357-7
35. Culbertson CT, Jacobson SC, Michael Ramsey J (2002) Diffusion coefficient measurements in microfluidic devices. Talanta 56:365–373. https://doi.org/10.1016/S0039-9140(01)00602-6
36. Van Den Bogaart G, Krasnikov V, Poolman B (2007) Dual-color fluorescence-burst analysis to probe protein efflux through the mechanosensitive channel MscL. Biophys J 92:1233–1240. https://doi.org/10.1529/BIOPHYSJ.106.088708
37. Fields AP, Cohen AE (2011) Electrokinetic trapping at the one nanometer limit. Proc Natl Acad Sci USA 108:8937–8942. https://doi.org/10.1073/PNAS.1103554108
38. Palo K, Mets Ü, Jäger S et al (2000) Fluorescence intensity multiple distributions analysis: concurrent determination of diffusion times and molecular brightness. Biophys J 79:2858–2866. https://doi.org/10.1016/S0006-3495(00)76523-4
39. Korlann Y, Dertinger T, Michalet X et al (2008) Measuring diffusion with polarization-modulation dual-focus fluorescence correlation spectroscopy. Opt Express 16:14609–14616. https://doi.org/10.1364/OE.16.014609
40. Dertinger T, Pacheco V, Von Der Hocht I et al (2007) Two-focus fluorescence correlation spectroscopy: a new tool for accurate and absolute diffusion measurements. Chem Phys Chem 8:433–443. https://doi.org/10.1002/cphc.200600638

Chapter 15

Mapping Labile Fe^{2+} and Fe^{3+} in Living *Arabidopsis thaliana* Roots: Imaging and Quantification

Carine Alcon and Tou Cheu Xiong

Abstract

Iron (Fe) is involved in numerous key physiological processes due to its transfer or gain electron capacity. However, in excess, Fe is toxic to cells. The understanding of mechanisms governing its homeostasis requires imaging techniques capable of localizing and quantifying Fe and its redox state in situ. The Perls-DAB histochemical staining is a widely used method for visualizing Fe at a cellular level. Other advanced approaches, such as micro X-ray fluorescence and X-ray absorption near-edge structure spectroscopy, also provide information on the spatial distribution and oxidation state of Fe. However, all of these methods require sample fixation before observing Fe localization, which can introduce biases in the results. Recent development in synthetic fluorescent probes now enables the real-time visualization of ferrous (Fe^{2+}) and ferric (Fe^{3+}) forms in living roots, offering a robust and user-friendly approach to study Fe redox distribution *in planta*. This live imaging approach provides access to Fe dynamics and enables semi-quantitative analyses, thereby opening new avenues for Fe homeostasis research. This chapter provides detailed protocols for applying these methods in Arabidopsis roots.

Keywords Iron, Redox, Fluorescent probes, *Arabidopsis thaliana*, Fluorescence microscopy, Quantitative fluorescence analysis

1 Introduction

Iron (Fe) is an essential micronutrient required for plant growth and development. As a transition metal, it is inherently redox-active, a property that underlies its central role in both photosynthesis and cellular respiration [1]. Fe bioavailability in soils is often limited, forcing plants to develop sophisticated acquisition mechanisms to meet their metabolic requirements. These mechanisms fall into two categories: Strategy I (reduction of Fe^{3+} to Fe^{2+} prior to uptake) or Strategy II (chelation of Fe^{3+} by phytosiderophores) [2]. These processes are both tightly regulated and spatially localized. In dicotyledon species such as Arabidopsis, Fe acquisition mechanisms involve the reduction of Fe^{3+} to Fe^{2+} prior to its uptake by epidermal root cells. Once absorbed, Fe is transported from the epidermis to the vascular

Steffen Vanneste and Inge Verstraeten (eds.), *Root Development*, Methods in Molecular Biology, vol. 3086,
https://doi.org/10.1007/978-1-0716-5603-7_15,

cylinder mainly as Fe^{3+}-citrate complexes. Its distribution within the root is both heterogeneous and dynamic [3], making Fe homeostasis challenging to characterize. Fe homeostasis ultimately depends on the balance between Fe transport, compartmentalization, storage, and remobilization, all of which rely on the redox interconversion between Fe^{2+} and Fe^{3+}. This complexity cannot be captured by traditional biochemical methods, which rely on tissue homogenization and therefore lose spatial information. Imaging techniques have therefore become essential for accurate characterization of the in situ distribution and chemical speciation of Fe, uncovering the heterogeneity at the organ, tissue, or cell scale and allowing quantification. Among these methods, the Perls–DAB histochemical staining technique is widely used to visualize Fe accumulation with cellular resolution [4]. Perls staining relies on the reaction of ferric iron (Fe^{3+}) with potassium ferrocyanide in acidic conditions, producing an insoluble Prussian blue precipitate. To increase sensitivity, this method is often combined with diaminobenzidine (DAB), which forms a dark brown precipitate when oxidized by hydrogen peroxide in the presence of Fe and endogenous peroxidase activity. Despite being simple and accessible, this approach has limitations in terms of quantification and distinguishing the Fe redox state.

Synchrotron-based micro x-ray fluorescence (μ-XRF) offers a powerful, non-destructive method for the spatial and quantitative mapping of Fe and other micronutrients in plant tissues with micrometer-scale resolution. As the emitted X-ray fluorescence is element-specific, μ-XRF can precisely localize Fe [5]. When coupled with X-ray Absorption Near-Edge Spectroscopy (XANES), μ-XRF can also yield information on oxidation state (Fe^{2+} vs. Fe^{3+}) [6]. Cryofixation and cryosectioning are generally required to preserve the elemental distribution. Despite its high resolution and chemical specificity, this technique is limited by restricted access to synchrotron facilities and cannot capture Fe dynamics in living tissues.

To overcome these limitations and allow real-time analysis of Fe behavior in living roots, synthetic fluorescent probes have been developed [7, 8]. Several Fe fluorescent dyes have been generated to detect Fe in vitro. Two main strategies have been employed to design turn-OFF or turn-ON read-out probes, the latter being more suitable for microscopy and live-cell science. Among all turn-ON iron probes, only a few have been applied in biological systems, and their application in plant cells remains largely unexplored, which limits their broader adoption. This kind of biosensor allows the detection of the labile Fe pool, which consists of Fe ions weakly bound to ligands. The use of Fe probes has recently been demonstrated in plant cells [9]. The two chemical sensors 4-(4-methylpiperazine-1)-7-nitrobenz-2-oxa-1,3-diazole (MP-NBD) [10] and SiRhoNox-1 [11] have been combined simultaneously for Fe^{3+} and Fe^{2+} ions detection in plant root, respectively. These organic molecules enter cells and can label different cell layers, allowing in vivo monitoring of both Fe species by producing fluorescence signals that specifically reflect the Fe redox state. The distinct fluorescence spectral properties of the two probes allow the simultaneous detection of both labile Fe^{2+} and Fe^{3+}. This new method reveals the heterogeneous distribution of Fe redox in cells [9] and enables dynamic changes to be monitored in response to environmental or genetic perturbations.

This chapter focuses on the detailed application of Fe fluorescent probes for imaging labile redox Fe^{2+} and Fe^{3+} in plant roots that we developed in Alcon et al. [9]. We also provide a semi-automated procedure for quantifying the fluorescence intensity of each cell layer, offering a powerful methodological framework for studying Fe homeostasis in living plant roots.

2 Materials

2.1 General

1. Wild-type Columbia-0 *Arabidopsis thaliana* seeds
2. Seeds sterilization solution: 70% Ethanol containing 0.01 % Triton X-100
3. Ethanol, absolute
4. Sterile 12 × 12 cm Petri dishes
5. MES-KOH buffer: Weigh 1.952 g of 2-(*N*-morpholino)-ethane sulfonic acid (MES) and dissolve it in 1 L of distilled water. Adjust pH to 6.0 with 1 M KOH solution and store at room temperature.
6. 0.5× MS medium with agar: Mix 2.2 g of Murashige and Skoog (MS), 10 g of sucrose, 0.5 g of MES, and 1 L of distilled water. Adjust pH to 5.7 using 1 M KOH. Add 8 g of agar and autoclave. Allow to cool down to ~60°C, and pour each time 50 mL of medium into each Petri-dish in a sterile hood.
7. Growth cabinet or growth chamber at 21°C in a 16-h light/8-h dark cycle with 65% relative humidity and a light intensity of 150 $\mu mol/m^2/s^1$.
8. Six-well plate
9. Double-sided tape
10. Forceps
11. Micro-pipette and corresponding pipetting tips
12. Microscopy slides and coverslips
13. Laser scanning confocal microscope (e.g., Leica SP8) (*see* **Note 1**), equipped with a HC PL APO 40×/1.10W CORR

CS2 water immersion objective with a 0.62 mm working distance (*see* **Note 2**).

2.2 Iron Synthetic Probes

1. SiRhoNox-1: dissolve Fe^{2+} probe (Goryo Chemicals, GC903-01) in DMSO at 1 mM and store aliquots at −20°C.
2. MP-NBD: dissolve Fe^{3+} probe in DMSO at 10 mM and store aliquots at −20°C.
3. Synthesis of MP-NBD: 4-chloro-7-nitrobenzofurazan (NBD-Cl); 1-methylpiperazine; chloroform ($CHCl_3$); dichloromethane (CH_2Cl_2); methanol (MeOH)diethyl ether (Et_2O); silica gel for column chromatography (SiO_2, 60 Å); Glass chromatography column; TLC plates (silica gel); Magnetic stirrer; Rotary evaporator; Cooling bath (ice).

2.3 Software

1. FIJI (version 1.54p) open source software https://imagej.net/software/fiji/downloads (*see* **Note 3**).
2. Plugin: Radial Profile Angle (http://questpharma.u-strasbg.fr/html/radial-profile-ext.html) (*see* **Note 4**).

3 Methods

3.1 MP-NBD Synthesis

MP-NBD is synthesized according to a previously reported procedure [10] with minor modifications described in Alcon et al. [9].

1. Dissolve 3.2 g of 4-chloro-7-nitrobenzofurazan in 100 mL of chloroform.
2. Add 2.07 mL of 1-methylpiperazine and stir the reaction mixture at room temperature until complete conversion, as monitored by TLC.
3. Evaporate the solvent under reduced pressure.
4. Purify the crude product by silica gel column chromatography using dichloromethane/methanol (95:5, v/v) as eluent.
5. Precipitate the purified product from chloroform by the addition of diethyl ether, cool to 0 °C, filter, and wash with diethyl ether to afford MP-NBD as an orange solid.
6. Dissolve MP-NBD as a 10 mM solution in DMSO and store at −20 °C.

3.2 Arabidopsis Root Staining with Chemical Iron Sensors

All the steps are summarized in Fig. 1.

1. Incubate the dry seeds in 1.5 mL Eppendorf tube containing the Seeds sterilization solution for 10 min under gentle shaking. The seeds are then washed four times in 100 % ethanol,

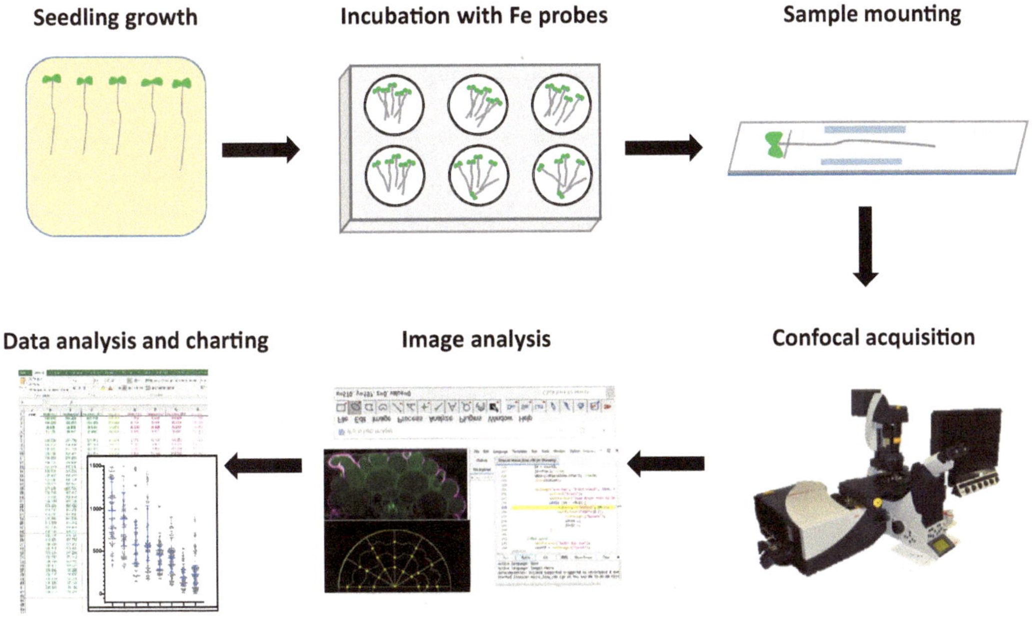

Fig. 1 Workflow for iron imaging of Arabidopsis roots. Seven-day-old seedlings were grown on 0.5x MS agar plates, positioned vertically. The seedlings were transferred to six-well plate containing 2 mL of MES-KOH buffer containing MP-NBD (10 μM) and SiRhoNox-1 (1 μM) for 3 h in the dark to allow the probes to penetrate. The seedlings were briefly washed in MES-KOH buffer and mounted on slides using double-sided tape as a spacer. A Z-stack of images was acquired by confocal microscopy (7–10 min per root zone) and was analyzed with Fiji software for signal intensity quantification. Data were subsequently exported for statistical analysis and charting

then dried in the sterile hood before being sown on 0.5× MS medium. After stratification at 4°C for two days, the plates are placed vertically in a growth culture chamber. The seedlings are grown for 7 days.

2. Transfer five 7-day-old seedlings to a six-well plate containing 2 ml of MES-KOH buffer (*see* **Note 5**).
3. Add the two dye solutions prior to adding the seedling in the MES-KOH buffer to get the following final concentrations: 10 μM MP-NBD and 1 μM SiRhoNox-1.
4. After 3 h of incubation at room temperature and in the dark, rinse the seedling with MES-KOH buffer. If there are several genotypes or conditions, stagger the start of the incubations according to the time required for microscope acquisitions (*see* **Note 6**).

An additional counterstaining can be used to identify tissue or subcellular compartment (*see* **Note 7**).

3.3 Sample Mounting

1. Cut two thin strips of double-sided tape, and stick them along each border of the slide to make spacers preventing the root from being flattened (Fig. 1).
2. Mount the 7-day-old seedlings gently on the slide containing MES-KOH buffer between the two spacers, then put the coverslip gently while avoiding air bubbles (*see* **Note 8**).

3.4 Image Acquisition Setting

1. MP-NBD: excitation at 488 nm, emission between 500 and 550 nm
2. SiRhoNox: excitation at 633 nm, emission between 650 and 700 nm

   In the far-red wavelength range, an autofluorescent punctuated signal from unstained roots can be observed (Fig. 2, *see* **Note 9**).
3. Perform the acquisitions in sequential line mode
4. Choose the laser power to allow a good dynamic range of the signal and avoid saturation. The power value for each beam line is fixed for all experiments.
5. Set the detectors to 12-bit and the scan speed to 400 Hz.
6. Rotate the scan head to orient the root horizontally in the field of view and select the resolution format 1024 × 512.
7. Ensure that all the roots are visible in the image by selecting a zoom magnification between 0.75 and 1.
8. Perform a z-stack acquisition with optimal axial resolution to collect the signal from the epidermis to the central cylinder (approximate distance of 65 μm; Fig. 3) (*see* **Note 10**). Both 3D and orthogonal views reveal the Fe distribution and the variation in redox state across the different cell layers (Figs. 3 and 4).

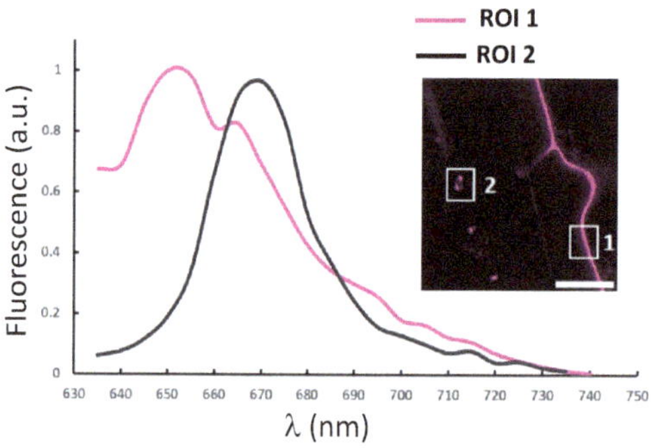

Fig. 2 Spectral emission of root stained with SiRhoNox-1. Normalized fluorescence emission spectra are recorded from two regions of interest (ROIs) indicated in the zoomed image upon excitation at 630 nm. The signal is collected from 635 to 740 nm with a bandwidth of 10 nm. The SiRhoNox-1 signal corresponds to ROI 1 (magenta line), while the autofluorescence corresponds to ROI 2 (dark line). The autofluorescence emission spectra are red-shifted compared to SiRhoNox-1. Scale bar = 20 μm

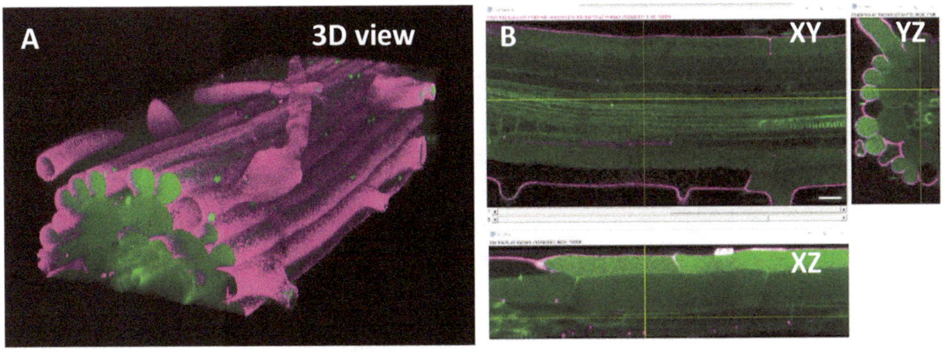

Fig. 3 3D visualization of Fe redox distribution in root of Arabidopsis. (**A**) 3D reconstruction created with LASX 3D viewer software. (**B**) Multi-dimensional view obtained with the orthogonal view plugin in Fiji, which dynamically projects XZ and YZ planes along the yellow cross. MP-NBD: green, SiRhoNox-1: magenta. Scale bar = 20 μm

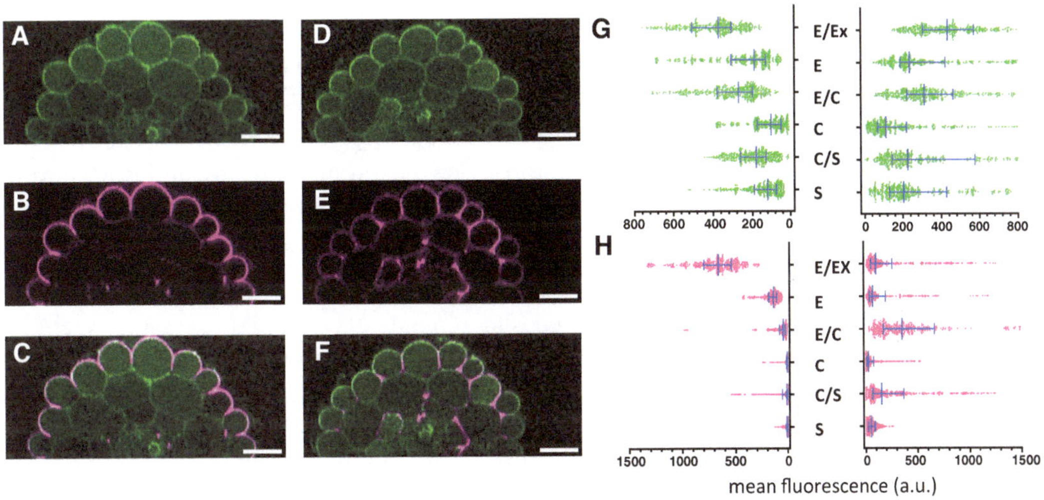

Fig. 4 Distribution of labile Fe in the different cell layers of the primary root of *Arabidopsis thaliana*. Representative orthogonal view of the young (**A**–**C**) and the mature differentiation zones (**D**–**F**) of the primary root stained with MP-NBD (Fe^{3+}, A and D), SiRhoNox-1 (Fe^{2+}, **B** and **E**), and a merged image of both probes (**C** and **F**). Quantification of MP-NBD (**G**) and SiRhoNox-1 (**H**) signals is shown for the root zones (young zone: left panel, mature zone: right panel). The MP-NBD signal is present in all cell layers, whether in the apoplast or inside the cells. SiRhoNox-1 signal is mainly localized in the epidermis apoplast. A notable result is the polarization of the labeling in the apoplast at the external apoplast side in the young zone, and at the cortex side in the mature zone (see Alcon et al. [9] for more details). Fe^{2+} can also be detected at the Casparian strip of few cells. Green LUT: MP-NBD; Magenta LUT: SiRhoNox-1. E/Ex epidermis/external, E epidermis, E/C epidermis/cortex, C cortex, C/S cortex/stele, S stele. Scale bar = 20 μm

9. Observe two root regions to account for the heterogeneous distribution of Fe: the young differentiation zone and the old differentiation zone (referred to as the mature zone) (Fig. 4).

These two zones are identified based on the apoplast-polarized staining of SiRhoNox-1 at the epidermis [9].

10. Select the field of view under the eyepieces in bright-field mode, as Far-Red light is not visible to the naked eye. The young differentiation zone corresponds to the region close to the elongating zone, where the epidermis bears mature root hair. The mature zone corresponds to the region of a stage VI lateral root primordium [12].

3.5 Image Processing and Quantification Analysis

All image analyses are conducted using Fiji software [13]. As no fully automatic image analysis tool is available for this type of dataset, we developed a semi-automatic method to segment the regions of the image corresponding to the different cell layers and interfaces between root cell types (*see* **Note 11**).

3.5.1 Orthogonal Image Generation

1. Open the z-stack image as split channels: (Drag and drop your confocal image file in Fiji bar menu and select "split channels" on the Bio-Format Import Options window).
2. Open the color Merge Channel tool (Image → Color → Merge Channels), and merge the MP-NBD channel as "C2 green channel" and SiRhoNox-1 channel as "C6 magenta channel").
3. Make an orthogonal view picture to quantify fluorescence from the different cell layers: draw a line vertically through the root and do Reslice (Image → stack → Reslice[/]). Set Slice count to 1.
4. Save the orthogonal view as Tiff image for further analysis.
5. Repeat these steps to generate another orthogonal view, ensuring that the selection is sufficiently distant from the previous one to include different cells.

3.5.2 Segmentation of the Different ROIs of the Root

1. Download and install the plugin "*Radial Profile Angle*" created by Philipp Carl (http://questpharma.u-strasbg.fr/html/radial-profile-ext.html) (*see* **Note 4**).
2. Open the orthogonal view file, find the center of the root (X, Y pixel coordinates), and report these coordinates in the macro below: replace "x_center" and "y_center" values with the X and Y coordinates. The default values are 256, which are shown in highlighted gray in the macro below.
3. Adjust the radius (default value: 220, highlighted gray in the macro below) so that it covers the entire root. In the orthogonal view, the root is considered as a circle, and only the upper half of this circle is analyzed (*see* **Note 10**).
4. Run the macro "*Piece of Pie*" in Fiji to generate the six pie-shaped ROIs: (Plugins → Macros → Startup Macros). Copy and paste the macro code below into the StartupMacros windows and then run the macro to generate six ROIs as shown in the Fig. 5 (*see* **Note 12**).

```
// === Macro Piece of Pie ===
run("ROI Manager...");
imgTitle = getTitle();
k = 1;
setBatchMode(true);
for (n = 0; n < 180; n = n + 30) {
p = 15 + n;
run("Radial Profile Angle", "x_center=256 y_
center=256 radius=220 starting_angle=" + p + "
integration_angle=15");
selectWindow(imgTitle);
roiManager("Add");
index = roiManager("count") - 1;
roiManager("select", index);
roiManager("Rename", "Pie" + k);
k++;
}
setBatchMode(false);
```

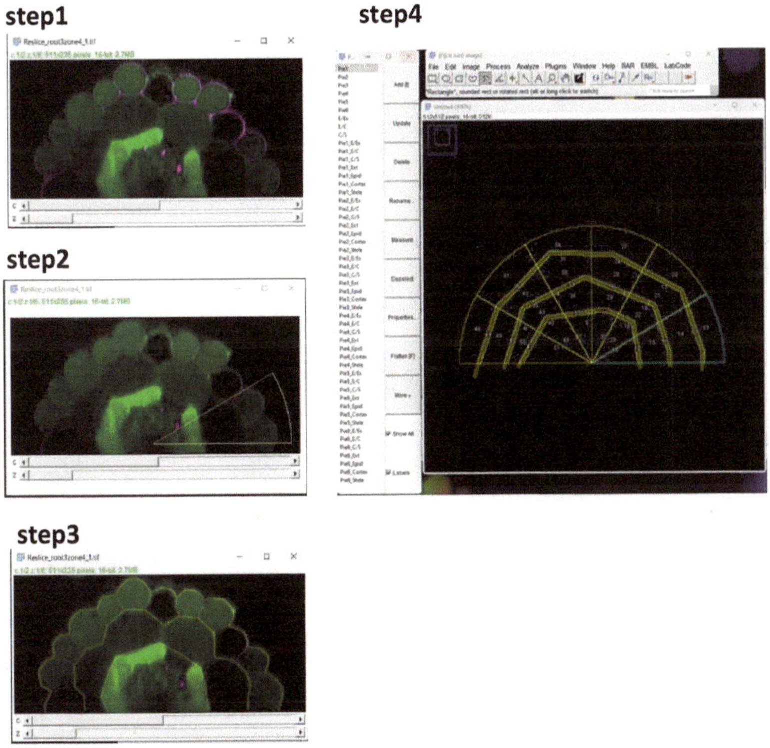

Fig. 5 Segmentation of orthogonal image of root stained with the two Fe probes. The different steps of the segmentation with Fiji macros are shown with screenshots. Step 1: use the "Reslice [/]" tool to generate the orthogonal view image from a z-stack acquisition. Step 2: run the "*Piece of pie macro"* to generate six ROIs. Step 3: draw three ROIs between cell layers with the line segmented tool. Step 4: run the second macro "*Cherry on the cake*" and measure the mean gray value for selected ROIs. The values are then exported in an excel file (or equivalent).

5. Use the segmented line selection tools with line width of 3 to mark each interface between the different cell layers (e.g., Epidermis/External; Epidermis/Cortex; Cortex/Stele). Accurate quantification depends on precise selection (*see* **Note 13**).
6. Rename each interface ROI as follows:
   Epidermis /External: E/Ex
   Epidermis/Cortex: E/C
   Cortex/Stele: C/S
7. Run the second macro "*Cherry on the cake*" below. Open a new macro window and copy and paste the macro code below (File →New) (*see* **Note 14**)

```
// === Macro Cherry on the cake ===
pieNames = newArray("Pie1", "Pie2", "Pie3", "Pie4",
"Pie5", "Pie6");
lineNames = newArray("E/Ex", "E/C", "C/S");
// === FUNCTION TO FIND ROI INDEX ===
function findROI(targetName) {
n = roiManager("count");
for (i = 0; i < n; i++) {
roiManager("select", i);
name = Roi.getName();
if (name == targetName) {
return i;
}
}
return -1;
}
// === PROCESS ===
setBatchMode(true);
originalTitle = getTitle();
getDimensions(width, height, channels, slices,
frames);
// For each Pie
for (p = 0; p < pieNames.length; p++) {
pieName = pieNames[p];
pieIndex = findROI(pieName);


if (pieIndex == -1) {
print("ATTENTION: Pie '" + pieName + "' Not find!");
continue;
}


print("=== Processing of " + pieName + " ===");
// === STEP 1: Extraction and Rename interface
(lines/pie) ===
delimROIs = newArray();
```

```
for (l = 0; l < lineNames.length; l++) {
lineName = lineNames[l];
lineIndex = findROI(lineName);


if (lineIndex == -1) {
print(" ATTENTION: Line '" + lineName + "' Not
find!");
continue;
}


// Intersection Pie + line = Interface
roiManager("select", newArray(pieIndex,
lineIndex));
roiManager("AND");


if (selectionType()!= -1) {
roiManager("Add");
newIndex = roiManager("count") - 1;
roiManager("select", newIndex);
roiManager("Rename", pieName + "_" + lineName);
delimROIs = Array.concat(delimROIs, newIndex);
print(" Interface " + pieName + "_" + lineName + "
create");
}
}


// === STEP 2: Create mask FULL SIZE to cut Pie
into zones ===
if (delimROIs.length == 3) {
// Create black image same size as original
newImage("mask_" + pieName, "8-bit black", width,
height, 1);


// Draw Pie in white
roiManager("select", pieIndex);
setForegroundColor(255, 255, 255);
run("Fill", "slice");


// Draw lines in black to split zones
setForegroundColor(0, 0, 0);
setLineWidth(6); // width to split properly
for (l = 0; l < delimROIs.length; l++) {
roiManager("select", delimROIs[l]);
run("Fill", "slice");
}


// Analyse particles to identify split zones
```

```
run("Select None");
countBefore = roiManager("count");


// Analyze Particles to detect white zones
run("Analyze Particles...", "size=100-Infinity add");


countAfter = roiManager("count");
nbZones = countAfter - countBefore;


print(" -> " + nbZones + " detected zones by
Analyze Particles");
// Table of zone names
zoneNames = newArray("Ext", "Epid", "Cortex",
"Stele");


// Rename zones
for (i = 0; i < nbZones; i++) {
roiIndex = countBefore + i;
if (roiIndex < roiManager("count")) {
roiManager("select", roiIndex);


// Rename ROIs with zone names
roiManager("Rename", pieName + "_" + zoneNames[i]);


print(" " + pieName + "_" + zoneNames[i] + "
create (index " + roiIndex + ")");
}
}
// close mask
close("mask_" + pieName);
selectImage(originalTitle);


totalROIs = 3 + nbZones;
print(" TOTAL for " + pieName + " : " + totalROIs
+ " ROIs (3 interfaces + " + nbZones + " zones)");


if (nbZones!= 4) {
print(" ATTENTION: Number of zones expected = 4,
found = " + nbZones);
print(" Check lines are crossing all Pies");
}
} else {
print(" ERREUR: Cannot create 3 interfaces " +
pieName);
}
```

```
print("");
}
setBatchMode(false);
roiManager("Show All");
```

8. The "*Cherry on the cake*" macro generates 42 additional ROIs in the ROI manager. A screenshot of the ROIs obtained after running the two macros is shown in Fig. 5. Double-check their names and positions. The Log file records every step of the macro. Each Pie should generate seven new ROIs.

3.6 Data Extract and Analysis

1. Delete the original ROIs (the Pie and the interface, meaning the first 9 ROIs) and save the newly generated 42 ROIs for further analysis.
2. Select all the 42 ROIs and measure the mean gray value.
3. The resulting data can then be exported to an Excel file (or equivalent) for charting and statistical analysis.

4 Notes

1. Epifluorescent wide-field microscopy is also an alternative for imaging. However, it is important to note that the metal-halide lamp used for fluorophore excitation induces a strong photobleaching of MP-NBD.
2. Long working-distance objectives are more efficient at collecting fluorescent signals from deeper regions and water-immersion objectives are more appropriate for the refractive index of the mounting medium.
3. The two macros developed in this study only work with Fiji version 1.54p or later.
4. To install the plugin, download the file "*Radial Profile Angle.jar*" from the following website: http://questpharma.u-strasbg.fr/html/radial-profile-ext.html. Then drop it in the *plugin* directory before restarting Fiji.app.
5. 10 mM MES-KOH buffer is used to maintain a pH equal to 6.0, which is close to the physiological pH.
6. A three-hour incubation is needed for complete penetration of the dyes into 7-day-old Arabidopsis roots. However, the incubation time can be adjusted depending on the sample.
7. The counterstain should exhibit fluorescence properties that do not overlap with MP-NBD and SiRhoNox-1. For example, propidium iodide can be used to label cell walls. This step is performed just before mounting the sample between slide

and coverslip. After incubation with the Fe probes, transfer the seedling to a solution of propidium iodide (2 μg/mL) for 10 min. For the confocal acquisition, add an additional channel with the following parameters: excitation at 561 nm and emission between 580 and 630 nm. Note that the diffusion of propidium iodide in the differentiation zone is restricted by the Casparian strip.

8. Avoid positioning the cotyledons between the slide and coverslip to keep the sample setup as flat as possible. When the coverslip becomes tilted due to the sample thickness, the fluorescent signal decreases significantly, and cellular structures may appear blurred.
9. Autofluorescence in the far-red spectral range can be distinguished from the SiRhoNox-1 fluorescence by using spectral acquisition or fluorescence-lifetime imaging.
10. Fluorescence signal from deeper tissues is reduced due to absorption of both excitation and emission photons by the sample thickness. For this reason, imaging only half the depth of the root along the Z-axis is sufficient and provides accurate data for quantification.
11. Manual selection with Image J tools may also be employed for quantification of fluorescence intensity values.
12. Segmenting the root into multiple Pie ROIs captures the heterogeneity in signal distribution and intensity within the root
13. Ensure that lines cross all Pie ROIs. Adjust the line width to a value between 2 and 5 for optimal results.
14. Before running the macro, ensure that the orthogonal image is open and that six "Pie" ROIs (Pie1–Pie6) and three "interface" ROIs (E/Ex, E/C, C/S) have been created and stored in the ROI Manager.

Acknowledgments

Tou Cheu Xiong has received research grant from the French Centre national de la Recherche Scientifique through the MITI interdisciplinary programs (Metallomix Grant 2021 DesciFer).

We acknowledge the Histocytology and Plant Cell imaging facility (PHIV-La Gaillarde) for technical support. We acknowledge the imaging facility MRI, member of the national infrastructure France-BioImaging (https://ror.org/01y7vt929) supported by the French National Research Agency (ANR-24-INBS-0005 FBI BIOGEN).

Conflict of Interest The authors have no conflicts of interest to declare that are relevant to the content of this chapter.

References

1. Briat J-F, Dubos C, Gaymard F (2015) Iron nutrition, biomass production, and plant product quality. Trends Plant Sci 20:33–40
2. Marschner H, Römheld V (1994) Strategies of plants for acquisition of iron. Plant Soil 165:261–274
3. Rogers EE, Guerinot ML (2002) FRD3, a member of the multidrug and toxin efflux family, controls iron deficiency responses in Arabidopsis. Plant Cell 14:1787–1799
4. Roschzttardtz H, Conéjéro G, Curie C, Mari S (2009) Identification of the endodermal vacuole as the iron storage compartment in the Arabidopsis embryo. Plant Physiol 151:1329–1338
5. Kopittke PM, Punshon T, Paterson DJ et al (2018) Synchrotron-based X-ray fluorescence microscopy as a technique for imaging of elements in plants. Plant Physiol 178:507–523
6. Terzano R, Mimmo T, Vekemans B et al (2013) Iron (Fe) speciation in xylem sap by XANES at a high brilliant synchrotron X-ray source: opportunities and limitations. Anal Bioanal Chem 405:5411–5419
7. Sahoo Crisponi (2019) Recent advances on Iron(III) selective fluorescent probes with possible applications in bioimaging. Molecules 24:3267
8. Hirayama T, Nagasawa H (2017) Chemical tools for detecting Fe ions. J Clin Biochem Nutr 60:39–48
9. Alcon C, Comte A, Curie C, Xiong TC (2024) Imaging of labile Fe^{2+} and Fe^{3+} in living *Arabidopsis thaliana roots*. Plant Physiol 195:2520–2523
10. Park M-J, Jung H-S, Kim Y-J et al (2014) High-sensitivity fluorescence imaging of iron in plant tissues. Chem Commun 50:8547–8549
11. Hirayama T, Tsuboi H, Niwa M et al (2017) A universal fluorogenic switch for Fe(II) ion based on N-oxide chemistry permits the visualization of intracellular redox equilibrium shift towards labile iron in hypoxic tumor cells. Chem Sci 8:4858–4866
12. Malamy JE, Benfey PN (1997) Organization and cell differentiation in lateral roots of *Arabidopsis thaliana*. Development 124:33–44
13. Schindelin J, Arganda-Carreras I, Frise E et al (2012) Fiji: an open-source platform for biological-image analysis. Nat Methods 9:676–682

Part III

Analysis of Transcript and (Phospho) Protein Levels

Check for updates

Chapter 16

Whole-Mount In Situ Hybridization Using Microwave Treatment

Liudmila Chelysheva and Jean-Christophe Palauqui

Abstract

In situ mRNA hybridization, which relies on the complementarity between antisense riboprobes and specific RNA populations, was originally used on thin sections with colorimetric dyes to visualize mRNA distribution. The use of fluorescent probes and confocal imaging enabled whole-mount in situ hybridization (WISH) and 3D reconstruction of gene expression domains. Here, we present a versatile protocol based on microwave treatment that results in high and homogeneous hybridization yields. When used with a cell wall marker and tissue clearing, it enables simultaneous visualization of gene expression and cell segmentation across various Arabidopsis organs, including leaves, shoots, roots, and embryos enclosed within integuments. This easy, rapid, robust, and low-cost method also accommodates different detection kits and cytological dyes, and it can be extended to Double WISH for the simultaneous detection of two distinct mRNA targets.

Keywords Arabidopsis, In situ hybridization, Gene expression, mRNA, Cell resolution, Fluorescence, Embryo, Shoot meristem, Leaf, Root

1 Introduction

Modern integrative biology requires fast, sensitive, and versatile tools to visualize gene expression within its spatial and temporal context. Among the most widely used approaches, transgenic reporter constructs coupled to fluorescent proteins allow tracking transcriptional dynamics in vivo. However, generating and analyzing reporter lines is labor-intensive: producing transgenic plants, selecting effective reporters, and introducing them into different genetic contexts require significant time and effort. Moreover, reporter expression may be biased due to incomplete promoter sequences or ectopic genomic insertion, leading to expression variability [1].

mRNA in situ hybridization, based on the complementarity between antisense riboprobes and endogenous cellular mRNAs, has long served as the reference method to reveal transcript accumulation across tissues and genetic backgrounds [2, 3]. Initially performed

Steffen Vanneste and Inge Verstraeten (eds.), *Root Development*, Methods in Molecular Biology, vol. 3086,
https://doi.org/10.1007/978-1-0716-5603-7_16,

on tissue sections with colorimetric detection, the method has continually evolved despite remaining technically demanding [4]. The development of fluorescent detection using tyramide signal amplification (TSA), combined with cell wall and nuclear markers and imaged by confocal microscopy, significantly improved cellular and subcellular resolution [5].

To avoid the constraining steps of embedding and sectioning, whole-mount in situ hybridization (WISH) protocols have been developed for permeable tissues or easily accessible organs such as roots, embryos, and meristems [6, 7]. Moreover, every new organ requires adapted tissue permeabilization to ensure a homogeneous penetration of probes and antibodies. Here, we present an improved and universal protocol—M2WISH (Microwave treatment for Whole-mount mRNA In Situ Hybridization)—that is simple, robust, and directly applicable to a wide variety of Arabidopsis organs, including embryos, roots, leaves, apical meristems, and floral meristems and also implemented for Brachypodium [8]. This method ensures high, homogeneous hybridization yields and unambiguous identification of both expressing and nonexpressing cells, greatly reducing the risk of false negatives. By combining alkaline phosphatase (AP) enzymatic amplification with dual colorimetric and fluorescent readout, M2WISH enables multi-scale analysis, from whole-organ visualization to cellular resolution after clearing, cell wall staining, and confocal imaging. The protocol can also be extended to double WISH, enabling co-expression analysis of transcripts from two distinct genes.

A key challenge in designing a WISH protocol that is broadly compatible with diverse tissues is to optimize tissue permeabilization, enabling uniformly access of riboprobes and antibodies to cells while preserving tissue integrity. Traditional protocols often employ enzymatic digestion of the cell wall to increase permeability [7]. However, this step requires organ-specific adjustments and frequently weakens the samples, particularly during high-temperature hybridization.

To overcome this limitation, we postponed the digestion step until after hybridization and introduced a more universal approach: microwave (MW) treatment during the dehydration step. MW treatment is known to enhance protein accessibility in chromosome preparations, while hot ethanol effectively delipidates the cuticle of aerial Arabidopsis tissues at all developmental stages [9, 10]. Both treatments provide efficient permeabilization without compromising tissue integrity.

Using the M2WISH protocol, we achieved consistent penetration of riboprobes across a broad range of organs, including embryos, roots, lateral roots, leaves, apical meristems, and floral meristems. The MW-based permeabilization results in improved accessibility to target RNAs while preserving fine cellular structures, thereby ensuring reliable and reproducible detection even in dense or naturally low-permeability tissues (Fig. 1).

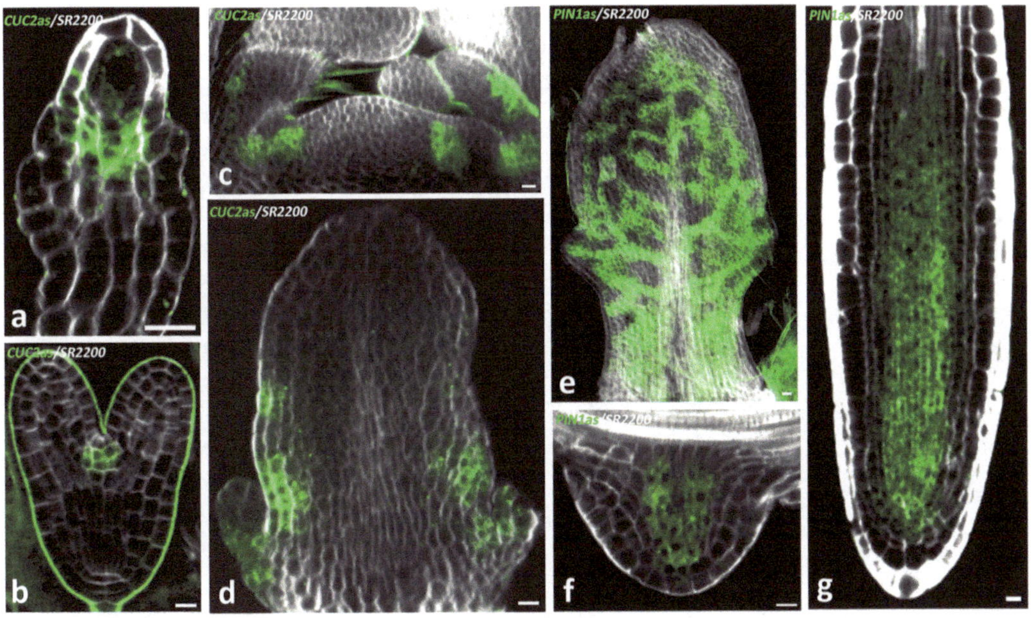

Fig. 1 *CUC2* and *PIN1* mRNA detection in ovules, embryo, SAM, leaves and roots. *CUC2* expression at the base of the nucellus in developing ovule (**a**), in embryo at heart stage (**b**), in SAM of 4-week-old, greenhouse-grown plants (**c**) and in the sinus of leaf serrations (**d**). *PIN1* expression along the pre-pattern of vascular network in a young leaf (**e**), in a lateral (**f**), and primary roots (**g**). Cell walls are stained with SR2200. Scale bar 10 μm

For probe labeling and detection, we implemented a cost-effective and highly reproducible strategy using DIG-UTP or FITC-UTP incorporation during in vitro transcription, followed by detection with anti-DIG or anti-FITC Fab fragments conjugated to alkaline phosphatase (AP).

To reveal AP activity, we employed the Vector® Blue Substrate Kit, which provides a blue colorimetric signal, allowing easy monitoring of the reaction in real time and a Far-Red fluorescent signal, ideal for confocal imaging due to minimal plant tissue autofluorescence in this spectral range. This dual detection system enables complementary levels of analysis: either a rapid macroscopic inspection via the colorimetric signal or high-resolution confocal imaging via fluorescence.

However, tissue thickness can limit fluorescence detection due to absorption and scattering. To resolve this, samples are incubated in ClearSee, a clearing reagent optimized for plant tissues [11]. After 1 week of clearing, strong and homogeneous fluorescence becomes visible throughout the entire z-stack, even in deeply embedded layers. A major advantage of ClearSee-treated samples is their long-term stability: tissues can be stored for months at room temperature without detectable signal loss, providing substantial experimental flexibility.

M2WISH can be extended to perform double hybridization, enabling simultaneous detection of two different transcripts within the same sample. This is achieved by preparing two riboprobes labeled with digoxigenin (DIG) and fluorescein (FITC).

Following co-hybridization, samples are incubated with a mixture of antibodies anti-DIG conjugated to AP and anti-FITC conjugated to POD (peroxidase). The detection proceeds in two sequential, but compatible steps: (I) POD detection using Alexa Fluor™ 488 Tyramide, producing a sharp and intense green fluorescent signal, (II) AP detection using Vector® Blue, producing both a colorimetric and Far Red fluorescent signal.

Importantly, this setup does not require enzymatic quenching between the two detection steps, simplifying the workflow and reducing sample loss. The lack of cross-reactivity between AP and POD ensures clean, nonoverlapping signals, enabling accurate colocalization analysis (Fig. 2). A minor but consistent observation is that AP detection induces additional cuticle staining regardless of the probe or organ. However, since this staining occurs only at the tissue surface, it does not interfere with interpreting internal transcript localization.

2 Materials

All solutions should be prepared using analytical-grade reagents and ultrapure water (18 MΩ·cm at 25°C, purified and deionized), unless otherwise stated. Dispose of all chemical and biological waste according to local institutional and governmental regulations.

2.1 Equipment

1. GeneScreen Plus Hybridization Transfer Membrane.
2. 20 mL glass vial with screw cap (*see* **Note 1**).

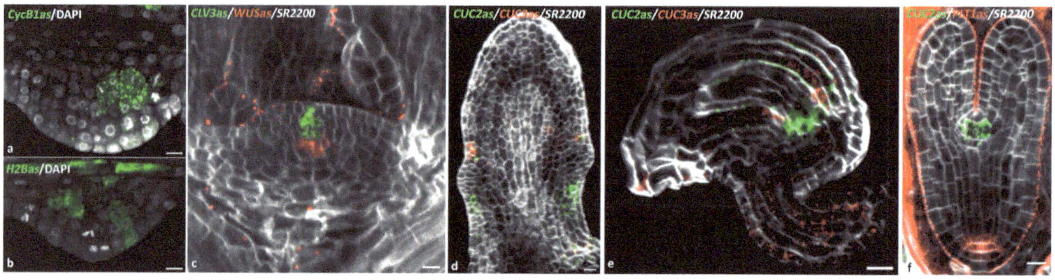

Fig. 2 Examples of combined counterstaining and co-detection in different Arabidopsis organs. Detection of *CYCB1;1* mRNA in the zone of dividing nuclei (**a**) and of *H2B* mRNA outside the zone of dividing nuclei (**b**) in a lateral root (nucleus counterstained with DAPI). Co-detection of *CLV3* and *WUS* mRNA in SAM (**c**). *CLV3* transcripts detected with Alexa Fluor™ Tyramide Reagent (POD activity) and *WUS* transcripts detected with Vector® Blue Substrate Kit (AP activity). *CUC2* and *CUC3* mRNA co-detection in leaf (**d**), and ovule (**e**). *CUC2* transcripts detected with Alexa Fluor™ Tyramide Reagent (POD activity) and CUC3 transcripts detected with Vector® Blue Substrate Kit (AP activity). *CUC2* and *PLT1* mRNAs co-detection in an embryo at the torpedo stage (**f**). *CUC2* transcripts detected with Alexa Fluor™ Tyramide Reagent (POD activity) and *PLT1* transcripts detected with Vector® Blue Substrate Kit (AP activity). Cell wall counterstained with SR2200. Scale bar 10 μm

3. Polypropylene 50 mL sterile conical tubes.
4. Sterile syringe filter with a pore size of 0.2 μm.
5. Untreated 24-round well multi-dish.
6. Polycarbonate cell culture inserts with 8-micron porosity in a 24-well plate.
7. Vacuum degassing system ensuring negative pressure at −0.09 MPa.
8. Fume hood.
9. Microwave oven, set to 180 W.
10. Dry bath incubator for 1.5 mL microcentrifuge tubes.
11. Water bath with horizontal shaking at 47°C and oven at 37°C.
12. Double-sided tape refill rolls (Overall Width = 18 mm; Overall Length = 32.2 m).
13. Confocal microscope, equipped with appropriate lasers and filters.

2.2 General Solutions and Buffers

1. 5 M NaCl: Dissolve 138.6 g of NaCl in 500 mL of distilled water and autoclave.
2. 0.5 M EDTA: Add 14.61 g of EDTA to 100 mL of distilled water, and add NaOH pellets one by one when stirring the solution. The EDTA salt will not go into solution until the pH reaches 8.0 (*see* **Note 2**).
3. 10×Phosphate-Buffered Saline (10× PBS): 1.37 M NaCl, 27 mM KCl, 100 mM Na_2HPO_4, and 18 mM KH_2PO_4 in water. Dilute tenfold in water to achieve 1×PBS.
4. 20×Saline Sodium Citrate (20× SSC): 3 M NaCl, 0.3 M Sodium Citrate in water. Set to pH 7.0.
5. 20% Sodium dodecyl sulfate (20% SDS): Dissolve 200 g of sodium dodecyl sulfate powder in 1 L of water.
6. 1 M Tris-HCl pH 7.5, 8.0, and 8.2: Dissolve 1059 g of Tris base in 800 mL of water, then adjust the pH to 7.5, 8.0, and 8.2 by adding 38% HCl. Top up to 1 L with water and then autoclave.

2.3 RNA Probe Synthesis

1. 1 μg of DNA template for your probe using 5′ gene-specific primers and 3′ gene-specific primers extended with the sequence of T7 promoter (*see* **Note 3**).
2. Digoxigenin-11-UTP (Roche, Cat No: 11209256910) and/or Fluorescein-12-UTP (Roche, Cat No: 11373242910).
3. Ribonucleotide Triphosphates (10 mM of rATP, rGTP, rUTP, and rCTP).

4. 20 U/μL T7 RNA polymerase.
5. RNasin® ribonuclease inhibitor.
6. Nuclease-free water.
7. RQ1 RNase-free DNase.
8. tRNA stock: Dissolve all the tRNA (from baker's yeast) powder in your packaging in water to a final concentration of 10 mg/mL. Aliquot in 1 mL volumes and store at −20°C.
9. 10 M Ammonium acetate in water.
10. Carbonate buffer: Mix 120 mM Na_2CO and 80 mM $NaHCO_3$ in nuclease-free water, adjust to pH 10.2.
11. 10% Acetic acid in water.
12. 3 M Sodium acetate in water.
13. Deioinized Formamide.

2.4 Hybridization, Washing, and RNA Probe Detection

1. Blot Incubation Buffer: 10 mM Tris-HCl pH 7.5, 15 mM NaCl, 1% BSA, 0.5% Triton™ X-100 in water (*see* **Notes 4 and 5**).
2. Blot wash buffer: 100 mM Tris-HCl pH 8.2; 0.1% TWEEN®20 (*see* **Notes 4 and 5**).
3. Anti-Fluorescein-AP, Fab fragments (Roche, Cat No: 11426338910) and Vector® Blue Substrate Kit, Alkaline Phosphatase (AP) (Vector Laboratories, Cat No: SK-5300).
4. PBS-T: Mix 5 mL of 10× PBS, 500 μL of 10% Triton™ X-100 with water. Top up to 50 mL with water (*see* **Notes 4 and 5**).
5. Anti-Digoxigenin-POD, Fab fragments (Roche, Cat No: 11207733910), and Kit Alexa Fluor™ 488 Tyramide SuperBoost™, Peroxidase (POD) (Thermo Fisher Scientific, Cat No: B40943) (*see* **Note 6**).
6. Pre-hybridization solution: Mix 20 mL of deionized Formamide, 5 mL of 20×SSC and 400 μL 10% TWEEN®20 in water. Top up to 40 mL with water (*see* **Notes 4 and 5**).
7. Salt Solution: Mix 60 mL of 5 M NaCl, 10 mL of 1 M Tris-HCl pH 8.0, 2 mL of 0.5 M EDTA, and add water up to 100 mL. Make 1 mL aliquots. Store at −20°C.
8. Hybridization solution: Mix 10 mL of deionized Formamide, 4 mL of 50% Dextran sulfate (sodium salt, M.W. ~500,000), 2 mL of tRNA stock, 1.5 mL of TWEEN® 20, 500 μL of 100× Denhardt's solution, 200 μL of Salt Solution, and 1.8 mL of sterile water (*see* **Note 7**). Store at −20°C.
9. 0.1× SSC, 0.5% SDS: Mix 200 μL of 20× SSC, 1 mL of 20% SDS in water. Top up to 40 mL with water (*see* **Note 4**).
10. 2× SSC, 50% Formamide: Mix 20 mL of deionized Formamide and 4 mL of 20× SSC in water. Top up to 40 mL with water (*see* **Note 4**).

11. TNE: Mix 500 µL of 1 M Tris-HCl pH 8.0, 100 µL of 0.5 M EDTA, and 100 µL of 5 M NaCl in water. Top up to 50 mL with water (*see* **Note 4**).
12. RNase A stock: Dissolve all the RNase A powder (from bovine pancreas) in your packaging in water to a final concentration of 10 mg/mL. Aliquot in 1 mL volumes and store at −20°C.
13. PBS-TB: Mix 50 mL of 10× PBS, 5 mL of 10% Triton™ X-100, and 5 g of Bovine Serum Albumin (BSA) in water. Top up to 500 mL with water. Aliquot in 20 mL volumes, store at −20°C (*see* **Note 5**).
14. For Two Probe Signal detection: Tyramide Conjugates Alexa Fluor™ 488 (Thermo Fisher Scientific, Cat No: B40953).

2.5 Sample Fixation, Permeabilization, and Digestion

1. PFA Fixative: Add 4 g Paraformaldehyde (PFA) powder under a fume hood to 100 mL 1×PBS. Stir the solution at 80°C in a water bath in a fume hood until completely dissolved (~1 h). Make 10 mL aliquots and freeze (*see* **Note 8**).
2. Ethanol/Water dilution series: Mix EtOH (absolute) with water to reach the following EtOH concentrations 10, 30, 50, 70%.
3. PBS-Tween: Mix 5 mL of 10× PBS, 500 µL of 10% TWEEN®20 in water in a total volume of 50 mL (*see* **Notes 4 and 5**).
4. Proteinase K stock: Dissolve all the Proteinase K powder (recombinant, PCR grade) in your packaging in water to a final concentration of 10 mg/mL. Aliquot in 1 mL volumes and store at −20°C.
5. 10% Glycine stock: Dissolve 10 g glycine powder in 100 mL of water, then aliquot by 3 mL. Store at −20°C. Dilute 400 µ L of 10% Glycine stock in 20 mL of 1×PBS to obtain a glycine working solution
6. Triethanolamine/Acetic anhydride working solution: Dilute 300 µL of Triethanolamine in 19 mL of sterile water, add 70 µL of 37% HCL, add 1 mL of Acetic anhydride (*see* **Note 9**).
7. RCL2®, formalin-free fixative (Exilone, Cat No: R1505J).
8. Enzymatic mix: Dissolve 250 mg of Macerozyme R10 (Duchefa, Cat No: M8002.0005), 250 mg of Cellulase RS (Duchefa, Cat No: C803.005), 125 mg of Pectolyase Y23 (Duchefa, Cat No: P9004.0001), and 375 mg of Pectinase (Serva, Cat No: 31660) in 50 mL of water. Filter the solution using a syringe filter with a pore size of 0.2 µm, then aliquot in 5 mL volumes. Store at −20°C.

2.6 Signal Visualization

1. ClearSee solution: Place 10 g of xylitol, 15 g of sodium deoxycholate, and 25 g of OmniPur® Urea in a 250 mL laboratory screw-top bottle. Add water to a final volume of 100 mL and stir (*see* **Note 10**).

2. Fluorescent brightening 28 stock: Dissolve Fluorescent Brightener 28 at 50 mg/mL in DMSO. Aliquot in 1 mL volumes. Store at −20°C protected from light.
3. Direct Red 23 stock: Prepare a 10 mg/mL Direct Red 23 solution in water, aliquot in 1 mL volumes. Store at 4°C protected from light.
4. DAPI stock: Prepare a 1 mg/mL 4′,6-diamidino-2-phenylindole (DAPI) solution by dissolving the contents of one vial (10 mg) in 10 mL of deionized water, then aliquot in 1 mL volumes. Store at −20°C protected from light.
5. SR 2200 (Renaissance Chemicals).
6. Citifluor Antifadent Mountant Solutions, AF1 (Agar Scientific).
7. VECTASHIELD® Antifade Mounting Medium (Vector Laboratories).

3 Methods

3.1 Probe Labeling by In Vitro Transcription from a PCR Amplified Fragment

1. 1 μg DNA template with T7 promoter for your probe.
2. Set up a reaction for in vitro transcription (*see* Table 1).
3. Incubate at 37°C for 1 h 30 min.
4. Add 2 μL of RNAse-free DNAse and 68 μL of nuclease-free water.
5. Incubate for 30 min at 37°C.

Table 1
Reaction mix for a single in vitro transcription reaction

DNA template (1 μg) up to	9 μL
Transcription optimized 5×Buffer	4 μL
10 mM rATP	1.5 μL
10 mM rGTP	1.5 μL
10 mM rCTP	1.5 μL
10 mM rUTP	0.9 μL
10 mM DIG (F)-UTP	0.6 μL
RNAse inhibitor	1 μL
T7 RNA polymerase	2 μL
Nuclease-free water	up to 20 μL

6. Precipitate the probe by adding 10 μL of 10 mg/mL tRNA stock, 20 μL of 10 M Ammonium Acetate, 240 μL of absolute EtOH, in a 1.5 mL tube, and incubate overnight at −20°C.
7. Centrifuge at 13,000 × *g* for 30 min at 4°C. Remove supernatant by decanting.
8. Wash the RNA pellet by adding 1 mL of 70% of EtOH, followed by 10 min centrifugation (13,000 × *g*) at 4°C. Remove supernatant as much as possible by decanting and pipetting.
9. Dry pellet by leaving the tube open at room temperature.
10. Hydrolyze pellet in 50 μL of carbonate buffer for 10 min at 60°C, thus probes of different length are hydrolyzed into 100–200 bp fragments.
11. Stop the reaction by adding 10 μL of 10% acetic acid, 12 μL of 3 M sodium acetate (pH 4.8), and 312 μL of cold absolute EtOH. Incubate overnight at −20°C.
12. Centrifuge at 13,000 × *g* for 30 min at 4°C. Remove supernatant by decanting.
13. Wash the RNA pellet by adding 1 mL of 70% EtOH, followed by 10 min centrifugation (13,000 × *g*) at 4°C. Remove supernatant as much as possible by decanting and pipetting.
14. Dry and then dissolve the pellet in 20 μL of nuclease-free water and 20 μL of deionised Formamide.

3.2 Dot-Blot Control of Labeling Efficiency

For Dig or FITC single probe detection:

1. Prepare dilution series (1:10, 1:100, 1:1000) of the labeled probes in nuclease-free water and apply 1 μL of each dilution to a small strip of positively charged nylon membrane.
2. Preincubate the membrane in 3 mL of Blot Incubation Buffer for 30 min at room temperature (RT).
3. Incubate membrane with Anti-Digoxigenin-AP, Fab fragments (or Anti-FITC-Fab-AP) diluted at 1/1250 in 3 mL of Blot Incubation Buffer for 15 min at RT.
4. Wash membrane three times in 3 mL of Blot Wash Buffer for 5 min.
5. Perform AP staining reaction with Vector® Blue Substrate Kit (Vector Laboratories SK-5300) according to the manufacturer's protocol: for 5 mL of Blot Wash Buffer, add in order: 80 μ L of Reagent 1, then 80 μL of Reagent 2, and 45 μL of Reagent 3. Incubate with a gentle agitation in the dark at RT.
6. Stop the detection process once the highest dilution is reached, or when there is a risk that background staining will obscure the detected dots.

7. Wash in water. If all three dilutions are detected on the membrane, we consider that the incorporation efficiency of DIG- or FITC-uracil is correct and that the probe can be used for hybridization.

For Dig and FITC dual detection:

1. Prepare a dilution series of both probes (1:10, 1:100, 1:1000) in nuclease-free water and apply 1 μL of each dilution to the same positively charged nylon membrane.
2. Preincubate the membrane in Blot Incubation Buffer for 30 min at RT.
3. Incubate membrane with Anti-Dig-Fab-AP and Anti-FITC-Fab-POD diluted 1/1250 in Blot Incubation Buffer for 15 min at RT.
4. Wash membrane three times in PBS-T for 10 min.
5. Perform POD staining reaction with Vector® VIP Peroxidase substrate kit according to manufacturer's protocol: for 5 mL of PBS, add 90 μL of Reagent 1, 75 μL of Reagent 2, 75 μL of Reagent 3, and 125 μL of hydrogen peroxide solution. Incubate with a gentle agitation in the dark at RT.
6. Stop reaction once the highest dilution appeared, or when there is a risk of background staining hiding the detected dots.
7. Wash membrane three times in Blot Wash Buffer for 5 min.
8. Perform AP staining reaction with Vector® Blue Substrate Kit according to the manufacturer's protocol.
9. Stop the reaction by washing in deionized water once the highest dilution appears. If all dilutions are detected on the membrane, we consider that the incorporation efficiency of DIG or FITC-uracil is correct and that the probe can be used for hybridization.

3.3 Samples Fixation and Permeabilization

1. Thaw a PFA Fixative aliquot and add 100 μL of 10% Triton X-100 (*see* **Note 5**).
2. Submerge the tissue in 7 mL PFA fixative, and apply vacuum for 1 h at RT. Stop the vacuum and put your samples overnight (O/N) at 4°C. For cell wall staining, SR2200 can be added directly to the PFA Fixative at a final concentration of 1/1000. Do not leave samples more than 2 days in fixative.
3. Transfer the tissue to 7 mL of 10% EtOH, and apply six pulses of 180 W microwaves for 30 s. Stir the sample by hand for a few seconds between each pulse. This stage is critical because the samples must be treated while avoiding boiling. Since the boiling point varies with ethanol concentration and the presence of samples in the vial, stop treatment when boiling occurs at higher ethanol concentrations. Leave the samples on the bench for 5 min.

4. Repeat this process for a series of EtOH/water dilutions, subsequently, 30, 50, and 70%. After this step, samples can be stored for up to 2 months at 20°C in 70% of EtOH.
5. Rehydrate by subsequently incubating the tissue each time for 5 min in 7 mL 70%, 50%, 30%, and 10% of EtOH, respectively, and finally, in PBS-Tween.
6. Digest in 5 mL of proteinase K stock for 10 min at RT.
7. Stop the digest with 5 mL glycine working solution for 2 min at RT.
8. Wash samples in 5 mL of Triethanolamine/Acetic anhydride working solution for 10 min at RT with a gentle shaking.
9. Fix the samples in 5 mL of the noncrosslinking fixative RCL2, for 15 min under vacuum at RT.
10. Wash three times in 5 mL PBS-Tween for 1 min at RT.

3.4 Hybridization

1. Distribute samples in cell culture inserts in a 24-well plate containing 400 μL of prehybridization solution per well and seal the plate with plastic film.
2. Put the 24-well plate in an additional plastic box and incubate the box with the samples in prehybridization solution for 60 min at 47°C, floating in a water bath. Do this for all steps at 47°C.
3. For one probe hybridization, dilute the RNA probe by halving with Nuclease-free water (*see* **Note 11**). Denature the RNA probe for 2 min at 80°C, and chill rapidly on ice and mix each sample with 400 μL of cold hybridization solution.
4. For simultaneous hybridization of two probes, dilute both RNA probes by halving with nuclease-free water. Denature RNA probes for 2 min at 80°C, chill on ice, and mix them with 400 μL of hybridization solution per well.
5. Hybridize O/N at 47°C in a water bath with gentle shaking.

3.5 Washouts

1. Wash with 400 μL of 0.1× SSC, 0.5% SDS for 30 min at 47°C.
2. Wash with 400 μL of 2× SSC, 50% formamide for 60 min at 47°C.
3. Wash in with 400 μL of TNE buffer for 5 min at 47°C.
4. Incubate with 400 μL of 20 μg/mL RNase A in TNE for 30 min in an oven at 37°C.
5. Wash with 400 μL of TNE for 5 min at 47°C.
6. Wash with 400 μL of 2× SSC, 50% formamide for 60 min at 47°C.
7. Wash with 400 μL of 0.1× SSC for 2 min at 47°C.
8. Wash with 400 μL of 1× PBS for 15 min at RT.

9. Leave the samples O/N at RT in 400 µL 1×PBS or proceed directly to the next step.

3.6 Sample Digestion

1. Perform a partial cell wall digestion with 400 µL of enzymatic mix for 10 min at RT.
2. Wash three times (5 min at RT) with 400 µL of 1× PBS.

3.7 One Probe Signal detection

1. Preincubate the digested samples in PBS-TB for 60 min at RT.
2. Incubate the samples with Anti-Dig-Fab-AP (1/1250 in PBS-TB) O/N at 4°C or 3 h at 37°C.
3. Wash three times in Blot Wash Buffer for 10 min at RT.
4. Perform AP staining reaction with Vector® Blue Substrate Kit according to manufacturer's protocol: For 5 mL of Blot Wash Buffer, add in order: 80 µL of Reagent 1, then 80 µL of Reagent 2, and 45 µL of Reagent 3. Mix well and distribute 500 µL per well.
5. Incubate with a gentle agitation in the dark at RT.
6. Watch for a signal and stop before the background staining risks hiding it. The staining process can take up to 2 days.
7. Stop the reaction by washing in water.
8. Incubate samples in ClearSee buffer for at least 5 days. Samples can be stored in Clearsee for months at 4°C.
9. Proceed to cell wall staining as described in **Sect.** 3.8.

3.8 Two Probe Signal Detection

1. Preincubate samples in PBS-TB for 60 min at RT.
2. Incubate samples with Anti-Dig-Fab-AP and Anti-Fluorescein-Fab-POD (each diluted at 1/1250 in PBS-TB) O/N at 4°C or 3 h at 37°C.
3. Wash three times in PBS-T for 15 min at RT.
4. Perform the fluorescent POD staining reaction with Tyramide Signal Amplification Kit according to [7]: Prepare Solution 1 by adding 1 µL of H_2O_2 to 200 µL of amplification buffer, prepare Solution 2 by adding 1 µL of Solution 1 to 100 µL of Amplification buffer. Prepare Solution 3 by adding 12 µL of Tyramide Conjugates Alexa Fluor™ 488 to 400 µL of Solution 2.
5. Incubate the samples in Solution 3 in the dark for 30 min at RT.
6. Wash three times in blot Wash Buffer for 10 min at RT.
7. Perform AP staining reaction with Vector® Blue Substrate Kit according to manufacturer's protocol: Per 5 mL of Blot Wash Buffer, add in order: 80 µL of Reagent 1, then 80 µL of Reagent 2, and 45 µL of Reagent 3. Mix well and distribute 500 µL per well.

8. Mix well before use, incubate with a gentle agitation in the dark at RT.
9. Watch the signal and stop when background staining appears.
10. Wash three times in Blot Wash Buffer for 10 min at RT.
11. Keep the samples in 1×PBS (*see* **Note 12**).
12. Proceed to cell wall staining (*see* Sect. 3.8).

3.9 Cell Wall Staining (see Note 13)

1. 1st option: Add Renaissance stain SR2200 directly to the PFA fixative (*see* Sect. 3.2, Step 1) in a 1/1000 dilution.
2. 2nd option: Add one of the cell wall dyes directly to the samples in the Clearsee buffer (*see* Sect. 3.6, Step 8). For Fluorescent Brightening, use a 50 µg/mL dilution, for Direct Red 23 use 0.01%, for SR2200 use a 1/500 dilution. Incubate O/N at RT.
3. 3rd option: Add one of cell wall dyes to samples in 1×PBS (*see* Sect. 3.7, Step 11). For Fluorescent Brightening use 50 µg/mL dilution. For Direct Red 80, use a 0.01% dilution, and for SR2200, use a 1/500 dilution. Incubate O/N at 4°C.

3.10 Nuclei Staining

1. Wash samples after cell wall staining three times in 1×PBS for 5 min.
2. Mount samples in VECTASHIELD® mixed with DAPI to a final concentration of 10 µg/mL.

3.11 Mounting Samples Between a Slide and a Coverslip

1. Use double-sided tape to make space between the slide and the coverslip, and prevent tissue damage.
2. Put the number of layers of double-sided tape you need, depending on the thickness of the samples.

   For *A. thaliana*, use one layer for roots and leaves, use two layers for developing seeds, and four layers for Shoot Apical Meristems and embryos within the seed.
3. Use spacers to define a 24 × 32 mm area on the slide. Place the samples in a drop of VECTASHIELD® and cover with a coverslip. Remove any excess liquid from the open borders.

3.12 Confocal Microscopy

Detection of the fluorescent probes and stains requires a suitable microscopic setup to detect them at optimal excitation and emission wavelengths:

1. Tyramide Conjugates Alexa Fluor™ 488 is excited at 488 nm and detected at 500 to 550 nm.
2. Vector Blue Substrate is excited at 633 nm and detected at 740–800 nm.
3. SR2200 is excited with a 405 nm UV laser and detected at 420–480 nm.

4. DAPI is excited by a 405 nm UV laser and detected at 420–480 nm.
5. Direct Red 23 is excited with a 561 nm laser and detected at 580–615 nm.

4 Notes

1. Remove silver seals from the vial screw cap to avoid sparkles during microwave treatment.
2. There are different forms of EDTA, both acid and salt. Please check the molecular weight of your EDTA to prepare a 0.5 M solution.
3. There are no restrictions on probe length because probes of different lengths are hydrolyzed into fragments of 100–200 bp suitable for hybridization at Step 10 of Sect. 3.1. The T7 promoter sequence should be added to the gene-specific reverse primer. The PCR conditions are standard and depend on primers' Tm for annealing temperature and sequence length for duration of the elongation step. The PCR template should be purified using a PCR clean-up kit.
4. All buffers are routinely prepared in 50 mL screw-cap tubes and can be kept on the bench for months.
5. Triton™ X-100 and TWEEN® 20 are difficult to dissolve in concentrated form. We prepare 50 mL of 10% solutions of both for further use.
6. We recommend purchasing one kit, which is sufficient for 50 microscope slides. You can purchase additional labeled tyramide conjugates as standalone reagents.
7. The hybridization solution should be prepared in advance, as mixing produces a lot of bubbles.
8. Aliquots can be stored at −20°C for years. Thawed aliquots cannot be reused.
9. The reaction begins when acetic anhydride is added, so the solution should be prepared just before adding it to the samples.
10. Mix the ClearSee solution on the magnetic stirrer with heating until everything is completely dissolved.
11. The amount of each probe placed in the hybridization buffer is adjusted. Several tests are often necessary to determine the quantity of probe required to obtain a signal for genes with relatively low expression levels. For example, 2.5 μL of

probe is required in 400 μL of hybridization buffer for *H2B*, whereas 10 μL is required for *CUC2/3*.

12. The ClearSee solution is not appropriate for Alexa fluorochromes.
13. When combining several fluorochromes, their spectral properties must be taken into consideration. To avoid fluorescence crosstalk, use markers with different fluorescence emission spectra. We favor DAPI staining (UV excitation and blue emission) and Direct red23 (561 nm) since SR2200 overlaps with the DAPI fluorescence spectrum.

Conflict of Interest

This work has benefited from the support of IJPB's Plant Observatory platforms: PO-Plants and PO-Cyto. The authors have no conflicts of interest to declare that are relevant to the content of this chapter.

References

1. Van Leeuwen W, Ruttink T, Borst-Vrenssen AWM et al (2001) Characterization of position-induced spatial and temporal regulation of transgene promoter activity in plants. J Exp Bot 52:949–959. https://doi.org/10.1093/jexbot/52.358.949
2. Gall JG (2016) The origin of in situ hybridization—a personal history. Methods 98:4–9. https://doi.org/10.1016/j.ymeth.2015.11.026
3. Gómez-Mena C, Roque EM (2018) Non-isotopic RNA *in situ* hybridization for functional analyses in *Medicago truncatula*. In: Cañas LA, Beltrán JP (eds) Functional genomics in *Medicago truncatula*. Springer New York, New York, NY, pp 133–144
4. Francoz E, Ranocha P, Pernot C et al (2016) Complementarity of medium-throughput in situ RNA hybridization and tissue-specific transcriptomics: Case study of Arabidopsis seed development kinetics. Sci Reps 6:1–10. https://doi.org/10.1038/srep24644
5. Yang W, Schuster C, Prunet N et al (2020) Visualization of protein coding, long noncoding, and nuclear RNAs by fluorescence in situ hybridization in sections of shoot apical meristems and developing flowers. Plant Physiol 182:147–158. https://doi.org/10.1104/pp.19.00980
6. Hejátko J, Blilou I, Brewer PB et al (2006) In situ hybridization technique for mRNA detection in whole mount Arabidopsis samples. Nate Prot 1:1939–1946. https://doi.org/10.1038/nprot.2006.333
7. Rozier F, Mirabet V, Vernoux T, Das P (2014) Analysis of 3D gene expression patterns in plants using whole-mount RNA in situ hybridization. Nat Prot 9:2464–2475. https://doi.org/10.1038/nprot.2014.162
8. Spiegelhalder RP, Berg LS, Nunes TDG et al (2024) Dual role of BdMUTE during stomatal development in the model grass *Brachypodium distachyon*. Development 151:dev203011. https://doi.org/10.1242/dev.203011
9. Chelysheva LA, Grandont L, Grelon M (2013) Immunolocalization of meiotic proteins in *Brassicaceae*: method 1. In: Pawlowski WP, Grelon M, Armstrong S (eds) Plant Meiosis. Humana Press, Totowa, NJ, pp 93–101
10. Truernit E, Bauby H, Dubreucq B et al (2008) High-resolution whole-mount imaging of three-dimensional tissue organization and gene expression enables the study of phloem development and structure in Arabidopsis. Plant Cell 20:1494–1503. https://doi.org/10.1105/tpc.107.056069
11. Kurihara D, Mizuta Y, Sato Y, Higashiyama T (2015) ClearSee: a rapid optical clearing reagent for whole-plant fluorescence imaging. Development 142:4168–4179. https://doi.org/10.1242/dev.127613

Check for updates

Chapter 17

Gene Expression Analysis from Specific Root Cell Types

Alessio Terenzi, Elena Salvi and Riccardo Di Mambro

Abstract

In plant roots, the diversity of tissues along the radial axis and of developmental stages along the longitudinal axis implies a highly dynamic and heterogeneous gene expression. Many master regulators of gene expression exhibit distinctive temporal expression patterns or are restricted to specific cell types, making whole-root molecular analyses insufficient to resolve their roles in developmental and physiological processes. Here, we report a protocol for fluorescence-activated cell sorting (FACS)-coupled RNA sequencing to analyze, reconstruct, and ultimately dissect the gene regulatory networks (GRNs) involved in root development and physiology, particularly well-suited for genes that are expressed transiently or in rare or restricted cell types. The method relies on selecting an appropriate cell-type-specific reporter line, which enables the isolation of labeled cells with high specificity and minimal contamination. Furthermore, we discuss the relative advantages of FACS-coupled RNA-seq compared with alternative gene-expression methodologies, including quantitative PCR (qPCR), bulk root RNA-seq, single-cell RNA-seq (scRNA-seq), and spatial transcriptomics, considering sensitivity, resolution, input requirements, and cost.

Keywords Plant root development, Cell-type-specific gene expression, Fluorescence-activated cell sorting (FACS), Bulk RNA-sequencing, Differential gene expression

1 Introduction

Plant roots are constituted by a complex organization of tissues and developmental zones, each with distinct physiological functions and transcriptional signatures [1].

In roots, gene expression patterns can vary dramatically in time and space, with genes expressed only in specific cell types and/or developmental stages.

Investigating the molecular basis of root development and physiology means utilizing methods that can resolve gene expression at high cellular resolution. Traditional approaches, such as bulk RT-qPCR or RNA-sequencing from entire or manually excised root segments [2, 3], though, result in averaging signals across

These authors "Alessio Terenzi and Elena Salvi" contributed equally to this work.

Steffen Vanneste and Inge Verstraeten (eds.), *Root Development*, Methods in Molecular Biology, vol. 3086,
https://doi.org/10.1007/978-1-0716-5603-7_17,

thousands of cells. This masks subtle transcriptional differences, particularly for rare cell populations, low-expression genes, or genes activated only during brief developmental windows, which could play paramount roles in root biology. The enrichment of defined cell types at a specific time can overcome this limitation. Fluorescence-activated cell sorting (FACS) is an established technology—successfully applied to characterize Arabidopsis root development [4–8]—based on flow-cytometry that physically isolates individual protoplasts from a heterogeneous population exploiting measurable optical cell features [9]. FACS provides single-cell resolution while pooling sorted isolated cells, thus combining cell-type specificity with bulking for improved sensitivity and robustness in downstream applications.

FACS relies on the preliminary selection of an appropriate cell-type specific fluorescent marker(s), which can be either dyes with a specific differential penetration in cells of interest or genetically encoded fluorescent reporters, such as transcriptional fusions (*promoter::fluorescent protein* fusions), translational fusions (*promoter:protein coding sequence:fluorescent protein* fusions) or publicly available promoter and enhancer trap lines [10, 11] (Fig. 1). The specificity, strength, and developmental timing of the reporter influence the quality of FACS-sorted populations: broad or leaky reporter expression reduce purity of sorted cell populations, whereas weak reporters may lead to insufficient fluorescence for robust FACS. Hence, choosing validated, cell-type- and time-restricted reporters is necessary to assess the success of specific cell enrichment and downstream analysis.

Nowadays, many methods that can be coupled to FACS are available for quantifying gene expression, each with strengths and limitations (Table 1). Given that these methods offer different resolutions and requirements, the choice depends on the biological question, required resolution, throughput, and budget (Table 1). When the study focuses on defined cell types or genes with putative low expression expected in a few very specific cells, FACS-bulk RNA-seq represents a strategic compromise. As only cells of interest are sequenced, FACS-coupled RNA-seq provides a high signal-to-noise ratio, which results in a more straightforward bioinformatic analysis compared to scRNA-seq.

Compared to qPCR, scRNA-seq or bulk RNA-seq, FACS-coupled RNA-seq is cost-efficient per cell type, avoiding the need to sequence or analyze thousands of single cells to identify the same population and allows the detection of low-abundance transcripts compared to scRNA-seq because a deeper coverage per gene grants the discovery of lowly expressed genes and alternative isoforms (Table 1). Potentially, a combination of different technologies applied to sorted cells can expand bulk RNA-seq gene expression investigation

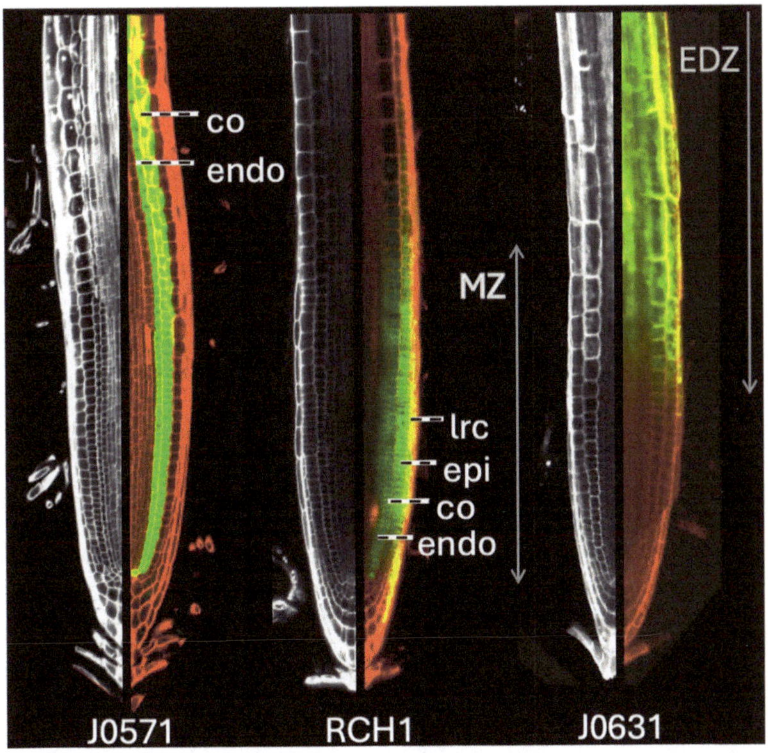

Fig. 1 Confocal microscopy images of *A. thaliana* root apical meristems (left halves) and examples of cell-type-specific fluorescent markers, available in public repositories, suitable for FACS (right halves). Enhancer trap line J0571, specifically expressed in the cortex and endodermis (co = cortex; endo = endodermis); pRCH1::GFP, specifically expressed in the meristematic zone, only in the lateral root cap, epidermis, cortex and endodermis tissue (LRC = lateral root cap; epi = epidermis; co = cortex; endo = endodermis; MZ); Enhancer trap line J0631, specifically expressed in the elongation/differentiation zone (EDZ = elongation/differentiation zone)

with other biological information (i.e., chromatin accessibility, protein markers, cell metabolomics, lineage information) [8, 12–14].

Here, we report a protocol that describes FACS-based cell-type enrichment with bulk RNA-seq to characterize the transcriptome of specific root cell populations marked by a fluorescent reporter. This protocol can be adapted to suit all plant species amenable to stable genetic transformation, for genetic investigations and root phenotypic characterization. We also outline a possible bioinformatic pipeline to analyze the results for differentially expressed genes (DEGs) identification and complementary downstream approaches to reconstruct, analyze, and dissect cell-type specific molecular mechanisms underpinning root development and physiology.

Table 1

Comparison of the FACS-RNA-seq technique with the main canonical methods for gene expression analyses in roots, including in situ hybridization, RT-qPCR, whole organ RNA-seq, scRNA-seq and spatial transcriptomics.

Method	Resolution	Advantages	Limitations	Relative cost[a]
In situ hybridization/ mRNA probes	Single cells within tissues and organs	Spatially precise; good for validation of other methods	Low throughput; semi-quantitative	Low–Medium
RT-qPCR	Gene-specific, bulk tissue/ whole organ	Sensitive and fast	Low throughput; flattens signals across diverse cell types; no cell-type resolution unless combined with FACS or Laser Capture Microscopy (LCM)	Low if few genes are tested; Medium if many genes are examined
Bulk root RNA-seq	Whole organ/ root tip / specific root region	Comprehensive organ transcriptome; cost-effective	Averages signals across diverse cell types; no cell-type resolution	Medium
FACS-RNA-seq	Defined marker^{+}, labeled cell populations	High cell-type specificity; good depth; compatible with fluorescent reporters	Requires protoplasting, which might induce cell stress; limited to known markers of cells of interest; requires a viable and effective fluorescent reporter line	Medium–High
scRNA-seq	Single cells	Unbiased discovery of cell types, cell states, developmental trajectories from pseudo-time analyses	Expensive; higher dropout rates compared to bulk techniques; limited spatial context; requires protoplasting, which might induce cell stress	High
Spatial transcriptomics	Spatial positions within tissue/organ	Preserves positional context; identifies expression "niches"	Expensive; resolution varies by platform; computationally intensive	High–Very High

[a]Absolute costs depend on country, platform, and sequencing depth

2 Materials

Analytical-grade reagents and ultrapure water obtained by deionization and further purification to a resistivity of 18 MΩ·cm at 25°C should be used for all solution preparations.

2.1 Preliminary Choice of the Marker and Validation of Its Expression

1. PC with internet connection and access to literature databases (e.g., Google Scholar, PubMed).
2. Access to online root cell-type databases (e.g., TAIR [15], EFP browser [16], Root Cell Atlas [17], https://bioit3.irc.ugent.be/plant-sc-atlas for *A. thaliana*).
3. Growth cabinets, chambers or greenhouses with a controlled environment (i.e., temperature, light, and humidity).
4. Confocal laser scanning microscope with appropriate lasers and filters for detection of the fluorescent reporters and dyes of interest (or, alternatively, fluorescent microscope).
5. Microscope slides: glass slides (25.4 76.2 mm, 1.0–1.2 mm thick) and coverslips (24 50 mm).
6. Surgical blades or scissors.
7. Tweezers.
8. Paper towel.
9. Microscopy mineral oil.
10. Microscope mounting solution (ddH_2O or PBS).
11. Optional: cell membrane or cell wall staining reagent (e.g., propidium iodide, FM4-64, FM1-43 dissolved in water).
12. Image analysis software (e.g., FIJI, IMARIS, or similar software).

2.2 Sample Preparation

1. Seeds: for experiments targeting a cell population representing approximately 1% of the total, 1.25 g of seeds should be used (approximately 38,500 seeds; *see* **Note 1**).
2. Seed sterilization solution: 25 mL commercial bleach (≤5% v/v), 15 μL of 10% Tween 20 (v/v) up to 50 mL with ultrapure water.
3. Sterile water.
4. 50 mL centrifuge tubes.
5. Nitex® Nylon mesh (10×10 cm squares), wrapped in aluminum foil in batches of up to five mesh sheets per packet, sealed with autoclave tape, sterilized, and oven dried prior to use (Fig. 2).
6. Murashige-Skoog plant medium: 4 g of Plant Agar (0.8% final concentration), 5 g of Sucrose (1% final concentration), 1.1 g of Murashige-Skoog Medium (0.22% final concentration),

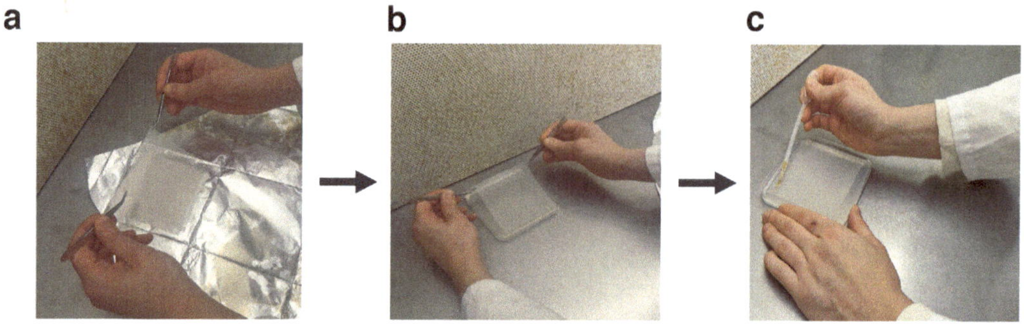

Fig. 2 Mesh handling and plating. (**a**) Pictures showing a sterile mesh sheet manipulated with sterile tweezers. (**b**) Individual mesh sheet positioned on an MS ½ medium plate. (**c**) Seeds plated onto the mesh-covered MS ½ medium using a sterile Pasteur pipette. Created in BioRender. Terenzi, A. (2025) https://BioRender.com/7yelqcr

50 mL of MES pH 5.8 (10% final volume), and up to 500 mL with ultrapure water, 12 × 12 cm square Petri dishes.

7. Parafilm.
8. L-shaped tweezers.
9. Ethanol 70% (v/v).
10. Laminar flow hood.
11. Growth chamber at 22°C at 60% relative humidity with a light intensity of 150 μmol/m^2/s under long-day conditions.

2.3 Sample Collection, Protoplasting, and FACS

1. Washing Buffer: 14.576 g Mannitol, 0.85 g of MES, 0.3 g of KCl, 0.3 g of $CaCl_2$, up to 200 mL ultrapure water. Adjusted to 5.7 pH with 1 M TRIS solution (*see* **Note 2**).
2. Enzyme Buffer: Prepare 1.5% w/v of Cellulase, 0.1% w/v of pectolyase in Washing Buffer on the same day of the experiment) (*see* **Note 3**).
3. Sterile scalpels.
4. 60 mm Ø Petri dishes.
5. Sterile Pasteur pipettes.
6. Orbital shaker.
7. 1000 μL tips and pipettes.
8. 70 μm cell strainers.
9. 50 mL centrifuge tubes.
10. 15 mL centrifuge tubes.
11. Centrifuge equipped with a swinging bucket.
12. 40 μm cell strainers.
13. Polystyrene tubes.
14. Wet ice.

15. Fluorescent-Activated Cell Sorter (FACS) with 100 μm nozzle size and 15.0 psi shelter pressure.

2.4 RNA Extraction and RNA Quality Check

1. Commercial RNA extraction kit.
2. Spectrophotometer (e.g., NanoDrop®), fluorimeter (e.g., Qubit™), or microfluidic electrophoresis analyzer (e.g., BioAnalyzer®).
3. Midori Green®.
4. Electrophoretic chamber.
5. Agarose.
6. Tris-acetate-EDTA (TAE) 50× solution: 242 g of Tris base, 57.1 mL of glacial acetic acid, 100 mL and up to 1 L with ultrapure water.

2.5 RNA Sequencing and Bioinformatic Analysis

1. Access to a workstation or high-performance computing environment (HPCC).
2. Operating system compatible with standard bioinformatics software (Linux or macOS recommended).
3. The latest available version of the reference genome assembly and gene annotation files for the species under study.
4. RNA-seq quality control and preprocessing software (e.g., `FastQC`, `MultiQC`, `Trimmomatic`).
5. Transcript quantification and alignment quality software (e.g., `Salmon` or `Kallisto`, `RSeQC`).
6. Statistical analysis tools (`R` environment and `R` packages https://www.r-project.org/).
7. Differentially expressed genes (DEG) identification package (e.g., for R, `DESeq2` from Bioconductor).
8. Access to gene annotation databases (e.g., NCBI, TAIR, or species-specific resources); `topGO` (Bioconductor package).
9. Enrichment analysis resources (gene ontology and pathway annotation databases such as https://pantherdb.org).

2.6 Data Interpretation and Downstream Functional Validation

1. Access to online root cell-type databases (such as TAIR [15], EFP browser [16], Root Cell Atlas [17], https://bioit3.irc.ugent.be/plant-sc-atlas for *A. thaliana*; see Sect. 2.1.2).
2. Literature and functional annotation databases of candidate genes.
3. Reference genome assembly and gene annotation files for the species under study.
4. Seeds of the wild-type and mutants (loss and gain-of-function), inducible or overexpression lines of candidate genes, if available in seed stock centers.

5. Transcriptional and translational reporter lines for candidate genes, if available in literature or in seed stock centers.
6. Growth cabinets, chambers, or greenhouses with a controlled environment.
7. Reagents for RT-qPCR and a list of reference genes for normalization.
8. Molecular cloning vectors suitable for reporter generation, genome editing, and overexpression; cloned promoters for constitutive, inducible, or cell-type-specific expression, including promoters previously validated for the sorted cell population.
9. CRISPR–Cas9 genome editing systems, including Cas9 expression constructs and guide RNA scaffolds.
10. Tools and online resources for guide RNA design (e.g., https://eu.idtdna.com).
11. Plant transformation systems and protocols appropriate for stable genetic transformation.
12. Reagents and primers for genotyping, validation of genome edits, and confirmation of overexpression or gene editing.
13. Instruments, materials, and reagents for fluorescence or confocal microscopy (see Sect. 2.1).
14. Equipment and reagents for developmental, morphological, and physiological assays relevant to the function of the target cell population under study.

3 Methods

In this section, we present a step-by-step protocol that comprises the choice and validation of a suitable fluorescent marker, sample preparation, protoplasting of root cells, FACS of protoplasted root cells, choice of bulk RNA-seq strategy, bioinformatic analysis pipeline to obtain a transcriptome profile and a DEG analysis from cell population(s) of interest, and suggestions for downstream functional characterization of candidate genes.

3.1 Preliminary Choice of the Marker and Validation of Its Expression

Identification and validation of the marker is an essential step that involves analyzing in time and space the expression dynamics of the marker in homozygous lines. We recommend—and describe below—using confocal microscopy rather than a fluorescent microscope to get the best possible resolution.

1. Compile a candidate list of potential markers that display high specificity and robust expression in the cell type/population

of interest with no or minimal expression outside that cell type (*see* **Note 4**).

2. Choose a marker line expressing the brightest available fluorophore, such as high-quantum-yield Turbo Green Fluorescent Protein (turboGFP). For dynamic analysis, the fast-maturing VENUS is an optimal choice (*see* **Note 5**).
3. Grow the selected marker lines using the same conditions as the ones planned for the FACS-RNA-seq analysis (i.e., age, light, photoperiod, temperature, humidity, hormonal treatments, drug treatments, stress conditions, etc.) (*see* **Note 6**).
4. To resolve the expression domain of the chosen marker, prepare samples for confocal microscopy using 25.4×76.2 mm glass microscopy slides. Counterstain roots with a drop of 10 μg/mL propidium iodide, or equivalent cell wall/cell membranes dyes, directly on the microscopy slide to visualize root tissue organization and cell boundaries.
5. Place roots in the dye drop on the glass slide with tweezers. Excising the root at approximately 0.5 cm from the root tip with surgical blades or scissors can ease this step. Place a coverslip over the sample and blot excess PI solution with a paper towel.
6. Image samples using a confocal laser scanning microscope equipped with a minimum 20× objective (40× is recommended) to allow visualization of individual root cells. Use the smallest pinhole setting to optimize the signal-to-noise ratio.
7. In the confocal acquisition software, select the appropriate laser line(s) to excite the cell wall/membrane dye and the fluorophore of the reporter line.
8. Adjust the dye's photomultiplier (PMT) gain settings to obtain a clear outline of individual root cells.
9. Set the PMT gain of the channel of the fluorophore carried by the marker line to get a strong and nonsaturated signal.
10. Image a minimum of 40 roots to assess variability in the expression of the fluorophore across the marker line population.
11. Verify that reporter expression is restricted to the cell population of interest by imaging also other tissues and regions of the root (*see* **Note 7**).
12. Assess that the fluorescent intensity of the selected marker is homogeneous (*see* **Note 8**). Check whether detectable subcellular states or subpopulations within the marker-positive cells are detectable (*see* **Note 9**).

13. Finally, for the following experiments, select the reporter line that gives the best fluorescent intensity and the highest specificity in the cell population of interest.

3.2 Sample Preparation

For this section of the "Methods," all steps can be conducted at room temperature (*see* **Note 10**).

1. Weigh an appropriate number of seeds (*see* **Note 11**).
2. Add 50 mL of seed sterilization solution to the seeds in a 50 mL centrifuge tube.
3. Shake for 5 min.
4. Centrifuge at 410 × *g* for 1 min.
5. Remove the supernatant solution in a laminar flow hood.
6. Wash seeds five times using sterile water. For each washing step, shake for a few seconds to remove all bleach residuals, then centrifuge at 410 × *g* for 1 min to spin down the seeds.
7. Stratify seeds in sterile water for 3 to 5 days in the dark at 4°C.
8. Prior to sowing, place the sterile 10 × 10 Nitex® Nylon mesh on pre-poured ½ MS plates using sterile tweezers (Fig. 2). Note that drying of the Nylon mesh takes 5 days, so this needs to be planned.
9. Remove some of the excess water from the 50 mL centrifuge tube containing the seeds using a sterile Pasteur pipette. Using the same pipette, collect the seeds from the tube and dispense them on the mesh, forming two rows per plate, 350 seeds per line (*see* **Note 12**), being careful to put the seeds in contact with the mesh and the medium (Fig. 2).
10. Seal each plate with Parafilm and place them in a growth chamber at 22°C at 60% RH under long-day conditions (*see* **Note 13**). Repeat for the required number of plates.

3.3 Sample Collection and Protoplasting

1. Excise the required length of the root apex using sterile scalpels (*see* **Note 14**) and collect the roots in a Ø 60 mm Petri dish filled with 7 mL of pre-heated Enzyme Buffer (*see* **Note 3**). Each Petri dish can accommodate roots collected from up to ten square Petri plates (Fig. 3).
2. Incubate the Ø 60 mm Petri dishes containing the harvested roots for 1 h on an orbital shaker at 70 rpm (Fig. 3). Every 20 min, gently mix using a 1000 μL pipette tip (without pipetting).
3. After the enzymatic incubation, gently pipette up and down the suspension using a 1000 μL tip until the solution becomes cloudy, indicating the release of protoplasts.

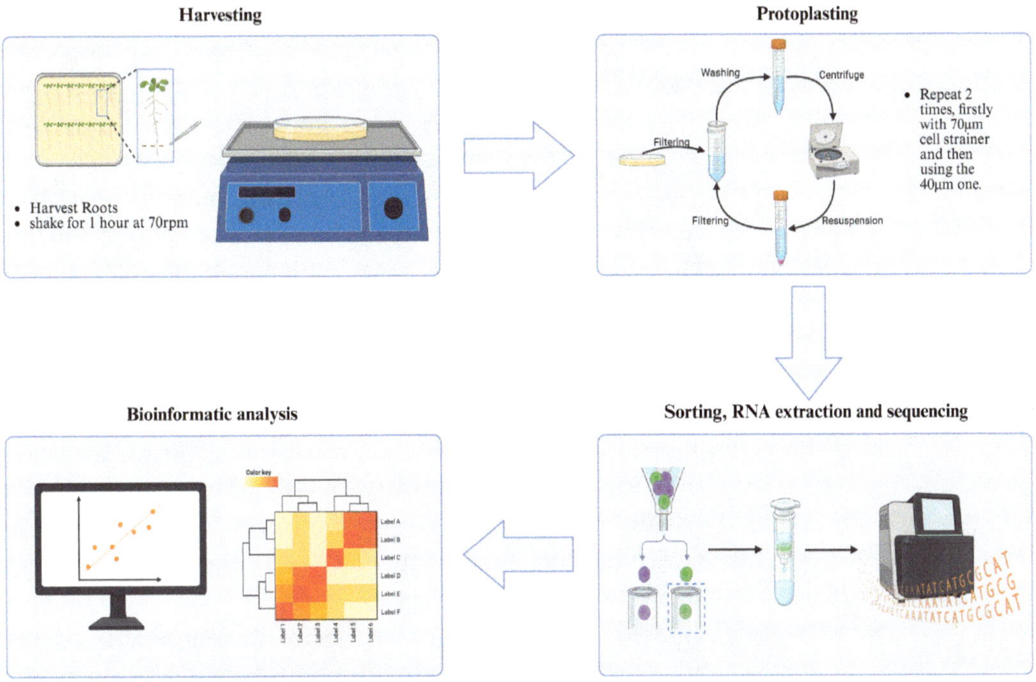

Fig. 3 Schematic overview of the experimental workflow. Following seed plating and plant growth, root tissues are harvested and incubated in an enzymatic solution to release protoplasts. Protoplasts are concentrated through sequential washing, centrifugation, resuspension, and filtration steps, and subsequently sorted by FACS. RNA is extracted from the sorted positive population and subjected to sequencing. Finally, raw sequencing data are processed to obtain a DEG dataset. Created in BioRender. Terenzi, A. (2025) https://BioRender.com/b2krdms

4. Place a 70 μm cell strainer onto a 50 mL centrifuge tube and pre-wet the membrane with 1 mL Washing Buffer (*see* **Note 15**) (Fig. 3).
5. Carefully pour the entire protoplast suspension from one Petri dish through the 70 μm cell strainer into the tube using a sterile Pasteur pipette (Fig. 3).
6. Rinse the empty Petri dish with 2 mL Washing Buffer and pour the resulting collected liquid through the 70 μm cell strainer to the 50 mL centrifuge tube.
7. Transfer the protoplast suspension to a 15 mL centrifuge tube using a sterile Pasteur pipette and centrifuge at 200 × *g* for 6 min (*see* **Notes 16**). With this step, a visible pellet is obtained (Fig. 3).
8. Discard the supernatant without perturbing the pellet by decanting or pipetting.

9. Gently resuspend the pellet in 1 mL of Washing Buffer and add three more mL of Washing Buffer to the 15 mL centrifuge tube.
10. Place a 40 μm cell strainer into a new 50 mL centrifuge tube and pre-wet the membrane with 1 mL Washing Buffer.
11. Carefully pour all the protoplast suspension from one 15 mL centrifuge tube through the 40 μm cell strainer into the new 50 mL centrifuge tube using a sterile Pasteur pipette (Fig. 3).
12. Rinse the first 15 mL centrifuge tube with 1 mL of Washing Buffer, pouring it over the 40 μm strainer into the centrifuge tube, then collect the residual protoplasts with one more mL of Washing Buffer.
13. Transfer the protoplast suspension from the 50 mL centrifuge tube to a new 15 mL centrifuge tube using a sterile Pasteur pipette (Fig. 3).
14. Centrifuge the suspension at 200 × *g* for 6 min to obtain a visible pellet (Fig. 3).
15. Discard the supernatant with a sterile Pasteur pipette, leaving approximately 1 mL of Washing Buffer.
16. Resuspend the protoplast pellet in the residual Washing Buffer by tapping.

3.4 Protoplasts Isolation by Fluorescence-Activated Cell Sorter (FACS)

We recommend conducting a pilot experiment to identify potential cellular morphological features distinctive to the population of interest. These data allow refinement of the sorter gating strategy for the experiment, ensuring that cell selection is based not only on fluorescence but also on physical characteristics.

1. Transfer the protoplasts solution into an appropriate tube for the sorting step (Fig. 3; *see* **Note 17**).
2. Use forward (FSC) and side scatter (SSC) parameters to assess cell size and internal complexity within the sorted population, identifying distinctive cellular morphological features of the protoplasts of interest. For more details and protocols regarding the FACS set [18] (*see* **Notes 17 and 18**).
3. Sort protoplasts directly in a tube in wet ice containing the first buffer of the RNA extraction kit, unless otherwise specified in the RNA extraction kit protocol in use (*see* **Note 19**).

3.5 RNA Extraction and RNA Quality Check

1. Extract RNA according to the manufacturer's instructions (Fig. 3; *see* **Note 20**).
2. Quantify extracted RNA and determine its purity by Ultraviolet–Visible spectrophotometry (e.g., Nanodrop®), calculating the 260 nm/280 nm (pure RNA: 2.0), and 260 nm/230 nm (ideal RNA: 2.0–2.2) ratios (*see* **Note 21**). In parallel, electrophoresis

with an RNA-specific dye such as Midori Green® provides a complementary, quick first evaluation of RNA integrity and overall quality (*see* **Note 22**).

3. Use microcapillary electrophoretic RNA separation to determine RNA integrity number (RIN) and total RNA input, which are key parameters for library preparation [19] (*see* **Note 23**).

3.6 Recommendations on Bulk RNA-Sequencing Strategy and Bioinformatic Analysis

1. Select the single-end sequencing or paired-end sequencing according to the experimental aims (*see* **Notes 24 and 25**).
2. Ensure you meet all technical requirements for library preparation and sequencing specified by the external sequencing facility, including RNA quality, concentration, and sample format.
3. Ship the extracted RNA samples in dry ice to the sequencing facility.
4. Download the sequencing results in `FASTQ` format.
5. Check the quality of raw sequencing reads using software like `FastQC` (https://github.com/s-andrews/FastQC). With `FastQC` assess per-base sequence quality, GC content, adapter contamination, and sequence duplication (*see* **Note 26**).
6. To ease identification of issues and to visualize quality metrics during the quality check step across samples (Sect. 3.5, Step 6), use an appropriate visualization tool such as the software `MultiQC` (https://github-com.translate.goog/MultiQC).
7. Trim adapter sequences and low-quality bases. This can be performed using software like `Trimmomatic` (https://github.com/usadellab/Trimmomatic): for `Trimmomatic`, indicative parameters include Phred scores below 20 and discard short reads of less than 20–30 bp after trimming.
8. Perform transcript-level pseudo-alignment using software like `Salmon` (https://combine-lab.github.io/salmon/) or `Kallisto` (https://github.com/pachterlab/kallisto) and the most recent version of the reference genome assembly and gene annotation. The `Tximport` package in R can be used to import the resulting transcript abundance estimates into R and summarize transcript-level data to gene-level counts for downstream differential gene expression analysis.
9. Check the quality of the alignment and absence of technical issues by assessing mapping rates, read distribution across genomic features, and strand specificity, also confirming consistency across biological replicates. In `R`, perform differential expression analysis on imported raw count matrices from previous steps, using packages like `DESeq2` (https://bioconductor.org/packages/release/bioc/html/DESeq2.html).

Normalize read counts using median of ratios normalization in `DESeq2`.

10. In `R`, perform exploratory data analysis, including principal component analysis (PCA), to check whether the replicates are consistent and whether there are batch effects or outliers.
11. Determine the comparison groups of interest between experimental conditions and accordingly identify DEGs using statistical models for count data, including covariates—like batch, genotypes, treatments—relevant for the experiment under analysis. Correct for multiple-testing (e.g., false discovery rate), then filter DEGs using thresholds and cut-offs for the adjusted *p*-value and fold-change appropriate for the biological question of the experiment.
12. Annotate the identified DEGs and the corresponding GOs referring to the latest gene models from functional databases (e.g., NCBI, TAIR), exploiting R or online resources such as Blast2Go [20].
13. Conduct a GO enrichment analysis to identify over-representation of any of the GO aspects (Molecular Function, Cellular Component, and Biological Process) of interest. This can be performed using tools like `topGO` (https://bioconductor.org/packages/release/bioc/html/topGO.html).
14. Record all bioinformatic steps (Steps 7–15) and scripts, software versions, parameters, reference genome assemblies, and annotation versions used in a logbook for reproducibility.

3.7 FACS–RNA-Seq Data Interpretation and Candidate Gene Selection for Downstream Functional Validation

This section suggests general strategies for interpreting the FACS-RNAseq results and selecting candidate genes (Steps 1–2), then illustrates possible downstream analyses to validate the target candidate genes (Step 3–4) and characterize the function of the identified candidate DEGs (Steps 5–7).

1. Explore the resulting DEG datasets and the GO enrichment analyses, searching for known genes expected to be enriched from former experimental results, similar published bulk or single cell or spatial transcriptomic datasets, to evaluate whether the identified DEGs are consistent with the identity and developmental state of the FACSed cell population.
2. Discern core cell-identity genes from genes that are responsive to treatments, genotypes, and conditions (when relevant in the experimental conditions). The genes that are consistently enriched in the FACSed cell population across all experimental conditions can be considered core cell-identity genes, whereas genes that show expression changes only upon treatments, within a genetic background, or under certain

environmental conditions can be considered condition-responsive genes (*see* **Note 27**).

3. Select candidate genes for downstream characterization by integrating expression level, gene expression fold-change, expression specificity across publicly available bulk, single-cell, or spatial transcriptomic datasets, specificity to the cell type of interest and expression consistency across biological replicates/conditions, predicted molecular function, and potential relevance to root development or physiology given by their functional annotations.
4. Prioritize candidates integrating DEG and gene ontology (GO) enrichment results to identify enriched biological processes, regulatory pathways, or cellular functions.
5. Assess the expression domain of the selected candidate genes. Several approaches can be applied to the same cell populations used in the FACS-RNAseq experiment.
6. We recommend validating the expression of the candidate genes in the subpopulation of cells of interest by analyzing transcriptional and translational reporter lines of the gene of interest, although we suggest using more than one approach. If not available, clone and generate those reporter lines, and examine their spatio-temporal expression across developmental stages and treatments.
7. Browse seed-stock databases—if available for the species under study—for mutants (both loss and gain of function) mapped in the candidate genes or constitutive/inducible overexpression lines of candidate genes and acquire those seeds for downstream phenotypic analyses in the organ/tissue of the cell population of interest (*see* **Note 28**).
8. Design targeted genetic knockouts (KO) using the CRISPR-Cas9 technology if mutants are not available (*see* **Note 29**). Strategies to generate cell-type-specific knockouts [21, 22] where the same/related promoters used for FACS marker expression or inducible promoters drive guideRNA(s) for tissue-specific CRISPR-Cas9-based genome editing of candidates can be exploited (*see* **Notes 30 and 31**).
9. Isolate homozygous lines devoid of the Cas9 enzyme, validate that they are KO and characterize the phenotype in the cell population of interest. Phenotypic characterization might include microscopy, growth assays, and measurements and characterization of physiological processes relevant to the cell population of interest or the developmental/biological process under investigation.
10. Plan an overexpression strategy for candidate genes, choosing among constitutive, inducible, or cell-type-specific promoters

based on the biological question and potential pleiotropy. Upon validation of the overexpression lines, characterize their phenotypes using relevant developmental, morphological, and physiological assays. Use these phenotypes to compare with those from the KO approaches to infer gene function (*see* **Note 32**).

4 Notes

1. It is recommended to perform a pilot experiment using approximately 0.25 g of seeds distributed in ten plates to estimate the size of the target cell population.
2. Washing buffer can be aliquoted into 50 mL centrifuge tubes and stored at −20°C. It can be thawed the day before and kept at +4°C.
3. Before use, heat the solution to 60°C for 10 min to optimize the enzyme activity, and allow it to cool down at room temperature [23, 24]. The indicated enzyme concentrations are optimized for species with relatively simple root architectures, such as *Arabidopsis thaliana*, *Cardamine hirsuta*, and *Capsella rubella*. For species with more complex root systems, both enzyme concentration and digestion time may need to be adjusted.
4. You can review preliminary experimental evidence or consult literature, querying transcriptomic databases that ideally provide single-cell resolution (such as the EFP browser [16], Root Cell Atlas [17], https://bioit3.irc.ugent.be/plant-sc-atlas for *A. thaliana*). You can also search publicly available promoter-, enhancer-trap lines, or similar. For example, transactivation *GAL4-UAS* marker lines with tissue-specific GFP expression are available from the Haseloff and Poethig collections [10].
5. If a fluorescent reporter is not available, design a cloning strategy for the marker of choice as either a transcriptional or translational fusion, depending on whether the expression domain of the gene or protein differs, to better reflect the expression domain that matches the cell type/population of interest. Select the fluorescent fluorophore for cloning, considering what was discussed in the main text regarding the brightness and fast maturation of the fluorophore. Possibly, if this does not affect the functionality of the translational fusion, select a fluorophore harboring a nuclear localization signal (NLS) to "concentrate" and enhance the fluorescence detection for downstream FACS.

6. If the planned FACS-RNA-seq experiments include mutant or overexpression lines, generate homozygous lines by crossing those backgrounds with the marker line.
7. If multiple independent, homozygous lines of the selected marker are available (and in case of mutants and overexpressor lines harboring the marker), verify expression consistency across lines.
8. To quantify the fluorescent intensity and perform statistical analyses, several imaging software programs are available. Software like FIJI, IMARIS, or proprietary microscope software can be used. If using FIJI, you can draw lines to select the region of interest encompassing the marker expression domain, and use the *Measure* tool to obtain the $\sum$ number of pixels in the channel (raw intensity density—RawIntDen), then normalize the results for the chosen area. Repeat this step for a region where the fluorescent marker is not expressed to subtract the background. Perform appropriate statistical analyses to compare different lines.
9. This step is important because all fluorescent cells will be sorted and pooled together for downstream RNA-seq analysis. Transcriptional heterogeneity arising from distinct cell states or subpopulations within the sorted cell population may be overlooked because no downstream bioinformatic analysis is usually applied to resolve transcriptional diversity within the marker-positive pool. Conversely, if fluorescent heterogeneity is observed in the marker line during marker validation and corresponds to biologically meaningful cell populations or cell states of interest, this heterogeneity can be exploited during FACS by applying sorting gates based on fluorescent intensity to separate distinct cell subpopulations.
10. All materials must be thoroughly cleaned with 70% ethanol before use in the laminar flow hood. Wearing gloves is not recommended during seed sterilization and handling due to electrostatic interference. Operator's hands must be washed thoroughly and, if possible, disinfected with 70% ethanol before seed manipulation and any other step performed in the laminar flow hood. On the contrary, gloves are required throughout the harvesting and protoplasting procedures.
11. Quantity should be calculated, including an additional 25% surplus, accounting for potential losses due to contamination events.
12. For Arabidopsis, two rows per plate provide an optimal arrangement. For other species, the number of rows should be adjusted according to root growth characteristics and root length.

13. Growth conditions must be adjusted according to the biological questions under investigation. The timing of sampling and experimental execution should also be determined by the specific biological process under investigation.
14. If the region of interest is the root meristem and the neighboring regions, it is recommended to cut no more than 0.3 mm from the root tip to ensure greater specificity and to limit a dilution effect of the cell population of interest.
15. Cell strainer pre-wetting facilitates the efflux of protoplast suspension through the membrane.
16. At the end of the centrifugation step, the rotor deceleration must be set to 0 to prevent protoplasts from resuspending.
17. Using a high-speed cell sorter, such as CytoFLEX SRT operating at a nozzle frequency of 32.55 kHz and set to Sort Mode: Purity (Drop Envelope 1). This setup prioritizes high purity by sorting only the target cell located within the central 50% of each droplet, avoiding contamination from leading or trailing droplets at the expense of a minimal overall yield loss.
18. A protoplast population derived from a nonfluorescent background line should be used as a blank control to set and calibrate the instrument prior to sorting.
19. Validation of marker-positive sorted cells by confocal microscopy or an equivalent detection approach is recommended, particularly when implementing the protocol for the first time. Viability dyes such as DAPI, PI, or equivalent markers may be used to identify living cells and discard dead cells, thereby minimizing the sorting of off-target cell populations.
20. Choose the most suitable RNA extraction kit considering the nature of the starting material (sorted protoplasts) and the downstream application. The kit must be compatible with RNA isolation from sorted protoplasts, and low-input formulations are needed when working with small cell populations. If the final experimental aim includes profiling small regulatory RNAs, the kit must efficiently recover these. Overall, it is critical that the chosen kit allows isolation of an RNA yield compatible with the sequencing facility's input requirements.
21. The 260 nm/280 nm ratio assesses contamination from proteins and phenolic compounds: if lower than 1.8, it indicates a likely protein, phenol, or peptide contamination; if higher than 2.1, it may indicate low concentration, baseline noise, or residual solvents. The 260 nm/230 nm ratio reflects contamination from organic compounds and salts. Lower than 1.8 suggests the likely presence of a significant contamination. A low 260/230 ratio indicates incomplete cleanup that can strongly inhibit enzymatic reactions.

22. qPCR analysis provides an additional method to verify that the correct cell population was sorted and to assess RNA quality by assessing the expression of known marker genes, including the reporter itself. However, this validation is performed at the expense of RNA consumption.

23. RIN provides a numerical assessment of RNA integrity on a 1–10 scale, with lower values indicating stronger degradation. RIN is determined from microcapillary electrophoretic RNA separation results. The algorithm considers multiple electrophoretic features, including the overall RNA ratio, the height of the 28S peak, the fast area ratio, and the marker height, to generate a comprehensive integrity score [25].

24. Single-end sequencing is generally sufficient for gene-level expression analyses, whereas paired-end sequencing improves alignment accuracy and facilitates analysis of transcript structure and alternative splicing, albeit at increased cost.

25. Poly(A)-selected RNA sequencing is recommended when samples have a RIN of 7 or higher. It provides efficient mRNA profiling and enables gene expression quantification. If RIN is compromised or a broader transcriptomic coverage is required, total RNA sequencing combined with ribosomal RNA depletion is preferable, as it detects both polyadenylated and nonpolyadenylated transcripts.

26. In `FastQC`, assess per-base sequence quality by inspecting Phred score distributions across read positions to identify quality drops at read ends. To evaluate GC content, compare the observed GC distribution to the theoretical distribution expected for the organism under investigation and library type. Inspect "Adapter Content" and "Overrepresented Sequences" modules to verify the presence of known adapter sequences. To identify potential PCR amplification biases or low library complexity, examine sequence duplication levels.

27. When assessing if the observed transcriptional changes reflect shifts in cell identity, developmental phases, or physiology, keep in mind that FACS–RNA-seq averages expression across the marker-positive population and that protoplasting and sorting induce cell stress. To identify transcriptional changes due to cell stress, inspect your dataset searching for expression of known stress- and protoplasting-responsive genes (e.g., heat shock, oxidative stress, wound-, or osmotic-stress-induced genes) and evaluate enrichment of stress-related Gene Ontology (GO) terms. Reduced expression of cell-type-specific markers or enrichment of cell-death-associated genes may also indicate cell damage. If available, you can additionally compare expression profiles with published datasets from nonprotoplasted tissues and assess whether stress-associated

genes are uniformly upregulated across replicates. All these must be considered when interpreting DEG results to distinguish biological effects from technical artifacts.

28. Unless the candidate gene shows a very narrow and specific expression pattern, the phenotypes given by the mutation might be recorded not only in the organ harboring the cell population of interest, but also in other organs.
29. Functional redundancy and/or compensatory mechanisms may mask phenotypes in targeted knockouts, particularly in gene families. Hence, perform a BLAST analysis and check in the literature whether other homologs are present in the genome of the species under investigation, whether they might be expressed in a similar pattern and, in case, acquire also those mutants if available to perform similar phenotypic characterization and to generate double or higher order mutants.
30. Tissue-specific KO (TSKO) restricts gene disruption to the population of cells of interest to minimize pleiotropic or systemic effects, unless the chosen promoter is also expressed in other organs apart from the one of interest [21].
31. Based on the RNA-seq datasets, additional cell-type-specific promoters can be used to generate TSKOs and inducible TSKOs upon validation of the candidate promoter [21, 22].
32. When applicable, secondary transcriptomic or targeted expression analyses in KO or overexpression backgrounds can be performed to identify downstream regulatory circuits.

Acknowledgments

This work was supported by Ministero dell'Università e Ricerca (grant number P2022B4C5H) to RDM.

Conflict of Interest The authors have no conflicts of interest to declare that are relevant to the content of this chapter.

References

1. Svolacchia N, Salvi E, Sabatini S (2020) Arabidopsis primary root growth: let it grow, can't hold it back anymore. Curr Opin Plant Biol 57:133–141. https://doi.org/10.1016/j.pbi.2020.08.005
2. Kyndt T, Denil S, Haegeman A et al (2012) Transcriptome analysis of rice mature root tissue and root tips in early development by massive parallel sequencing. J Exp Bot 63:2141–2157. https://doi.org/10.1093/jxb/err435
3. Protocol: Basic Bulk RNA-Seq Work Flow|PGRP Roots. https://sites.lsa.umich.edu/pgrp-roots/protocol-basic-rna-seq-work-flow/. Accessed 28 Dec 2025
4. Birnbaum K, Shasha DE, Wang JY et al (2003) A gene expression map of the Arabidopsis

root. Science 302:1956–1960. https://doi.org/10.1126/science.1090022
5. Sozzani R, Cui H, Moreno-Risueno MA et al (2010) Spatiotemporal regulation of cell-cycle genes by SHORTROOT links patterning and growth. Nature 466:128–132. https://doi.org/10.1038/nature09143
6. Di Mambro R, Svolacchia N, Dello Ioio R et al (2019) The lateral root cap acts as an auxin sink that controls meristem size. Curr Biol 29:1199–1205.e4. https://doi.org/10.1016/j.cub.2019.02.022
7. Ortiz-Ramírez C, Guillotin B, Xu X et al (2021) Ground tissue circuitry regulates organ complexity in maize and Setaria. Science https://doi.org/10.1126/science.abj2327
8. Serrano-Ron L, Perez-Garcia P, Sanchez-Corrionero A et al (2021) Reconstruction of lateral root formation through single-cell RNA sequencing reveals order of tissue initiation. Mol Plant 14:1362–1378. https://doi.org/10.1016/j.molp.2021.05.028
9. Iyer-Pascuzzi AS, Benfey PN (2010) Fluorescence-activated cell sorting in plant developmental biology. In: Hennig L, Köhler C (eds) Plant developmental biology: methods and protocols. Humana Press, Totowa, NJ, pp 313–319
10. Eurasian Arabidopsis Stock Centre (uNASC). https://arabidopsis.info/. Accessed 28 Dec 2025
11. Machin FQ, Beckers M, Tian X et al (2019) Inducible reporter/driver lines for the Arabidopsis root with intrinsic reporting of activity state. Plant J 98:153–164. https://doi.org/10.1111/tpj.14192
12. Petersson SV, Lindén P, Moritz T, Ljung K (2015) Cell-type specific metabolic profiling of *Arabidopsis thaliana* protoplasts as a tool for plant systems biology. Metabolomics 11:1679–1689. https://doi.org/10.1007/s11306-015-0814-7
13. Dong A, Chan ITC, Cheung TH (2023) ATAC-seq protocol for the profiling of chromatin accessibility of *in situ* fixed quiescent and activated muscle stem cells. STAR Protoc 4:102376. https://doi.org/10.1016/j.xpro.2023.102376
14. Shi X, Fan W, Mehrpouyan M et al (2024) Flow cytometry analysis of protein expression using antibody-derived tags followed by CITE-Seq. Cytometry Part A 105:62–73. https://doi.org/10.1002/cyto.a.24792
15. TAIR-Home. https://www.arabidopsis.org/. Accessed 28 Dec 2025
16. Jamie Waese NP, Pasha A, Sullivan A, The BAR and other Data Analysis Tools for Plant Biology. In: https://bar.utoronto.ca/. https://bar.utoronto.ca/. Accessed 28 Dec 2025
17. Root Cell Atlas. https://rootcellatlas.org/. Accessed 28 Dec 2025
18. Clark NM, Fisher AP, Sozzani R (2018) Identifying differentially expressed genes using fluorescence-activated cell sorting (FACS) and RNA sequencing from low input samples. Methods Mol Biol 1819:139–151. https://doi.org/10.1007/978-1-4939-8618-7_6
19. Hu R, Islam MN, Varghese RS, Ressom HW (2024) Short-read RNA-Seq. Methods Mol Biol 2822:245–262. https://doi.org/10.1007/978-1-0716-3918-4_17
20. Conesa A, Götz S, García-Gómez JM et al (2005) Blast2GO: a universal tool for annotation, visualization and analysis in functional genomics research. Bioinformatics 21:3674–3676. https://doi.org/10.1093/bioinformatics/bti610
21. Decaestecker W, Buono RA, Pfeiffer ML et al (2019) CRISPR-TSKO: a technique for efficient mutagenesis in specific cell types, tissues, or organs in Arabidopsis. Plant Cell 31:2868–2887. https://doi.org/10.1105/tpc.19.00454
22. Wang X, Ye L, Lyu M et al (2020) An inducible genome editing system for plants. Nat Plants 6:766–722. https://doi.org/10.1038/s41477-020-0695-2
23. Köster T, Marondedze C, Meyer K, Staiger D (2017) RNA-binding proteins revisited–the emerging Arabidopsis mRNA interactome. Trends Plant Sci 22:512–526. https://doi.org/10.1016/j.tplants.2017.03.009
24. Zhang Z, Boonen K, Li M, Geuten K (2017) mRNA interactome capture from plant protoplasts. J Vis Exp 56011. https://doi.org/10.3791/56011
25. Schroeder A, Mueller O, Stocker S et al (2006) The RIN: an RNA integrity number for assigning integrity values to RNA measurements. BMC Molecular Biol 7:3. https://doi.org/10.1186/1471-2199-7-3

Chapter 18

Single-Cell RNA-Sequencing in Nodulated Roots of *Medicago truncatula*

Sophia Müller, Kelvin Adema, Amber van Seters, Jana Witmer, Wouter Kohlen and Carolin Grones

Abstract

This section details the optimized methods for single-cell RNA sequencing (scRNA-seq) in *Medicago truncatula* roots and nodules. We describe a detailed experimental workflow, comprising plant growth and bacteria application, protoplast isolation, quality control, and library processing according to 10× Genomics®, suitable for both long- and short-read sequencing. This protocol can serve as a starting point for establishing robust, high-quality scRNA-seq workflows to study nodulation and root development in legumes.

Keywords *Medicago truncatula* , Single-cell RNA sequencing, Protoplast, Nodule

1 Introduction

The symbiotic process of nodulation in legumes has long captured the interest of plant biologists, both for its agronomic value by enabling atmospheric nitrogen fixation and reducing fertilizer inputs and for its striking developmental plasticity [1,2]. In *Medicago truncatula*, infection by compatible rhizobia bacteria (e.g., *Sinorhizobium meliloti)* triggers the formation of indeterminate nodules whose meristem arises from differentiated root tissues that undergo extensive developmental reprogramming [3]. Early cellular events in the Medicago root during nodule formation include the initiation of cell divisions in the pericycle, which later on expand into the endodermis and inner cortex, with the future meristem emerging from the third inner cortex layer [4]. Single-cell RNA sequencing (scRNA-seq) offers the resolution needed to capture and dissect early molecular events at the cell-type and even cell-stage levels to establish a nodule in legumes [5–9]. The general workflow to obtain single-cell transcriptomes includes tissue collection, isolation of single cells (here referred to as protoplasts) or nuclei, enrichment, cell compartmentalization (e.g.,

Steffen Vanneste and Inge Verstraeten (eds.), *Root Development*, Methods in Molecular Biology, vol. 3086,
https://doi.org/10.1007/978-1-0716-5603-7_18,

using microfluidics) and processing, library generation, sequencing, and subsequent computational processing, including quality control, clustering, and visualization (reviewed in detail in [10,11]). A major bottleneck for plant scRNA-seq, however, is the efficient isolation of intact single cells. Protoplast isolation from intact plant tissue requires the enzymatic digestion of the plant cell walls. Yet digestion efficiency varies with cell-type- and stage-specific cell wall composition and typically requires extensive optimization. Furthermore, high yields of viable cells are critical to ensure that the resulting transcriptional data are reproducible and free from preparation-induced artifacts.

This chapter provides a detailed protocol, including practical guidance, alternative methods, and troubleshooting tips, for processing *Medicago truncatula* roots and nodules to obtain large numbers of viable protoplasts suitable for scRNA-seq, with an emphasis on speed, efficiency, and preservation of transcriptome integrity.

2 Materials

Prepare all solutions in Milli-Q water unless otherwise stated, and store at room temperature.

2.1 Seed Sterilization and Plant Growth

1. Spore-elements β-solution: Mix 0.0354 g/L $CuSO_4 \bullet 5H_2O$, 0.462 g/L $MnSO_4 \bullet H_2O$, 0.974 g/L $ZnSO_4 \bullet 7H_2O$, 1.269 g/L H_3BO_3, and 0.398 g/L $Na_2MoO_4 \bullet 2H_2O$ in water.
2. Fåhraeus (FA) medium: Mix 0.12 g/L $MgSO_4 \bullet 7H_2O$, 0.10 g/L KH_2PO_4, 0.15 g/L $Na_2HPO_4 \bullet 2H_2O$, 1 mL/L 15 mM Fe-Citrate, 1 mL/L 125 mM $Ca(NO_3)_2$, and 2.50 mL/L Spore-elements β-solution in water. Adjust the pH to 6.7, add 0.9% Daishin agar, and autoclave (*see* **Note 1**).
3. FA + carbendazim medium: Prepare FA medium as described above. After autoclaving, cool the medium to approximately 50°C and add filter-sterilized carbendazim to a final concentration of 1 g/L.
4. Sterile 120 × 120 mm square Petri dishes.
5. Sterile 100 × 100 mm autoclaved filter paper.
6. ~300 seeds of *Medicago truncatula* Jemalong A17.
7. Concentrated (95%–98%) H_2SO_4.
8. 12% NaOCl.
9. Autoclaved water.
10. Sterile tweezers.
11. Micropore tape.
12. Light-protecting plastic covers or aluminum foil pouches to protect the roots.

2.2 Rhizobial Culture and Application

1. A culture of *Sinorhizobium meliloti* 2011 (Sm2011).
2. Yeast Extract Mannitol (YEM): Mix 5 g/L mannitol, 5 g/L sodium gluconate, 0.5 g/L yeast extract, 0.2 g/L $MgSO_4 \times 7$ H_2O, 0.1 g/L NaCl, 0.5 g/L K_2HPO_4, 1 mL/L 16% (w/v) $CaCl_2$ in water and autoclave. For a solid variant, add 1.5% agar prior to autoclaving. After autoclaving, let the medium cool to approximately 50°C, then add filter-sterilized tetracycline to a final concentration of 10 μg/mL.
3. Liquid Fåhraeus (FA) medium: Mix 0.12 g/L $MgSO_4 \bullet 7H_2O$, 0.10 g/L KH_2PO_4, 0.15 g/L $Na_2HPO_4 \bullet 2H_2O$, 1 mL/L 15 mM Fe-Citrate, 1 mL/L 125 mM $Ca(NO_3)_2$, and 2.50 mL/L Spore-elements β-solution in water. Adjust to pH 6.7 and autoclave.
4. OD meter.
5. Cuvettes for OD meter.

2.3 Generating Protoplasts

The following enzyme buffer was optimized for isolating Medicago root protoplasts and will likely require further optimization and adjustment for other tissues and species. Buffers are best prepared fresh on the day of the experiment.

1. Enzyme buffer: Mix 400 mM D-sorbitol, 10 mM MES (pH 5.7), 10 mM $CaCl_2$, 1.5% (w/v) Cellulase R10, 0.5% (w/v) Macerozyme R10 in water. Heat the enzyme buffer to 55°C for 10 min, let it cool to room temperature, and add 5% (v/v) Viscozyme and 1% (w/v) bovine serum albumin (BSA).
2. Wash buffer: Mix 400 mM D-sorbitol, 10 mM MES (pH 5.7), 10 mM $CaCl_2$, 0.1% (w/v) BSA in water.
3. Staining solution: Mix 1 mL wash buffer, 8 μg/mL fluorescein diacetate (FDA), and 2 μg/mL propidium iodide (PI). Keep the solution on ice and shield from light.
4. Scalpel with razor blade (e.g., blade size 22).
5. Six-well plates.
6. Soft-tip tweezers.
7. 15 mL or 50 mL conical centrifuge tubes.
8. RNase/DNase-free 2 mL tubes (*see* **Note 2**).
9. RNase/DNase-free filtered wide-bore pipette tips (200 μL and 1000 μL) (*see* **Note 3**).
10. Cell strainers with mesh sizes of 100, 70, and 40 μm, where the 40 μm and 70 μm strainers are mini formats compatible with 2 mL tubes (e.g., pluriStrainer Mini).
11. Hemocytometer.
12. Vacuum pump.

13. Orbital shaker (30–85 rpm).
14. Swinging-bucket centrifuge with adaptors for conical centrifuge tubes.
15. Tabletop centrifuge.
16. Light microscope with differential interference contrast optics.

2.4 Library Preparation for Single-Cell Transcriptomics

1. Chromium Next GEM Single Cell 3′ Reagent kits (V3.1).
2. Nuclease-free water.
3. SPRIselect Reagent Kit (Beckman Coulter cat. B23318).
4. Glycerin solution: 50% (v/v) glycerol in water.
5. Ethanol, Pure.
6. High Sensitivity DNA Kit (Agilent, 5067-4626).
7. Qubit dsDNA HS Assay Kit (Thermo Fisher Scientific, Q32854).
8. RNase/DNase-free 1.5 mL (*see* **Note 2**).
9. RNase/DNase-free PCR Tubes 0.2 mL 8-tube strips.
10. Emulsion-safe pipette tips (*see* **Note 4**).
11. Magnets for 1.5 mL microcentrifuge and PCR tubes.
12. Chromium Controller (10× Genomics) with associated consumables, including the manufacturer-recommended vortex adapter, gasket, and Chromium NextGEM chip (e.g., Chromium NextGEM Chip G, 10× Genomics, cat. no. 1000127 or 1000120), and a compatible 96-well PCR plate (e.g., Eppendorf twin.tec® Trace PCR Plate 96 LoBind, Eppendorf cat. no. 0030129822).
13. PCR machine (ideally with Deep Well adapter).
14. Vortex.
15. Tabletop centrifuge.
16. Temperature-controlled heat block.
17. Agilent 2100 Bioanalyzer.
18. Qubit Fluorometer.
19. Optional: If more than two samples are processed simultaneously, it is recommended to use an 8-channel pipette with a 20–200 μL volume range.

3 Methods

Carry out all experimental procedures at room temperature, unless specified otherwise.

3.1 Plant Growth and Material Generation

1. In a sterile flow hood, prepare Petri dishes with FA medium and let the medium solidify. Place one clean and autoclaved filter paper on each plate, using sterile tweezers. Leave the plates open without a lid for 5 min to allow excess water to evaporate.
2. Surface-sterilize the seeds by placing them in a 50-mL tube and submerging them in concentrated H_2SO_4. Incubate for 7 min, while gently inverting the tube occasionally to ensure thorough mixing.
3. Carefully discard the H_2SO_4 and rinse the seeds with 40 mL of autoclaved water for 1 min, then discard the rinse. Repeat this water-wash step five additional times.
4. Submerge the seeds in 40 mL of 12% NaOCl and incubate for 7 min, again gently inverting the tube occasionally.
5. Carefully discard the 12% NaOCl and rinse the seeds with 40 mL of autoclaved water for 1 min, discarding the rinse afterward. Repeat this water-wash step five more times.
6. Keep the seeds in autoclaved water for 1–3 h and protect them from light by covering the tube with aluminum foil.
7. Prepare Petri dishes with FA+carbendazim medium and let the medium solidify. Place one clean, autoclaved filter paper on each plate using sterile tweezers. Leave the plates open without a lid for 5 min to allow excess water to evaporate.
8. Transfer the seeds to filter-covered plates containing FA+carbendazim medium, cover with aluminum foil, and place at 4°C for 20–24 h for stratification (*see* **Note 5** and Fig. 1A).
9. Allow seeds to germinate at room temperature for 15–20 h, covered with aluminum foil. Ensure that plates are placed lid-down to allow uniform root growth in response to gravity.
10. Place the plates in a sterile flow hood and crack seed pots using sterile tweezers. Add 40 mL of autoclaved water to the plate to help seed coat removal.
11. Place eight to ten seeds on a filter-covered plate with FA medium using sterile tweezers (see Fig. 1B).
12. Close the plates with micropore tape on three sides, leaving one side open to allow gas exchange (*see* **Note 6**).
13. Cover the seeds with a light-protecting plastic cover (or aluminum foil) to protect the roots from excess light.
14. Place plates vertically in growth cabinets at 22°C and a light/dark cycle of 16 h/8 h.
15. Let the seedlings grow for 5–6 days. Treat or harvest roots for further processing.

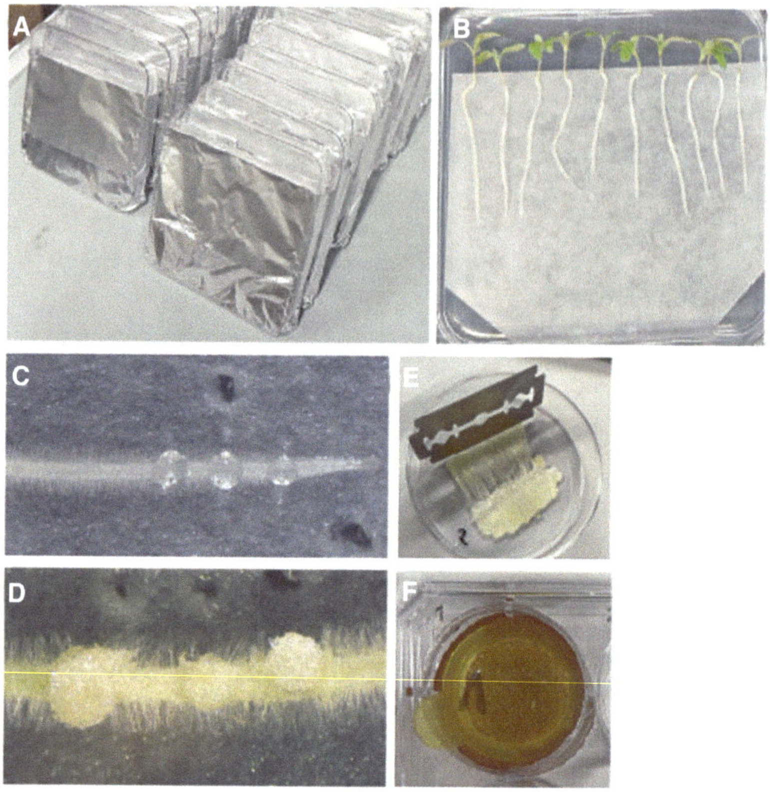

Fig. 1 In vitro growth conditions and root processing for single-cell transcriptomics of *Medicago truncatula*. (**A**) In vitro growth conditions for *Medicago truncatula* seedlings used for single-cell transcriptomic analysis following bacterial treatment. Plates were covered with foil to prevent excessive light exposure to the roots. (**B**) Representative plate showing seedling distribution and root length at the time of infection. (**C**, **D**) Root infected with bacteria in the susceptible zone, with (**C**) showing bacterial droplets at the time of inoculation and (**D**) showing formed nodules 96 h after infection. (**E**) Representative image illustrating the cutting of root tissue into small pieces to accelerate cell wall digestion during protoplast isolation. (**F**) Representative image of root fragments in digestion buffer. Root pieces were placed in a cell strainer during digestion, which can help reduce background cell debris in the resulting protoplast suspension

3.2 Bacteria Application

1. For bacteria treatment: Grow *Sinorhizobium meliloti* 2011 (Sm2011) on YEM agar for 2 days at 28°C to obtain colonies.
2. Pick a colony and transfer to liquid YEM and grow for 2 days at 28°C. Take 500 μL of this culture and dilute it with 4.5 mL liquid YEM. Grow it overnight (approximately 18 h) at 28°C to an OD_{600} of 0.6–0.9.
3. Dilute the culture to an OD_{600} of 0.02 using liquid FA medium (no antibiotics are added).

4. Apply $3 \times 2\,\mu L$ of this bacterial suspension to the root of 5–6 day old seedlings within the susceptible zone (5–15 mm above the root tip; see Fig. 1C, D). For mock treatment, replace the bacterial suspension with an equal amount of liquid FA medium.
5. Mark the positions of the bacterial spots on the filter (e.g., with a needle), carefully close the plate with micropore tape, and cover the roots with a light-protective plastic cover. Return the plates to the growth room until further processing.

3.3 Protoplast Isolation and Purification

1. Harvest the roots by cutting the infected region (marked during infection; see Fig. 1C, D). Immediately transfer the collected material into a droplet of enzyme buffer and keep all harvested tissue submerged until collection is complete to prevent drying (*see* **Note 7**).
2. Cut the harvested material with a razor blade into smaller pieces of approximately 1 mm (see Fig. 1E).
3. Transfer the harvested tissue to a 100 μm cell strainer and into a six-well plate (*see* **Note 8** and Fig. 1F).
4. Carefully add enzyme solution (*see* **Note 9**).
5. Incubate the mixture for 15 min under gentle vacuum (e.g., approximately 200–400 mbar) to enhance enzyme infiltration, then release the vacuum slowly (can take up to 5 min) and incubate the tissue for 2 h at 85 rpm in the dark (*see* **Notes 10 and 11** and Figs. 2 and 3). During incubation, gently compress the tissue two to three times with soft-tipped tweezers to promote protoplast release.

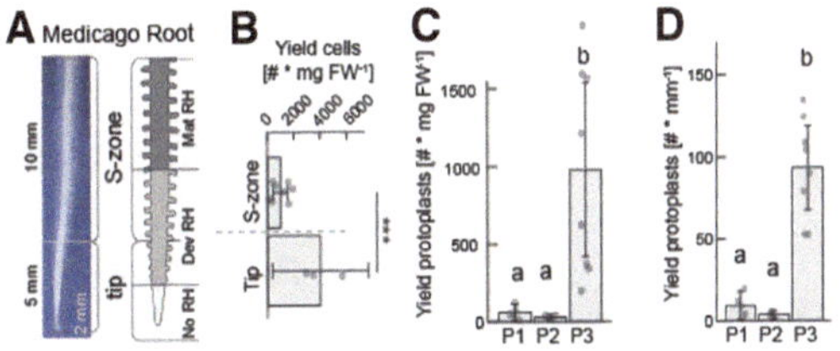

Fig. 2 Comparison of protoplast isolation efficiency across three digestion buffers. (**A**) Schematic of a Medicago root showing sampled regions: the 2–5 mm root tip lacking root hairs (No RH) and the 10 mm susceptible zone (S-zone) above the tip, where root hairs are developing (Dev RH) or mature (Mat RH). (**B**) Protoplast yield comparison between root tip and S-zone. (**C**, **D**) Protoplast yields obtained using isolation methods P1–P3, expressed as protoplasts per mg fresh weight (FW; **C**) or per mm root length (**D**), excluding root tip samples. Bars represent means with 95% CI ($n \geq 2$). Different letters indicate significant differences (ANOVA with Tukey post hoc test, $p \leq 0.05$). Asterisks denote significance between treatments (***$p \leq 0.001$)

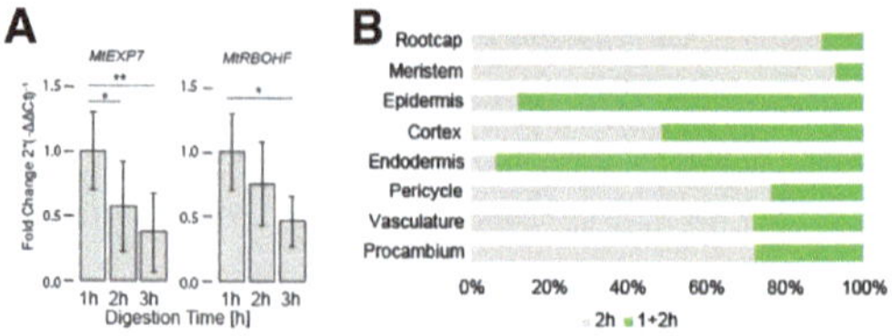

Fig. 3 Effect of enzymatic digestion duration on recovery of distinct cell types. (**A**) Expression of the epidermal markers *MtEXP7* and *MtRBOHF* in protoplast suspensions generated using digestion times of 1, 2, or 3 h. Fold-change expression values are normalized to the housekeeping gene *MtUBQ*. Bars represent the mean ± 95% CI ($n \geq 2$). Asterisks denote significant differences between treatments (*$p \leq 0.05$, **$p \leq 0.01$, ***$p \leq 0.001$). (**B**) Cell type representation in single-cell datasets generated from Medicago root protoplasts obtained after a 2-h digestion, and a combined sample containing protoplasts from both 1- and 2-h digestions. The bar graph shows the comparison of how many cells (in percent) were obtained for each cell type. A higher percentage after the 2-h digestion indicates that most cells captured in the combined dataset originate from the longer digestion, suggesting that extended incubation is more effective for that cell type. Conversely, a higher percentage in the combined 1 + 2-h digestion sample indicates that including both short and long digestions improves capture for that cell type. Only the cortex cell type appears to be independent of digestion duration, showing equal contribution from both conditions

6. Pre-wet a 100 μm cell strainer with 1 mL of wash buffer within a new Petri dish.
7. Transfer the enzyme-protoplast mixture with a 1000 μL wide-bore tip to the pre-wetted 100 μm cell strainer. Tilt the dish to the side so that cells collect at the bottom, and circle a few times to maximize protoplast recovery.
8. Keep the protoplast suspension on ice or cooled to 4°C from this step onwards to minimize enzyme activity and reduce the risk of protoplast lysis (*see* **Note 12**).
9. Transfer the filtered protoplasts to a 50 mL conical centrifuge tube with a 1000 μL wide-bore tip and gentle aspiration. Keep the tube at an angle while gently releasing the solution along the walls.
10. Centrifuge at 100 × *g* for 3 min (4°C) in a swinging bucket centrifuge. Acceleration and brake level should be set to low (*see* **Note 13**).
11. Carefully pipette off the supernatant without disturbing the protoplast pellet at the bottom of the tube.
12. Pre-wet a 40 and a 70 μm cell strainer with some wash buffer and place each into an RNase/DNase-free 2 mL tube.
13. Resuspend the protoplast pellet in 1000 μL wash buffer and transfer to the 70 μm cell strainer.

14. Remove the strainer and centrifuge the tube at 100×*g* for 3 min (4°C) using a tabletop centrifuge.
15. Carefully pipette off the supernatant without disturbing the protoplast pellet.
16. Place a pre-wetted 40 μm cell stainer on top of a new RNase/DNase-free 2 mL tube.
17. Resuspend the protoplasts in 50 μL of pre-chilled wash buffer. Strain the protoplast solution through the pre-wetted 40 μm cell strainer (*see* **Note 14**).
18. Aspirate the protoplast suspension using a pipette with a wide-bore tip and adjust the aspirated volume to obtain the final precise suspension volume.
19. Take 10 μL protoplast suspension and mix with 10 μL staining solution in a new RNase/DNase-free 2 mL tube. Incubate in the dark for 3 min (at room temperature). Add 10 μL of the stained protoplast solution to a hemocytometer and count the number of intact, round cells under a DIC light microscope (*see* **Notes 15 and 16** and Figs. 4 and 5).
20. Determine cell concentration and viability. Adjust to a concentration of 700–1200 cells/μL for use in Sect. 3.4.
21. Optional: To ensure that all cell types are represented in the protoplast suspension, a separate preliminary test can be performed in which an independently prepared protoplast suspension is snap-frozen in liquid nitrogen and later used for RNA extraction to assess the expression of cell-type-specific marker genes by qRT-PCR (*see* **Note 17**).

3.4 Single-Cell Transcriptomics Library Preparation and Sequencing

1. Proceed according to the instruction from either 10× Chromium Next GEM 3′ v3.1 rev E Demonstrated protocol (CG000315 Rev E, for Dual index libraries) or 10× Chromium

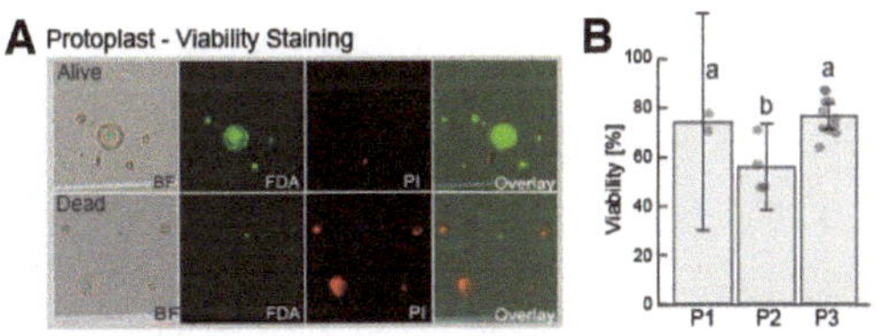

Fig. 4 Assessment and quantification of protoplast viability. (**A**) Fluorescence microscopy images illustrating protoplast morphology and viability. Brightfield (BF) images (left) show cell shape; fluorescein diacetate (FDA, middle) marks viable cells, and propidium iodide (PI, right) marks dead cells. The top row shows an intact, FDA-positive viable protoplast; the bottom row shows PI-positive dead protoplasts with irregular shapes. (**B**) Quantification of cell viability across three different protocols (P1–P3), with P3 representing our optimized protocol for protoplast isolation

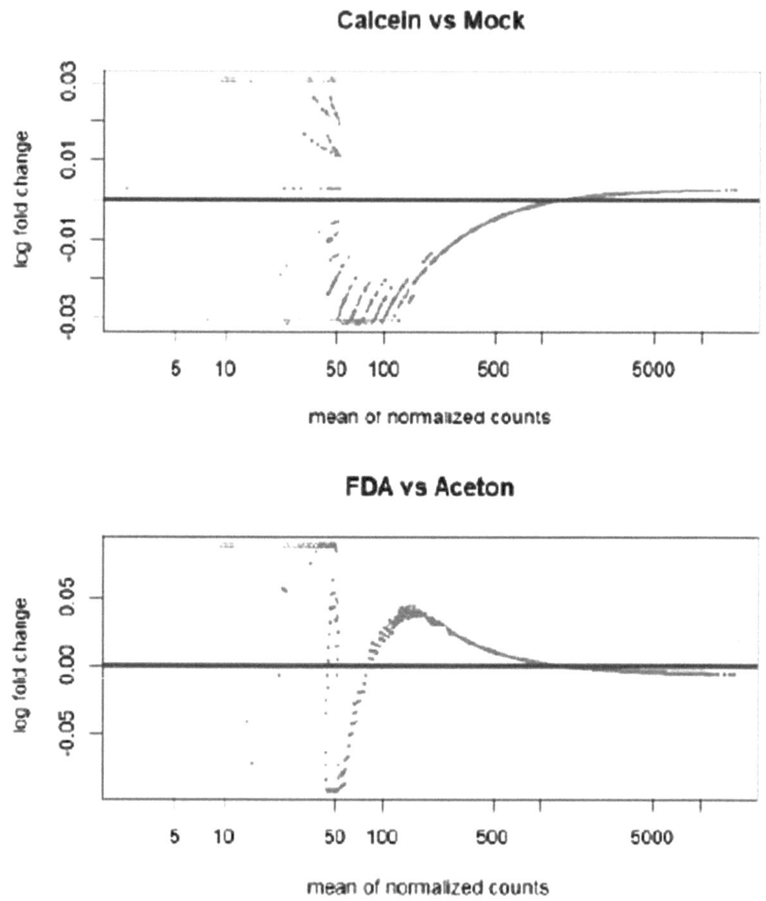

Fig. 5 Assessment of transcriptional effects of cell viability staining. Whole transcriptome RNA-seq data were generated from Arabidopsis root protoplast suspensions treated for 15 min with 18 μM calceinFM, mock (DMSO), 100 μ M fluorescein diacetate (FDA), or acetone (mock for FDA). Each treatment was performed in three biological replicates. MA plots show the $\log_2$ fold-change in gene expression between calceinFM and DMSO (top panel) or FDA and acetone (bottom panel). Genes are ordered along the *x*-axis by their mean normalized expression, with lowly expressed genes on the left and highly expressed genes on the right. No genes were found to be differentially expressed at a statistically significant level (adjusted *p*-value >0.05)

Next GEM 3′ v3.1 Rev D Demonstrated protocol (CG000204 Rev D, for Single index libraries) (*see* **Note 18**).

2. Use 15,000–20,000 cells as input per sample, resulting in an estimated recovery rate of approximately 10,000 cells and a multiplet rate below 8%.
3. Proceed by following the manual instructions from 10× Genomics. Adjust the number of PCR cycles needed to

construct the cDNA library according to the estimated number of recovered cells.

4. Indexed NGS libraries were generated according to the manufacturer's instructions (*see* **Note 19** and Fig. 6).

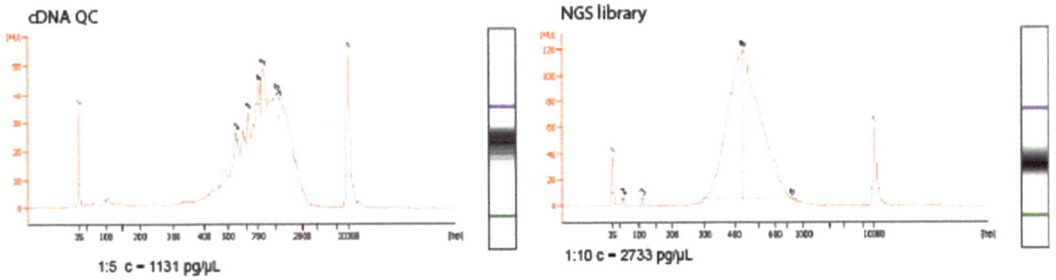

Fig. 6 Library fragment size profiles obtained using the 10× Chromium platform. Fragment size distributions of cDNA (left panel) and NGS library (right panel), corresponding to protocol Steps 2.4 and 3.7 for 10× Genomics

A

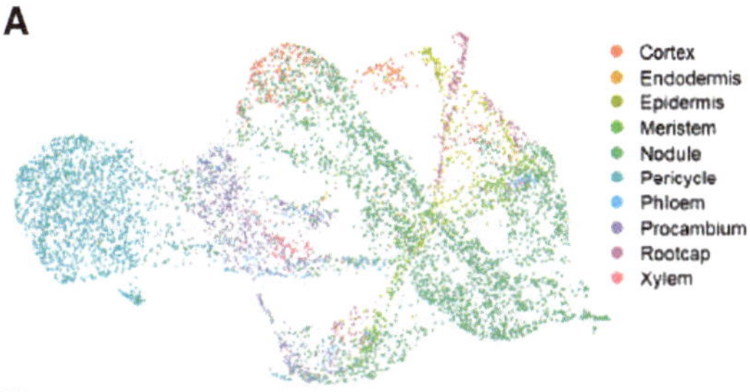

B

Sample	Mock	24 h	48 h	72 h	96 h
median_UMI	5639	7999	7347	12868	10292
median_genes	2525	3220	2861	3348	3395
mean_UMI	6622	10120	12814	17002	13806
mean_genes	2624	3405	3271	3505	3580
n_cells	848	1583	1433	1858	1978

C

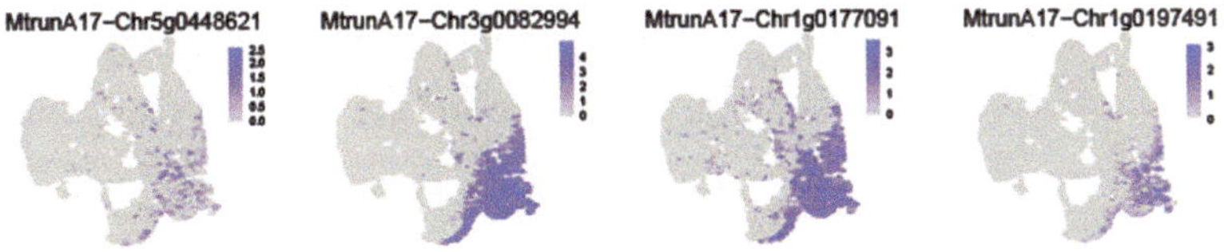

Fig. 7 Integrated single-cell RNA-seq dataset from the susceptible zone of Medicago roots under mock and bacterial treatment. (**A**) Uniform Manifold Approximation and Projection (UMAP) visualization after integration of single-cell RNA-seq datasets from the susceptible zone of Medicago roots treated with mock or bacteria for 24, 48, 72, and 96 h. Colors indicate the annotated cell types. (**B**) Summary of general sequencing and cell-level statistics for each sample used in panel (**A**). **(C)** UMAP showing the expression patterns of nodule marker genes across the integrated dataset

5. Libraries are sequenced using paired-end 100 bp reads with an intended sequencing depth of 30–75 Gb/sample (*see* **Note 20**).
6. Optional: Combine short- and long-read sequencing to enhance scRNA-seq resolution. Long-read platforms such as Oxford Nanopore enable full-length transcript reconstruction, improved isoform detection, and more accurate analysis of alternative splicing and gene family members. To generate long-read data from the same single-cell experiment, process the amplified cDNA library into a Nanopore-compatible library [12] (*see* **Note 21**).
7. As a reference, we provide a summary of general quality control parameters for single-cell datasets after bacterial application (*see* **Note 22** and Fig. 7).
8. In the case of incomplete cell type recovery (e.g., low capture of endodermis cells), single nuclei RNA-seq could be considered as a complementary approach (*see* **Note 23**).

4 Notes

1. To increase nodulation efficiency, nitrogen concentrations in the plant growth medium must be strictly limited. To ensure this, we utilize a low-nitrogen FA medium throughout the stages of plant growth and bacterial infection. The liquid version of this low-nitrogen medium also serves as the diluent for the bacterial inoculum and mock inoculum. This maintains consistency in nitrogen availability across plates and prevents nutrient imbalances that could introduce confounding variables into the data.
2. We recommend using RNase/DNase-free reaction tubes to minimize sample contamination and RNA degradation, which could compromise scRNA-seq data quality. Optionally, and particularly during library preparation, RNase/DNase-free LoBind tubes can be used to further improve RNA recovery and stability by reducing adsorption of nucleic acids to the tube walls.
3. We recommend using wide-bore tips for all pipetting steps during protoplast isolation to minimize mechanical damage to protoplasts.
4. The manufacturer recommends using only validated emulsion-safe plastic consumables during sample and library preparation, as some plastics can destabilize gel bead-in-emulsions (GEMs). Following this guidance, Rainin LTS low-retention tips were used throughout the workflow.

5. Representative images depicting critical steps of the protocol are provided in Fig. 1.
6. Closing the plates completely with micropore tape causes the roots to dry out.
7. We recommend keeping the duration of material harvest as short as possible. To facilitate this, assistance from colleagues can help reduce harvesting time. Additionally, using an indicator of the required length (e.g., indicating the range of 5–15 mm resembling the length of the application zone) is useful to ensure consistent harvesting across all samples.
8. Enzymatic digestion can be performed in various types of vessels. In this study, we used six-well plates as they prove practical for the following washing steps; however, glassware such as Erlenmeyer flasks or glass Petri dishes can also be employed, as cells are less prone to adhere to glass surfaces.
9. Our digestion buffer was optimized to maximize the efficiency of protoplast isolation from both the root meristem and the susceptible zone, including root hairs (Fig. 2A). The effectiveness of enzymatic cell-wall digestion varies across cell types and developmental stages, as reflected by the lower protoplast yield obtained from the Medicago root tip compared to the susceptible zone (Fig. 2B). To assess buffer performance and compatibility, we benchmarked three protocols. P1 differed from our buffer by containing 2% macerozyme and using a high-calcium W5 (154 mM NaCl, 125 mM CaCl2, 5 mM KCl, 10 mM MES, pH 5.7) wash buffer (total digestion time: 2 h). P2 reduced macerozyme to 0.5% (w/v) and used PBS+ (PBS supplemented with 0.04% BSA) for washing (total digestion time: 2 h). P3 corresponds to the enzyme and wash buffers described in this manuscript (total digestion time: 2 h). Our optimized protocol (P3) yielded significantly more protoplasts per milligram of fresh weight and per millimeter of root length, indicating that it also enhanced protoplast release from more mature cells in the susceptible zone compared with the alternative protocols (Fig. 2C, D).
10. The optimal activity temperature of the enzyme in the digestion buffer is 50–60°C [13]. However, performing the digestion at such high temperatures would impose severe heat stress on the plant tissue and is therefore not recommended. Instead, the digestion temperature should be kept as close as possible to the plants' growth temperature.
11. Digestion time should be tailored to the tissue types and cell populations of interest. In our experiments, the expression of epidermal cell markers dropped significantly in protoplast suspensions generated with digestion times longer than 1 h (Fig. 3A). In contrast, extending digestion to 2 h increased

protoplast concentration and recovery of also inner root tissues in single-cell datasets, including pericycle, vasculature, and meristematic cells (Fig. 3B). To balance representation across tissue layers, we opted to process the material in two batches: half of the sample was digested for 2 h, and the remaining half for 1 h. The resulting protoplast suspensions were collected simultaneously, pooled at a 1:1 ratio, and processed together through the straining and washing steps as a single combined sample.

12. Keeping the protoplast suspension on ice and performing the centrifugation steps at low temperature is optional. Alternatively, all steps can be carried out at room temperature to avoid temperature-induced transcriptional perturbations.

13. To minimize bursting of fragile protoplasts, we recommend using gentle centrifugation conditions (100–250 × *g*) with low acceleration and deceleration settings.

14. Droplet-based systems such as 10×Genomics rely on microfluidics, which require the protoplasts to pass through narrow channels that can accommodate only particles smaller than 40 μm in diameter (recommended by the technology provider). Loading excessive debris, oversized protoplasts, or large protoplast clusters can clog the system, ultimately leading to sample loss or incomplete capture. However, this approach may also exclude larger protoplasts and certain cell states. As an alternative, nanowell-based single-cell platforms (e.g., BD Rhapsody) can be used, as they are less sensitive to cell-size constraints.

15. Protoplast viability can be assessed by examining cell morphology: live protoplasts appear evenly round, whereas dead protoplasts are often burst or irregularly shaped (Fig. 4A). Additionally, staining with cell viability dyes (e.g.,FDA for live cells and PI for dead cells) provides robust quantitative viability estimates (Fig. 4B). Using our optimized protocol (P3), protoplasts consistently exhibited high viability (>70%), which meets commonly recommended thresholds for downstream single-cell platforms. As an alternative clean-up step, viable protoplasts can be enriched by sorting FDA-positive cells using flow cytometry or by magnetic levitation (e.g., LeviCell, Levitas Bio), approaches that have been successfully applied to enrich viable protoplasts from Medicago roots in our hands, but have not yet been combined with subsequent workflows for scRNA-seq.

16. Several cell viability stains are available for labeling viable cells. In our study, we primarily used calceinFM and FDA, observing that FDA produced more robust staining and brighter fluorescence. Although long-term exposure to FDA has been

reported to induce cell death [14], bulk transcriptome analysis of protoplasts treated with FDA or calceinFM for 15 min revealed no significant transcriptional changes under our conditions (Fig. 5).

17. Isolation of RNA from low-input material, such as protoplast suspensions, can be challenging. We have successfully obtained sufficient total RNA for cDNA synthesis and qRT-PCR using the E.Z.N.A. Plant RNA Kit (Omega Bio-Tek) by following the manufacturer's instructions.
18. Keep the processing time as short as possible after protoplast generation and before loading. Smaller meristematic cells tend to aggregate and form clumps, which can obstruct the 10×Chromium Controller. To mitigate this, we recommend minimizing the number of samples processed per chip, as protoplast viability declines the longer the workflow takes, and processing multiple samples in parallel can introduce additional delays (e.g., during cell counting, viability assessment, and dilution adjustments).
19. Bioanalyzer traces, generated using the Agilent® DNA High Sensitivity Kit, should display a broad peak ranging from ~300–2000 bp for the cDNA and ~300–700 bp for the NGS library (Fig. 6). Lower molecular weight product (35–150 bp) may be present. This is normal and does not affect sequencing or application performance.
20. Because the optimal sequencing depth depends on genome complexity and size, as well as the number of captured cells, a multi-step approach can be used to achieve the desired coverage. First, a shallow sequencing run of approximately 10 million reads is performed, followed by a pre-analysis to assess quality metrics (e.g., number of cells, UMIs per cell, etc.). Based on these results, an additional sequencing run of the same sample can then be performed to reach the optimal depth.
21. For detailed instructions on generating a Nanopore-compatible library, please refer to the protocol provided by Oxford Nanopore Technologies (https://nanoporetech.com/document/single-cell-transcriptomics-with-cdna-prepared-using-10x).
22. In the single-cell datasets shown in Fig. 7, reads were mapped to the *Medicago* genome (version 5, annotation version 1.9). The UMAP displays cell types annotated in the merged dataset from Medicago roots (susceptible zone) following Mock or bacterial treatment (24–96 h) (Fig. 7A). Using our optimized protocol, we achieved ~40%–50% mapped reads, with ~13%–15% classified as usable. Additional quality metrics—including UMIs per cell, genes per cell, and total high-quality cell counts—are summarized in Fig. 7B. Analysis of nodulation

marker gene expression confirmed that our protocol effectively isolated these target cells (Fig. 7C).

23. The choice between using nuclei or protoplasts as input depends on the experimental setup and the underlying biological question. For example, nuclei isolation offers greater flexibility, as it is compatible with frozen starting material. In addition, the efficiency of nuclei isolation is not influenced by cell wall composition, making it effective across different cell types and developmental stages. In our experience, working with nuclei was challenging due to issues such as RNA leakage, ambient RNA contamination, and the need for extensive data filtering.

Acknowledgments

This work was supported by the Dutch Research Council NWO-VIDI (VI.Vidi.193.119) to WK. C.G. was supported by the Dutch Sector Plan for Biology (Sectorplan Biologie) through Wageningen University & Research.

Conflict of Interest The authors have no conflicts of interest to declare that are relevant to the content of this chapter.

References

1. Roy S, Liu W, Nandety RS et al. (2020) Celebrating 20 years of genetic discoveries in legume nodulation and symbiotic nitrogen fixation. Plant Cell 32:15–41
2. Jhu M-Y, Oldroyd GED (2023) Dancing to a different tune, can we switch from chemical to biological nitrogen fixation for sustainable food security?. PLoS Biol 21:e3001982
3. Kohlen W, Ng JLP, Deinum EE et al. (2018) Auxin transport, metabolism, and signalling during nodule initiation: indeterminate and determinate nodules. J Exp Bot 69:229–244
4. Xiao TT, Schilderink S, Moling S et al. (2014) Fate map of *Medicago truncatula* root nodules. Dev Camb Engl 141:3517–3528
5. Limpens E, Moling S, Hooiveld G et al. (2013) Cell- and tissue-specific transcriptome analyses of *Medicago truncatula* root nodules. PLOS ONE 8:e64377
6. Cervantes-Pérez SA, Zogli P, Amini S et al. (2024) Single-cell transcriptome atlases of soybean root and mature nodule reveal new regulatory programs that control the nodulation process. Plant Commun 5:100984
7. Frank M, Fechete LI, Tedeschi F et al. (2023) Single-cell analysis identifies genes facilitating rhizobium infection in *Lotus japonicus*. Nat Commun 14: 7171
8. Pereira WJ, Boyd J, Conde D et al (2024) The single-cell transcriptome program of nodule development cellular lineages in *Medicago truncatula*. Cell Rep 43:113747
9. Ye K, Bu F, Zhong L et al. (2024) Mapping the molecular landscape of *Lotus japonicus* nodule organogenesis through spatiotemporal transcriptomics. Nat Commun 15: 6387
10. Seyfferth C, Renema J, Wendrich JR et al. (2021) Advances and opportunities in single-cell transcriptomics for plant research. Annu Rev Plant Biol 72:847–866
11. Chen C, Ge Y, Lu L (2023) Opportunities and challenges in the application of single-cell and spatial transcriptomics in plants. Front Plant Sci 14:1185377
12. Lebrigand K, Magnone V, Barbry P et al. (2020) High throughput error corrected Nanopore single cell transcriptome sequencing. Nat Commun 11:4025.

13. Ming X, Wan M-C, Zhang Z-D et al. (2025) FX-cell: a method for single-cell RNA sequencing on difficult-to-digest and cryopreserved plant samples. Nat Methods 22:2551–2562

14. Šranková M, Dvořák A, Martínek M et al. (2022) Antiproliferative and cytotoxic activities of fluorescein—a diagnostic angiography dye. Int J Mol Sci 23:1504

Chapter 19

Enrichment of Arabidopsis Plasma Membrane Proteins by Sequential Differential Centrifugation

Gijs Wilbrink, Roman Pleskot and Peter Grones

Abstract

The plasma membrane (PM) is the primary interface between plant cells and their environment, and its resident proteins mediate key processes such as extracellular signal perception and downstream cellular reprogramming. Yet, PM proteins are typically underrepresented in total protein extracts, and existing enrichment strategies are often laborious and require extensive optimization. Here, a simple and robust workflow is described for enriching PM proteins from *Arabidopsis thaliana* seedlings using total microsomal membranes obtained by differential centrifugation as starting material. Sequential low- and high-speed spins are used to isolate total microsomal membranes and progressively deplete contaminating organelles, thereby increasing the relative abundance of PM proteins. Coupled with the rich genetic toolkit available in Arabidopsis, this protocol provides an accessible platform for systematic characterization of the PM proteome.

Keywords Plasma membrane, Extraction, Protein enrichment, Differential centrifugation, Arabidopsis

1 Introduction

The plasma membrane (PM) delineates the interface between the intracellular and extracellular environments. It is heavily involved in the regulation of essential cellular processes, including endocytosis, cell proliferation, and transport of metabolites and ions [1–3]. These functions are mediated by a diverse repertoire of transmembrane proteins, embedded in the lipid bilayer or plasma membrane-associated proteins [2, 3]. Initially, the PM was considered to exhibit a homogeneous distribution of proteins and lipids [4]. However, accumulated evidence indicates that PM proteins are confined to specific nanodomains, enabling differential cellular responses and precise regulation of signal transduction, highlighting the PM as an important subject of research [5–7].

Membrane proteins are intrinsically challenging to study due to their low solubility and abundance relative to cytosolic proteins

Steffen Vanneste and Inge Verstraeten (eds.), *Root Development*, Methods in Molecular Biology, vol. 3086,
https://doi.org/10.1007/978-1-0716-5603-7_19,

[1–3]. Numerous methods have been developed to isolate membrane proteins through fractionation, such as free-flow electrophoresis and isoelectric focusing [8, 9]. In free-flow electrophoresis, membrane fractions are continuously introduced into a separation chamber and resolved according to the surface charge difference using an appropriate buffer system [8]. Isoelectric focusing analogously exploits charge heterogeneity but achieves separation within a gradient column positioned between cathode and anode compartments [9].

In the 1980s, a simpler, less labor-intensive alternative was introduced: membrane isolation using an aqueous polymer two-phase system [2, 10]. This method usually employs polyethylene glycol (PEG) and dextran, which form two immiscible aqueous phases after centrifugation. The PM has a high affinity for the PEG-rich upper phase, while the other organelles predominantly associate with the dextran-rich bottom phase, thus allowing efficient separation [2, 10]. Here, we describe a modified variant of this approach.

After homogenizing the sample, ultracentrifugation is performed to separate the microsomal fraction from the cytosolic fraction. Subsequently, phase separation using a PEG-dextran gradient is achieved through low-speed centrifugation, during which the upper phase is further enriched in PM proteins. Finally, an enriched PM fraction is isolated by ultracentrifugation, and its enrichment is verified by Western blot analysis using different antibodies recognizing proteins from different compartments. The enriched PM fraction can then be subjected to downstream analyses, such as comparative mass spectrometry.

2 Materials

1. 7-day old *Arabidopsis thaliana* (L.) Heynh. seedlings (*see* **Note 1**).
2. 10 mM vitamin C stock: Dissolve 17.6 mg in 10 mL of Milli-Q (MQ) H_2O.
3. 200 mM Phenylmethylsulfonyl fluoride (PMSF) stock: Dissolve 34.8 mg PMSF in 1 mL of isopropanol.
4. 150 mM benzamide stock: Dissolve 181.8 mg in 10 mL of MQ H_2O.
5. 1.5 mM pepstatin A stock: Dissolve 1 mg in 1 mL 10% (v/v) acetic acid in MeOH.
6. 11.7 mM leupeptin stock: Dissolve 5 mg in 1 mL of MQ H_2O.
7. 767 μM aprotinin stock: Dissolve 5 mg in 1 mL of MQ H_2O (*see* **Note 2**).
8. Homogenization buffer: Add 34.23 g sucrose, 2.98 g HEPES, 1.85 g KCl, 5.08 g $MgCl_2 \times 6H_2O$ to 200 mL of milliQ (MQ)

H_2O in a 500 mL graduated cylinder or glass beaker. Mix well and adjust the pH to 7.5 with 1 M NaOH. Then add 440 μL of vitamin C stock, 288 μL PMSF stock, 1.36 mL benzamide stock, 125 μL pepstatin A stock, 25 μL leupeptin stock, and 25 μL aprotinin stock. Adjust with MQ H_2O to the final volume of 250 mL.

9. Phosphate buffer A: Dissolve 6.81 g of KH_2PO_4 in MQ H_2O to a total volume of 1 L. The pH of the solution is 4.5.
10. Phosphate buffer B: Dissolve 8.90 g of K_2HPO_4 in MQ H_2O to a total volume of 1 L. The pH of the solution is 9.0.
11. 50 mM Phosphate buffer pH 7.8: Mix 900 mL of phosphate buffer A, and 100 mL of phosphate buffer B in a 1 L graduated cylinder or glass beaker.
12. 5 mM Phosphate buffer pH 7.8: Dilute 50 mL of 50 mM phosphate buffer, in 450 mL of MQ H_2O to a 500 mL graduated cylinder or glass beaker.
13. 1 M KCl.
14. 0.5 mM EDTA: Dissolve 93 mg EDTA in 10 mL MQ H_2O. Adjust the pH to 8.
15. 20% (w/w) Dextran (Mr 500 000): Dissolve 13 g of dextran in 50 g of MQ H_2O in a 100 mL graduated cylinder or glass beaker.
16. 40% (w/w) Polyethylene Glycol (PEG; Mr 3350): Add 14 g of PEG, and 21 g of MQ H_2O to a 50 mL graduated cylinder or glass beaker.
17. 1 M Sucrose: Dissolve 17.16 g of sucrose in MQ H_2O to a final volume of 50 mL.
18. Separation tube A solution: Add 4.96 g of 20% dextran into a 50 mL glass conical tube, and add 40% PEG up to a total weight of 7.44 g. Add 48 μL of 1 M KCl, 3.5 mL of 1 M sucrose, 1.4 mL of 5 mM phosphate buffer pH 7.8, and MQ H_2O up to a total weight of 14 g.
19. Separation tube B solution: Prepare, analogously to separation tube A (by adding dextran, PEG, KCl, sucrose, and phosphate buffer to a total weight of 14 g), into a 50 mL glass conical tube. Afterward, add 5 mM phosphate buffer, pH 7.8, up to a total weight of 16 g.
20. Storing buffer: Dissolve 42.8 g of sucrose, 2.6 g of MOPS, 10 mL of 0.5 mM EDTA in 400 mL of MQ H_2O, mix, and adjust to pH 7.5 with 1 M KOH. Adjust to the final volume of 500 mL with MQ H_2O.
21. 15- and 50-mL conical tubes.
22. Mortar and pestle (*see* **Note 3**).

23. Liquid nitrogen.
24. Ice.
25. Spinning wheel.
26. Miracloth, 15 × 15 cm squares.
27. Microcentrifuge tubes (2.0 mL).
28. Table-top microcentrifuge with temperature control (*see* **Note 4**).
29. Ultracentrifuge with a fixed-angle rotor fitted with thick-wall polycarbonate tubes (mm) (Beckman Coulter).
30. Bradford or TBA reagent.
31. Spectrophotometer with plate reader.
32. Syringe and long needle for transfer.
33. Sodium dodecyl sulfate-polyacrylamide gel (SDS-PAGE) and electrophoresis equipment.
34. Tris-Glycine-Saline (TGS) running buffer (0.025 M tris(hydroxymethyl) aminomethane [Tris] base, 0.192 M glycine, 0.1% [w/v] SDS).
35. Nitrocellulose membrane for protein transfer.
36. Rapid blotting system (e.g., TransBlot Turbo).
37. TBST buffer (Tris Buffer Saline with 0.1% [v/v] polyoxyethylenesorbitan [Tween 20]).
38. Skim milk or Bovine Serum Albumin (BSA).
39. Reagents and equipment for protein immunoblot transfer.
40. Antibodies against different subcellular proteins (*see* **Note 5**).
41. Chemiluminescence detection kit.
42. Imaging system for Western Blots (e.g., ChemiDoc).

3 Methods

3.1 Total Protein Isolation from Arabidopsis Seedlings

A mortar and pestle are used to homogenize whole Arabidopsis seedlings for total protein isolation. To minimize the degradation of proteins in the samples, work on ice, and pre-chill all solutions on ice during the full procedure (Fig. 1).

1. Pre-chill a mortar and pestle in liquid nitrogen, add plant material (*see* **Note 6**) and grind to a fine powder (*see* **Note 7**, Fig. 1a, b).
2. Add the ground plant material to a 50 mL tube and add 50 mL of homogenization buffer per 10 g of ground plant material.

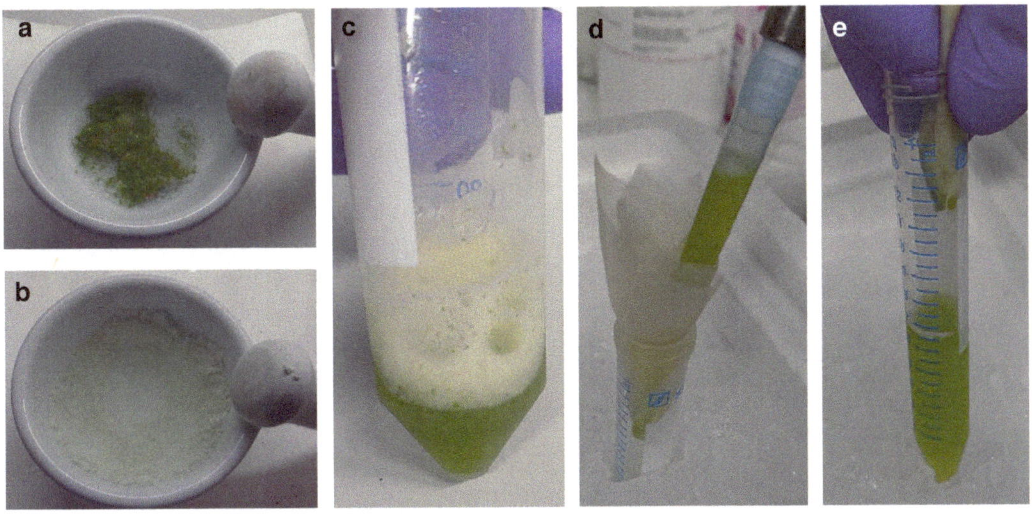

Fig. 1 Grinding of Arabidopsis seedlings and filtering of protein homogenate. (**a**, **b**) Six- to 7-day-old wild-type (Col-0) Arabidopsis seedlings are typically used for an experiment. Before (**a**) and after fine grinding (**b**). (**c**) Dissolved ground material in the homogenization buffer. (**d**) Two layers of Miracloth are placed into a 15-mL conical tube for filtering the big debris. (**e**) Squeezing the Miracloth with one finger against the side of the tube to pass the homogenate through

3. Incubate at a spinning wheel at 4°C for 30 min.
4. After dissolving the ground plant material in the homogenization buffer, place the Miracloth above the opening of a 15 mL falcon tube, put the cell homogenate onto it and filter out the cell debris by gentle squeezing (*see* **Note 8**).
5. Take an aliquot of approximately 50 μL of cell extract for immunoblot analysis and keep it in a microcentrifuge tube on ice.

3.2 Isolation of Microsomal Fraction

Large organelles, such as nuclei, chloroplasts, or mitochondria, are pelleted from the total protein extract by low-speed centrifugation. The supernatant, containing cytosolic, soluble, and microsomal proteins, is consequently ultra-centrifuged to pellet the microsomal fraction.

1. Transfer the filtered homogenate from Step 4 (i.e., from the 15-mL Falcon tube) to pre-chilled 2 mL microcentrifuge tubes.
2. Centrifuge the homogenate in a pre-chilled centrifuge at 9300 × *g* for 10 min, at 4°C (*see* **Note 4**). This step removes any unfiltered cell debris as well as big organelles, such as nuclei, chloroplasts, or mitochondria.
3. Transfer the supernatant containing cytosolic, soluble, and microsomal proteins to pre-chilled ultracentrifugation cuvettes.

Ensure to balance the centrifuge with the homogenization buffer (*see* **Note 9**).

4. Centrifuge at 190,000 × *g* for 60 min at 4°C in the pre-chilled ultracentrifuge.

3.3 Phase Separation

During membrane extraction with nonionic detergents or polymer-based systems, proteins and lipids partition into distinct aqueous phases according to their hydrophobicity, charge, and association with specific membrane nanodomains. This phase separation can be exploited to enrich the plasma membrane-derived fraction (*see* **Note 10**) by driving these components into one phase while soluble or non-target proteins partition into the other.

1. Gently remove the cuvettes from the rotor, avoiding shaking to avoid disturbing the pellet, which contains the microsomal fraction consisting of PM, ER, Golgi, and/or tonoplast proteins.
2. Carefully decant the supernatant, containing the cytosolic fraction, from the cuvettes, avoiding disturbing the pellet. Take an aliquot of approximately 50 μL of the supernatant and keep it in a microcentrifuge tube on ice. Remove as much of the supernatant as possible by wiping the insides of the cuvettes with a paper tissue.
3. Homogenize the pellet in an adequate amount of pre-chilled 5 mM phosphate buffer (*see* **Note 11**) and take an aliquot of approximately 20 μL of resuspended pellet in a microcentrifuge tube on ice.
4. Transfer the solution into separation tube A solution (*see* **Note 12**), close the tube by applying parafilm on top of the separation tube A, and mix 40× by arm movement from above the head to the waist.
5. Without adding anything, mix separation tube B in the same way as separation tube A (*see* **Note 13**).
6. Leave the tubes overnight at 4°C.
7. Check the concentration of all extraction fraction aliquots using an appropriate protein concentration assay (e.g., Bradford, TBA).

3.4 Enrichment of Plasma Membrane Fraction

The microsomal fraction obtained in the previous step is purified using a two-phase system, which removes contamination by organellar proteins and enriches for plasma membrane proteins.

1. Centrifuge both separation tubes (A and B) at 840 × *g* for 5 min at 4°C, using the table-top centrifuge.

2. After centrifugation, discard the top fraction of the separation tube B.
3. Transfer the top fraction of separation tube A (containing the enriched plasma membrane fraction) to the bottom of separation tube B using a syringe and long needle.
4. Apply parafilm on top of the separation tube B and mix 40× by arm movement from above the head to the waist. After mixing, centrifuge at 840×*g* for 5 min at 4°C using a tabletop centrifuge.
5. Repeat Steps 1–4 two more times to acquire a higher yield and purity of the plasma membrane (*see* **Note 14**).
6. After centrifugation, transfer the top fraction of the separation tube B to the pre-chilled ultracentrifugation cuvettes and fill them with storage buffer. The ideal ratio of top fraction: storage buffer = 1:3.
7. Centrifuge at 190,000×*g* for 60 min at 4°C.
8. Immediately after centrifugation, remove the supernatant by decanting, then wipe the inside of the cuvette with a tissue.
9. Homogenize the pellet in 100 μL of storage buffer.
10. Determine the concentration of the enriched PM fraction analogous to the method described above (Sect. 3.3, Step 7: Bradford, TBA).

3.5 Validation of Enrichment by Western Blotting

Immunoblot analysis can be used to verify whether the protein of interest is enriched in the plasma membrane fraction by probing fractionated samples with a specific antibody. By comparing signal intensity between total, cytosolic, and PM-enriched fractions, it is possible to assess the efficiency of the fractionation and confirm PM localization of the target protein (Fig. 2). The enriched PM fraction can then be subjected to downstream analyses such as mass spectrometry (*see* **Note 18**).

1. Adjust the concentration of each extracted fraction to 2 μg protein/μL by adding storage buffer.
2. Add an appropriate amount of Laemmli loading buffer to the sample and heat the sample at 70°C for 5 min (*see* **Note 15**). After heating, centrifuge the sample at 3000×*g* for 30 s to bring down condensates.
3. Load equal protein amounts (for example, ~30 μg) of the boiled protein samples from each fraction (e.g., total extract, microsomal, and PM) on 12% (w/v) polyacrylamide gels. Run the gel in TGS running buffer until the Laemmli loading dye reaches the bottom of the gel. Consequently, the proteins are transferred to the nitrocellulose membrane.

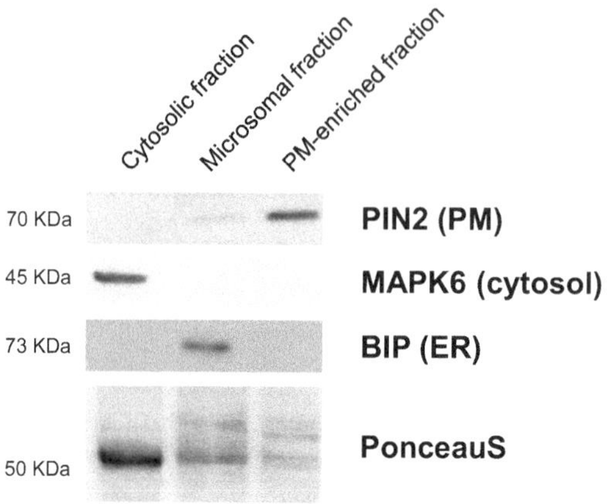

Fig. 2 Representative immunoblot using antibodies against different subcellular marker proteins. The PM marker PIN2 protein is heavily enriched in the PM-enriched fraction compared to the microsomal or the cytosolic fraction. The cytosolic marker MPK6 is only present in the cytosolic fraction, indicating the lack of contamination in the other fractions. The membrane marker for endoplasmic reticulum BIP is mostly present in the microsomal fraction, but heavily reduced from the PM-enriched fraction. As proof of concept, these results demonstrate the feasibility of PM protein extraction and enrichment. Collectively, these results provide proof of concept that this approach enriches for PM proteins while reducing contamination by organellar proteins in Arabidopsis seedlings

4. The membrane is blocked with 5% (w/v) skim milk or BSA in TBS-T for 60 min at room temperature (*see* **Note 16**) with gentle shaking.
5. Afterward, the membrane is incubated with primary antibodies diluted in 5% skim milk/BSA in TBS-T buffer at an appropriate concentration for 60 min at room temperature (*see* **Note 17**) with gentle shaking, subsequently washed 5× with an adequate amount of TBST buffer.
6. The membrane is incubated with secondary antibodies diluted in 5% skim milk/BSA in TBS-T buffer at an appropriate concentration for 60 min at room temperature with gentle shaking, and afterward washed 5× with an adequate amount of TBS-T buffer.
7. Washed membranes are developed using a chemiluminescence detection kit, and the signal is detected using an imaging system (e.g., ChemiDoc).

4 Notes

1. Grow the sterilized Arabidopsis seeds on 0.8% (w/v) agar plates, with half-strength Murashige and Skoog salts and 1% sucrose for 7 days under constant light at 22°C.

2. All the above-mentioned chemicals can be prepared beforehand as stock solutions and stored at −20°C for several months.
3. For a typical experiment, we use a porcelain mortar with a capacity of 50 mL and a porcelain pestle with a flat bottom.
4. Use a table-top microcentrifuge placed within a 4°C deli case fridge.
5. We usually use antibodies from Agrisera (Vännäs, Sweden) against proteins from different cellular compartments: PM-localized PINFORMED2 (PIN2; AS20 4521), endoplasmic reticulum molecular chaperone (BIP; AS09 481), cytoplasmic MAPK-type protein kinase (MAPK6; AS12 2633).
6. Roughly 2 g of fresh tissue is sufficient for use. This PM enrichment protocol also works well with a higher input amount, if scaling up is necessary. A larger volume of the homogenization buffer is then recommended.
7. Pour liquid nitrogen into a mortar and place the pestle in it to ensure the pestle is cooled as well. As soon as the liquid nitrogen has evaporated, add the plant material and start grinding it. Frequently add liquid nitrogen to keep the plant material cool. Repeat this procedure until a fine powder has been obtained.
8. Miracloth is used to remove tissue debris before the first centrifugation step. For efficient filtration, the Miracloth layers are packed snugly into the tube with a finger or pen, and the homogenate is then forced through by squeezing or twisting rather than relying on gravity.
9. Ultra centrifuge balancing is crucial, as even small imbalances at very high rotational speeds can generate significant vibrations, leading to reduced performance, inaccurate results, and premature wear or failure of components. Proper balancing ensures smooth operation, protects sensitive bearings and rotors from excessive stress, and enhances the safety of both the equipment and the operator. Additionally, well-balanced ultra-centrifuges deliver more reliable and reproducible experimental outcomes, which are essential for high-precision scientific and industrial applications.
10. During this extraction procedure, the peripheral membrane proteins (PM) are preferentially retained.
11. The total volume of the resuspension buffer should not exceed 2 μL so as not to dilute the microsomal fraction too much. Once the pellet is detached from the side of the cuvette and floating in the buffer, it can be transferred to a pre-chilled Potter-Elvehjem homogenizer and homogenized by moving

the pistil up and down. Afterward, the pellet can be transferred into the separation tube A.

12. For separation tubes A and B, glass tubes with a volume of 50 mL must be used, as plastic tubes will not result in correct separation and gradient formation.
13. Separation tube B must be mixed in the same way as separation tube A: mixing 40× by arm movement above the head, to the waist.
14. We recommend repeating Steps 1–3 of the isolation of plasma membrane fraction at least 2–3 times, as the purity of the plasma membrane in the final solution will increase significantly.
15. Avoid boiling samples at high temperatures (above 70°C), as this may cause aggregation of PM proteins and prevent successful separation on the acrylamide gel.
16. The membrane can also be blocked in the blocking solution overnight at 4°C.
17. The membrane can also be incubated in the primary antibody solution overnight at 4°C.
18. The enriched PM fraction can subsequently be used for mass spectrometry analysis [11].

Acknowledgments

The work in RP's lab is supported by the Czech Science Foundation grant number 22-35680M.

Conflict of Interest Authors declare no competing interests.

References

1. Marmagne A, Rouet M, Ferro M et al (2004) Identification of new intrinsic proteins in Arabidopsis plasma membrane proteome. Mol Cell Proteomics 3:675–691. https://doi.org/10.1074/mcp.m400001-mcp200
2. Schindler J, Nothwang HG (2006) Aqueous polymer two-phase systems: effective tools for plasma membrane proteomics. Proteomics 6:5409–5417. https://doi.org/10.1002/pmic.200600243
3. Garcia SaL PanisB, Swennen R, Carpentier SC (2014) Evaluation of four different strategies to characterize plasma membrane proteins from banana roots. Ciênc Agrotec 38:424–434. https://doi.org/10.1590/s1413-70542014000500002
4. Singer SJ, Nicolson GL (1972) The fluid mosaic model of the structure of cell membranes. Science 175:720–731. https://doi.org/10.1126/science.175.4023.720
5. Bücherl CA, Jarsch IK, Schudoma C et al (2017) Plant immune and growth receptors share common signalling components but localise to distinct plasma membrane nanodomains. eLife 6:e25114. https://doi.org/10.7554/elife.25114
6. Bahammou D, Recorbet G, Cassim AM et al (2024) A combined lipidomic and proteomic profiling of Arabidopsis thaliana plasma membrane. Plant J 119:1570–1595. https://doi.org/10.1111/tpj.16810

7. Jaillais Y, Bayer E, Bergmann DC et al (2024) Guidelines for naming and studying plasma membrane domains in plants. Nat Plants 10:1172–1183. https://doi.org/10.1038/s41477-024-01742-8
8. Bardy N, Carrasco A, Galaud J et al (1998) Free-flow electrophoresis for fractionation of Arabidopsis thaliana membranes. Electrophoresis 19:1145–1153. https://doi.org/10.1002/elps.1150190715
9. Griffing LR, Quatrano RS (1984) Isoelectric focusing of plant cell membranes. Proc Natl Acad Sci USA 81:4804–4808. https://doi.org/10.1073/pnas.81.15.4804
10. Widell S, Larsson C (1981) Separation of presumptive plasma membranes from mitochondria by partition in an aqueous polymer two-phase system. Physiol Plant 51:368–374. https://doi.org/10.1111/j.1399-3054.1981.tb05571.x
11. Iwasaki Y, Itoh T, Hagi Y et al (2020) Proteomics analysis of plasma membrane fractions of the root, leaf, and flower of rice. Intl J Mol Sci 21:6988. https://doi.org/10.3390/ijms21196988

Chapter 20

High-Throughput Sample Preparation for Plant (Phospho) Proteomics

Mark Roosjen and Dolf Weijers

Abstract

Mass spectrometry-based proteomics allows the unbiased identification and quantification of proteins and phosphopeptides in biological materials. The nature of walled plant cells requires specific protocols for effective and efficient protein isolation, and, in general, the plant sciences can benefit from more accessible, optimized proteomics workflows. Advances in MS instrumentation now allow the measurement of large numbers of samples, shifting constraints in proteomics toward the accurate, high-throughput preparation of samples. Here, we describe a high-throughput (phospho)proteomics protocol that enables processing of samples using different filter types in a 96-well format.

Keywords Plant proteomics, Phosphoproteomics, High-throughput, Suspension trapping

1 Introduction

Mass spectrometry (MS)-based proteomics is a fast and sensitive tool to quantify the dynamics of protein behavior in complex biological materials. Recent advancements in instrumentation speed and sensitivity allow for increased sample throughput, enabling the measurement of multi-variable and longitudinal biological datasets (e.g., time-series analysis, multiple treatments). Large sample cohorts, however, put constraints on the sample preparation techniques used in a proteomics workflow. Typically, plants are mechanically (e.g., grinding in liquid nitrogen) treated in combination with strong detergent-containing lysis buffers to release proteins. Lysis buffer components subsequently need to be removed to minimize their interference with subsequent liquid chromatography (LC)-MS/MS measurements. Often, protein precipitation techniques are employed to remove these interfering substances [1]. However, such approaches are cumbersome when performed on large sample sets and usually cause significant losses of materials. Other vesicle-based approaches, such as filter-aided sample preparation (FASP) [2], protein aggregation capture (PAC)

Steffen Vanneste and Inge Verstraeten (eds.), *Root Development*, Methods in Molecular Biology, vol. 3086, https://doi.org/10.1007/978-1-0716-5603-7_20,

[3], and suspension trapping (S-Trap) [4], allow for manipulating the protein material on matrices, making these approaches amenable to high-throughput sample preparation. Here, we describe a high-throughput approach for measuring shotgun proteomes and phosphoproteomes from plants using S-Trap in a 96-well plate-based format. The benefit of this approach is that peptides can be eluted directly onto phosphopeptide enrichment material, eliminating sample transfers and the usual C18 cleanup steps, thereby substantially reducing peptide loss during sample preparation. Moreover, this strongly decreases hands-on time and meanwhile increases phosphosite identifications compared to precipitation-based approaches or a recently published high-throughput amenable phosphoproteomics approach, GreenPhos [5] (Fig. 1). While we describe the protocol in a high-throughput format, this approach can also be performed on single S-Trap columns as described [6]. The protocol is adapted for high-throughput shotgun proteome, phosphoproteomics, or similar applications. We have successfully applied this approach to diverse plant species and tissue types.

Fig. 1 Overview of the centrifugation-based 96-well (phospho)proteomics protocol. Plant tissue is harvested, ground, and sonicated in a 96-well format. All subsequent downstream steps are accomplished using 96-well filter plates with centrifugation. Samples can then be acquired using high-end mass spectrometers, such as the timsTOF HT, allowing for short gradients that maximize throughput. In this setup, 96 samples can be measured in 4 measurement days. The S-trap protocol outperforms a precipitation-based protocol, such as the GreenPhos protocol, in recovering phosphopeptides

2 Materials

All solutions should be prepared fresh, in 15 or 50 mL Falcon tubes and in ultrapure water (MilliQ). Since strong acids and organic solvents are used throughout the protocol, users should take appropriate safety precautions and follow dedicated waste disposal regulations.

1. Lysis buffer: 100 mM Tris-hydrochloride (Tris-HCl) pH 8.0, 7 M Urea, 1% Triton-X, 5 mM Dithiothreitol (DTT).
2. S-Trap Binding/Wash buffer: 100 mM triethylammonium bicarbonate (TEAB) in 90% Methanol.
3. 0.5 M Iodoacetamide stock.
4. 10% Sodium dodecyl sulfate (SDS).
5. Trifluoracetic acid (TFA).
6. Formic acid (FA).
7. 1 M Triethylammonium bicarbonate buffer (TEAB).
8. 100% Acetonitrile (ACN).
9. Ammonium dihydrogen phosphate ($NH_4H_2PO_4$).
10. Methanol.
11. Pierce BCA protein assay kit.
12. 1.2 mL 8 Strip tubes and lids.
13. S-Trap 96-well (C02-96well-1).
14. Acroprep Advance 96-well 350 μL 1.2 μm Supor (8039 Pall).
15. Acroprep advance cell lysate clearance plate (8175 Pall).
16. 96-well full-skirted PCR plate.
17. Eppendorf Protein Lo-Bind 0.5/1/2 mL Deepwell plates.
18. Trypsin sequencing grade (11418033001 Roche).
19. SepPak C18 μElution plates.
20. PureCube Fe-NTA MagBeads.
21. Centrifuge for 96-well plates (e.g., Eppendorf 5810).
22. Vacuum centrifuge with 96-well adapter.
23. Sonicator with 96-well capability (e.g., QSonica Q700 Cuphorn).

3 Methods

3.1 Harvesting and Lysis

1. Harvest plant material and grind to a fine powder in liquid nitrogen (*see* **Note 1**).

2. Add the lysis buffer at a 1:1 (volume:volume) ratio to the ground plant material and resuspend until homogeneous (work with the minimal volume/maximum concentration possible).
3. Sonicate samples in a 4°C water bath for 5 min O/N using a cycle of 30 s ON and 30 s OFF at 100% amplitude.
4. After sonication, transfer lysed material to a 96-well cell lysate clearance plate placed above a 500 μL deepwell collection plate (*see* **Note 2**).
5. Centrifuge lysates at 1500 × *g* for 20 min.
6. After centrifugation, discard the cell lysate clearance plate and seal the deepwell plate with a silicone mat (*see* **Note 3**).
7. Determine the sample volume, add iodoacetamide to a final concentration of 20 mM, and incubate in the dark for ~30 min.
8. During alkylation, the protein concentration is measured using the BCA assay.

3.2 Protein Concentration Measurement

1. For the BCA assay, use the kit from Pierce (#23225). The kit is provided with BSA (2 mg/mL). For reference standard, see Table 1.
2. For reference dilution stocks, add 10 μL each to a well (*see* **Note 4**).
3. For the samples, make a 1:10 dilution by resuspending 5 μL sample in 45 μL of MilliQ, and add 10 μL to each well.
4. Mix component A and B in a 50:1 ratio (e.g., 50 mL A, 1 mL B).

Table 1
Dilution series of 2 mg/mL BSA standard

Vial	Volume of MilliQ (μL)	Volume and source of BSA (μL)	Final BSA concentration (mg/mL)
A	0	300 in stock	2
B	125	375 in stock	1.5
C	325	325 in stock	1
D	175	175 of B	0.75
E	325	325 of C	0.5
F	325	325 of E	0.25
G	325	325 of F	0.125
H	400	100 of G	0.025
I	400	0	0

5. Add 200 µL of the BCA mixture to each well containing the BSA references or samples.
6. Incubate for 30 min at 37°C.
7. Measure the OD at 562 nm in a microplate reader.
8. Calculate protein concentration using simple linear regression.

3.3 S-Trap Protein Clean-Up and Trypsin Digestion

1. Take 500 µg of protein per sample and proceed with S-Trap cleanup (*see* **Note 5**). See Table 2 for precalculated example amounts.
2. Choose a desired plate type based on the total volume after addition of SDS, formic acid, and S-Trap binding buffer. For example, for samples with a volume of 50 µL, a 1 mL deep-well plate is used.
3. Add 10% SDS stock solution to reach a final concentration of 5%, then mix by briefly vortexing.
4. Add 10% formic acid to a final concentration of 1% and mix by short vortexing.
5. Add six volumes of S-Trap Binding/Wash buffer and mix with a short vortex.
6. Transfer the whole sample (including any protein precipitates) to the S-Trap plate.
7. Place S-Trap plate above a 2 mL deepwell collection plate and centrifuge at 1500 × *g* for 2 min (*see* **Note 6**).
8. Visually confirm that all samples have passed through the column; if not, centrifuge again until all solutions have passed through.
9. Discard flowthrough.
10. Add 200 µL binding/wash buffer and centrifuge at 1500 × *g* for 2 min.
11. Repeat washes three times.
12. Perform an additional centrifugation step to remove all liquid.

Table 2
Volume amounts for suspension trapping preparation

Sample (µL)	10% SDS (µL)	10% Formic acid (µL)	S-Trap Binding/Wash buffer (µL)
10	10	2	132
50	50	10	660
100	100	20	1320
200	200	40	2640

13. Move the S-Trap plate to a new 96-well collection plate.
14. Add 100 μL of 5 μg trypsin (trypsin:protein, 1:100) (prepared in 50 mM TEAB) to each well and incubate O/N at 25°C with shaking.

3.4 Elution for Shotgun Proteomics

1. Add 100 μL of loading buffer (1% TFA in 80% ACN) to the S-Trap plate and centrifuge at 1500 × *g* for 2 min.
2. Repeat the last step for a second elution.
3. Discard the S-trap plate and vacuum centrifuge eluted peptides until a small volume (~10–20 μL) remains.
4. Proceed with peptide desalting and concentrating.

3.5 Phosphopeptide Enrichment

1. Resuspend PureCube Fe-NTA Magbeads thoroughly until homogeneous (*see* **Note 7**).
2. Transfer 100 μL of beads to a 96-well Acroprep Advance 1.2 μm Supor membrane plate.
3. Place the 96-well Acroprep Advance plate above a 2 mL deep-well plate and centrifuge at 200 × *g* for 2 min.
4. Add 200 μL of loading buffer (1% TFA in 80% ACN) and centrifuge at 200 × *g* for 2 min.
5. Repeat the last step for a total of three equilibrations.
6. Add 100 μL of loading buffer (1% TFA in 80% ACN) to the beads.
7. Add 100 μL of loading buffer (1% TFA in 80% ACN) to the S-Trap plate.
8. Place the S-Trap plate above the Acroprep Advance plate and the Acroprep Advance plate above a 0.5 mL deepwell collection plate.
9. Centrifuge at 100 × *g* for 10 min.
10. Add 100 μL of loading buffer (1% TFA in 80% ACN) to the S-Trap plate and centrifuge at 200 × *g* for 5 min.
11. Discard the flow-through and remove the S-Trap plate.
12. Add 200 μL of loading buffer (1% TFA in 80% ACN) and centrifuge at 1000 × *g* for 2 min.
13. Repeat the last step for a total of three washes, discarding the flow-through each time.
14. Place the Acroprep Advance plate on top of a new 96-well collection plate.
15. Elute phospopeptides with 100 μL 500 mM $NH_4H_2PO_4$ and centrifuge at 200 × *g* for 5 min.
16. Repeat the last step for an additional elution.

3.6 Peptide Desalting and Concentrating

1. For shotgun proteomics: Resuspend peptides with 0.1% formic acid to 100 μL, wherein peptides should be acidic (pH 2–3).
2. For peptide desalting, the SepPak C18 μElution plate from Waters is used.
3. Place the C18 plate on a 2 mL deepwell collection plate and add 50 μL of 100% Methanol.
4. Centrifuge at 1000 × *g* for 2 min.
5. Add 50 μL of 80% ACN and 0.1% FA, then centrifuge at 1000 × *g* for 2 min.
6. Discard the flowthrough.
7. Add 50 μL of 0.1% FA and centrifuge at 1000 × *g* for 2 min.
8. Repeat the last step for a second wash.
9. Add acidified shotgun proteomics samples or phosphopeptide-enriched samples and centrifuge at 200 × *g* for 10 min.
10. After peptide loading, wash with 100 μL of 0.1% FA and centrifuge at 1000 × *g* for 2 min.
11. Place the C18 well plate above an LC-MS-compatible 96-well plate.
12. Elute peptides with 30 μL of 80% ACN, 0.1% FA and centrifuge at 1000 × *g* for 2 min.
13. Vacuum-centrifuge samples at 45°C in V-AQ until dry.
14. Resuspend peptides to 7 μL with 0.1% FA.
15. Sonicate in a 4°C water bath for 1 min ON time using a cycle of 30 s ON and 30 s OFF at 100% amplitude to facilitate peptide resuspension.
16. Centrifuge plate at 1000 × *g* for 2 min.
17. Samples are ready for MS/MS and can be stored at −20°C.

4 Notes

1. The amount of starting material depends on the tissue type used. Typically, as of *Arabidopsis thaliana*, per biological replicate, for 5- to 7-day-old roots, ten to 15 square Petri dishes are grown and used; for 5 to 7-day-old whole seedlings, one square Petri dish is grown and used; for Arabidopsis suspension-cultured PSB-D cells, 10 mL culture is grown and used; for 14-day-old rosettes, one rosette is used. For high-throughput applications, 1.2 mL cluster tube strips are used with the bead-milling method. For manual grinding with a mortar and

pestle, cluster tubes are filled with up to 500 μL of powdered tissue.

2. Cell lysate clearance plate has a maximum capacity of 350 μL. If the volume exceeds the maximum capacity, repeat the loading step until all samples have passed. Note that receiver plate volume should also be chosen based on sample volume.
3. At this stage, the sample can be frozen and stored at −80°C
4. BSA reference standards and samples are usually measured in duplicates or triplicates.
5. If only proteomes are measured, an input of 50 to 100 μg is sufficient. Lowering the input for phosphoproteomics might be possible, but it should be confirmed whether sufficient identifications can be achieved. We have noted that this relies heavily on the type of tissue used and should therefore be empirically tested before conducting large-scale experiments.
6. S-Trap plate has a maximum capacity of 350 μL. If the volume exceeds the maximum capacity, repeat the loading step until all samples have passed. Discard flowthrough each time.
7. PureCube Fe-NTA Magbeads also work for small-scale phosphoproteomic experiments. PureCube Fe-NTA Magbeads can be replaced by other types of beads, which, however, need to be tested before conducting large-scale experiments.

Acknowledgments

MR and DW acknowledge the European Research Council (ERC AdG "DIRNDL; contract 833867), the Sector Plan Biology of the Dutch Ministry of Education, Culture and Science, and the Netherlands Organization for Scientific Research (NWO; CropXR) for funding proteomics research in their team. We thank Jiashu Chu for critically reading and commenting on the protocol.

Author Contributions MR and DW conceived the manuscript. MR wrote the initial draft and DW edited the final manuscript.

Conflict of Interest None declared.

References

1. Brajkovic S et al (2023) Getting ready for large-scale proteomics in crop plants. Nutrients 15(3):783. https://doi.org/10.3390/nu15030783
2. Wiśniewski JR, Zougman A, Nagaraj N, Mann M (2009) Universal sample preparation method for proteome analysis. Nat Methods 6(no. 5):359–362. https://doi.org/10.1038/nmeth.1322
3. Batth TS et al (2019) Protein aggregation capture on microparticles enables multipurpose proteomics sample preparation. Mol Cell

Proteom 18(no. 5):1027–1035. https://doi.org/10.1074/mcp.TIR118.001270

4. Zougman A, Selby PJ, Banks RE (2014) Suspension trapping (STrap) sample preparation method for bottom-up proteomics analysis. Proteomics 14(no. 9):1000–1006. https://doi.org/10.1002/pmic.201300553
5. Duan X et al (2024) GreenPhos, a universal method for in-depth measurement of plant phosphoproteomes with high quantitative reproducibility. Mol Plant 17(no. 1):199–213. https://doi.org/10.1016/j.molp.2023.11.010
6. Chen C-W, Tsai C-F, Lin M-H, Lin S-Y, Hsu C-C (2023) Suspension trapping-based sample preparation workflow for in-depth plant phosphoproteomics. Anal Chem 95(no. 33):12232–12239. https://doi.org/10.1021/acs.analchem.3c00786

FSC
www.fsc.org
MIX
Papier | Fördert gute Waldnutzung
FSC® C083411

Kolibri 360 GmbH
Ferdinand-Jühlke-Straße 7
99095 Erfurt, Deutschland
produktsicherheit@kolibri360.de